# RELATIVISTIC ATOMIC COLLISIONS

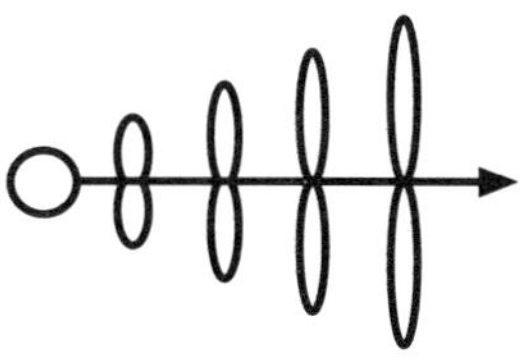

# RELATIVISTIC ATOMIC COLLISIONS


Jörg Eichler

Bereich Theoretische Physik
Hahn-Meitner-Institut
14109 Berlin
Germany

Fachbereich Physik
Freie Universität Berlin
14195 Berlin
Germany

Walter E. Meyerhof

Department of Physics
Stanford University
Stanford, CA 94305



ACADEMIC PRESS

San Diego Boston New York
London Sydney Tokyo Toronto

This book is printed on acid-free paper. ∞



Academic Press, Inc.
A Division of Harcourt Brace & Company
525 B Street, Suite 1900, San Diego, California 92101-4495

*United Kingdom Edition published by*
Academic Press Limited
24-28 Oval Road, London NW1 7DX

Library of Congress Cataloging-in-Publication Data

Eichler, Jörg, date.
Relativistic atomic collisions / Jörg Eichler, Walter E. Meyerhof.
p. cm.
Includes index.
ISBN 0-12-233675-5
1. Collisions (Nuclear physics) I. Meyerhof, Walter E. (Walter Ernst), date. II. Title.
QC794.6.C6E27 1995
539.7'23--dc20 94-48900
CIP

PRINTED IN THE UNITED STATES OF AMERICA
95 96 97 98 99 00 QW 9 8 7 6 5 4 3 2 1

# Contents

# II Elementary Atomic Processes

# IV Appendix

# Preface

This book is addressed to graduate students and professional scientists who are interested in atomic collisions at relativistic energies. Although normally one would not associate the investigation of atomic phenomena with relativistic collisions, these collisions bring some unique opportunities to atomic-structure and collision studies in experiment and theory, especially for ions of high atomic number $Z$. (1) Atomic collisions at relativistic energies allow the production of highly stripped ion beams up to uranium, with which atomic-structure studies of few electron ions can be undertaken. In such high-$Z$ systems, quantum-electrodynamic effects are emphasized. (2) At relativistic energies, the processes of inner-shell electron excitation, ionization, and capture become relatively simple, even for the highest-$Z$ ions, allowing a systematic investigation of these processes. The availability of relativistic heavy-ion storage rings is particularly helpful in these investigations. (3) Peripheral collisions of high-$Z$ nuclei at extreme relativistic energies can be viewed as virtual photon-photon collisions because of the Lorentz contraction of the nuclear Coulomb field in the direction of motion. The field of photon-photon collision physics thereby becomes accessible. The expected production of single and multiple electron-positron pairs presents new experimental and theoretical challenges and may assist in the production of new elementary particles of interest in high-energy physics.

This book concentrates on atomic collisional processes and leaves the consideration of atomic-structure studies to other reviews. It is assumed that the reader is familiar with nonrelativistic quantum mechanics and had an introduction to the Dirac equation. Part I of the book, Chaps. 2 – 5, is in the nature of a text and discusses the basic equations of special relativity, of electromagnetic fields, and of relativistic electron motion in a Coulomb potential and in the electromagnetic field of colliding nuclei. This part should be of general interest to readers who wish to become familiar with the theoretical treatment of relativistic collisions as well as with relativistic electron theory. Part II of the book, Chaps. 6 – 10, is more in the nature of a review and considers the basic atomic-collision processes, excitation and ionization, and capture without and with the emission of radiation. Electron-positron pair production, which is the signature of relativistic collisions, is treated by perturbative and nonperturbative methods. An outlook on extreme relativistic collisions is included, where multiple-pair production is expected. Part III of the book, Chaps. 11 – 13, presents experimental techniques which might be useful in the field. Appendices on units, a list of the book's tables and graphs, and a bibliography are collected in Part IV.

The book could not have been written without frequent meetings of the authors, which were facilitated by support from the NATO Collaborative Research Grant No. 910044 and from the International-Programs and Physics Divisions of the National Science Foundation (U.S.) through grants to Stanford University, Nos. INT-9013087 and PHY-9019293, respectively. One of us (J.E.) profited from a 1-month stay supported by the National Science Foundation through a grant to the Institute for Theoretical Atomic and Molecular Physics at Harvard University and Smithsonian Astrophysical Observatory, and also from extended visits to the University of Tsukuba, Japan, and to the Japan Atomic Energy Research Institute.

The authors are grateful for the cooperation of many professional colleagues, in particular A. Belkacem, F. Decker, D.P. Dewangan, R. Dörner, H. Gould, D. Ionescu, P. Mokler, G. Soff, T. Stöhlker, N. Toshima, and J. Ullrich. The able assistance of Ms. I. Lang in preparing the figures is greatly appreciated.

Jörg Eichler
Walter E. Meyerhof

# Chapter 1

# Introduction

A relativistic atomic collision is one in which an ion or atom impinges on an atomic system with a velocity that is comparable to the speed of light. It is assumed that during the entire collision the distance between the two nuclei involved remains large enough so that only electromagnetic interactions prevail and strong interactions do not play any role. In order to experimentally ensure a purely atomic collision, in which the nuclei never overlap, one has to accept only those events in a detector, in which the projectile emerges intact, except perhaps for the gain or loss of one or more electrons. Included in the definition is also a relativistic motion of the electrons involved. This book specializes in the range of collision energies[1] from about 100 MeV/u up to the highest energies that might be achieved in the near future and it covers atomic charge numbers up to $Z = 92$.

One of the fascinations of the field is the hope that a realm of new phenomena will be opened up. In the GeV/u ($10^9$ eV per atomic mass unit) energy range, it is possible to produce new atomic systems such as high-$Z$ few-electron atoms and bare high-$Z$ nuclei and to subject them to experimental study. In the TeV/u ($10^{12}$ eV per atomic mass unit) energy range, one hopes to create elementary particles or pairs of them by the intense high-energy pulse of electromagnetic radiation due to the passing heavy nuclei.

This book intends to set a framework, theoretical as well as experimental, for describing and measuring the basic elementary processes involved in relativistic atomic collisions without, however, venturing into the field of particle physics. In order to bring out the similarities and differences of relativistic atomic collisions with the corresponding nonrelativistic processes,

[1]As is customary, the phrase "collision energy" is used interchangeably for the projectile energy in MeV or in MeV/u.

we start in Sec. 1.1 with some general remarks on theoretical descriptions of atomic collisions and in Sec. 1.2 proceed to a brief qualitative discussion of nonrelativistic collisions as far as their relativistic generalization enters into the subsequent chapters. We then discuss in Sec. 1.3 the rough dependencies of the basic elementary processes on the nuclear charges and the collision energy. In Sec. 1.4, we give a brief outline of the book, and finally, in Sec. 1.5, we comment on the units used.

## 1.1 Theoretical treatments of atomic collisions

What is of interest in atomic collisions is the behavior of the *electrons* while the *nuclei* mainly serve as vehicles and as sources of time-varying electromagnetic fields.

If many-electron atoms take part in the collision, the dynamics become very complicated owing to the many degrees-of-freedom involved. To avoid complications arising from many-body effects, we mainly concentrate on theoretical discussions of a three-body ion-atom collision system, comprising a projectile nucleus, a target nucleus, and an electron. Such a system embraces the basic dynamics of a relativistic atomic collision. For example, for inner-shell electrons, if the electron-electron interaction is small compared to the electron-nucleus interaction, the one-electron model is a good approximation.

In principle, all particles involved, electrons and nuclei, must be described theoretically by quantum mechanics. This approach is called the "wave picture." However, because of the large mass ratio $M/m_e$ of the nuclei to the electrons, it is possible to treat the nuclear motion classically. This approach is called the "semiclassical approximation" (SCA) or the "impact parameter picture." In the following, we briefly discuss both descriptions.

*Quantum description of the nuclear motion*

In the wave picture, the nuclear motion is described by solving the stationary wave equation for a fixed energy. In this description, the incoming and outgoing waves for the relative nuclear motion extend over all space and are characterized by their asymptotic momenta. Energy and momentum are conserved for the collision system of nuclei and electrons.

Adopting the one-electron approximation, the electron whose transition one wishes to study, called the *active* electron, shares energy and momentum with the colliding nuclei. In a transition between initial and final states, a momentum $\Delta\mathbf{p}$ is transferred from the nuclear motion to the electron. In general, the larger the momentum difference, the smaller is the cross section. Since a transition normally also involves an energy difference $\Delta E$, also denoted as Q-value, a minimum momentum change is required to bring about the transition. This can be shown to be equal to [MeL58]

$$\Delta p_{\text{min}} = \frac{\Delta E}{v} , \tag{1.1}$$

where $v$ is the projectile velocity. The quantity $\Delta p_{\text{min}}$ plays an important part in all atomic-collision cross sections.

In order to implement the formalism, one generally must introduce a partial-wave expansion for the wave function of the relative nuclear motion. This is often impractical because a large number of partial waves is needed. Moreover, in a relativistic collision, the wave description of the nuclear motion requires the use of the Dirac equation which, however, is valid only for spin-$\frac{1}{2}$ particles. Since the nuclear spin usually is unimportant for atomic collisions, approximations must be introduced to render a quantum treatment manageable. For these reasons, the wave description is very rarely applied to relativistic atomic collisions.

*Classical description of the nuclear motion*

In energetic ion-atom collisions, because of the large masses of target and projectile nuclei, the relative momentum of the colliding atoms is very large compared with the average momentum of a bound electron. Therefore, the de Broglie wavelength, and hence the dimension of the wave packet associated with the relative momentum, is small compared to the Bohr radius of the electronic orbits. Furthermore, only a very small fraction $\Delta E$ of the initial kinetic energy of the relative motion is transferred to the atom in electronic transitions. This implies that, to a very good approximation, one can describe the trajectory of the relative motion by a predetermined elastic Rutherford trajectory, parametrized by the asymptotic relative velocity $\mathbf{v}$ and the impact parameter $\mathbf{b}$, see Fig. 3.1 for the general definition. For high projectile velocities and for impact parameters on the atomic scale (see Sec. 3.2.1) the deflection is negligible, so that the Rutherford trajectories may be replaced with straight-line trajectories. The approximations involved in assuming a predetermined classical nuclear trajectory entail that energy and momentum are not formally conserved within the impact parameter picture.

Of course, the motion of the electron in the field of the target nucleus and the moving projectile nucleus must be described quantum mechanically. The electrons then obey a time-dependent Schrödinger equation (for low nuclear charge numbers $Z$) or a time-dependent Dirac equation (for high-$Z$ nuclei).

Comparing both approaches, one observes that the relative nuclear motion in a quantum description is specified by energy and momenta (or angular momenta) and in a semiclassical description in space and time.

Both treatments yield identical results for quite general cases of ion-atom collisions. Within the approximations of the order of $m_e/M$, they are related to one another by Fourier transforms. The quantity $\Delta p$ defined in Eq. (1.1) occurs also in a formalism based on the impact parameter picture as a result of this Fourier transform. Usually, the latter method is mathematically more tractable and has a greater intuitive appeal. Therefore, it is used throughout this book.

## 1.2 Nonrelativistic versus relativistic collisions

Nonrelativistic collisions span a wide energy range, starting with chemical reactions at thermal energies ($\sim$ 0.1 eV) and reaching to tens of MeV/u in accelerator-produced collisions. The lower part of this range has no relativistic counterpart. These slow collisions are defined by the requirement that the speed of the colliding nuclei be small compared to the velocity of the electrons whose transitions are considered. If this is so, the electron wave functions can follow the nuclear motion and form quasi-molecular wave functions appropriate to the instantaneous (slowly time-dependent) internuclear separation $R(t)$. This approach is called the molecular-orbital model. It is not discussed any further.

*Coupled-channel calculations*

Although relativistic collisions are never truly slow, for high-$Z$ atoms the collision velocity can be comparable to the velocity of the innermost electrons (for a more accurate criterion, see Sec. 5.4.1). The corresponding situation occurs in the energy range of some keV/u for low-$Z$ atoms and can be treated successfully in nonrelativistic two-center coupled-channel calculations (see e.g. [FrL91]). A "channel" in this context specifies that the target as well as the projectile are in particular atomic eigenstates. In the SCA, the electron wave function (assuming *one* electron) obeys the time-dependent Schrödinger equation

$$\left(H - i\hbar\frac{\partial}{\partial t}\right)\Psi(\mathbf{r},t) = 0 \tag{1.2}$$

with

$$H = -\frac{\hbar^2}{2m_\mathrm{e}}\nabla^2 - \frac{Z_\mathrm{T}e^2}{r_\mathrm{T}} - \frac{Z_\mathrm{P}e^2}{r_\mathrm{P}}. \tag{1.3}$$

The charge numbers of target and projectile are $Z_\mathrm{T}$ and $Z_\mathrm{P}$, while $\mathbf{r}_\mathrm{T}$ and $\mathbf{r}_\mathrm{P}$ are the electron coordinates measured from the target nucleus and from the moving projectile nucleus, respectively. In many cases, as discussed above, it is justified to adopt a rectilinear trajectory $\mathbf{R}(t) = \mathbf{b} + \mathbf{v}t$ for the motion of the projectile nucleus with respect to the target nucleus, where $\mathbf{b}$ is the impact parameter (see Fig. 3.1 for a general trajectory) and $\mathbf{v}$ is the constant projectile velocity.

The salient point of the coupled-channel method is the expansion of the total wave function

$$\Psi(\mathbf{r}, t) = \sum_j a_j(t)\psi_j(\mathbf{r}_\mathrm{T}, t) + \sum_k a_k(t)\psi_k(\mathbf{r}_\mathrm{P}, t) \tag{1.4}$$

in terms of target wave functions

$$\psi_j(\mathbf{r}_\mathrm{T}, t) = \varphi_j(\mathbf{r}_\mathrm{T})\, e^{-i\epsilon_j t/\hbar} \tag{1.5}$$

and the projectile wave functions

$$\psi_k(\mathbf{r}_\mathrm{P}, t) = \varphi_k(\mathbf{r}_\mathrm{P})\, e^{-i\epsilon_k t/\hbar}\, e^{im_\mathrm{e}\mathbf{v}\cdot\mathbf{r}_\mathrm{T}/\hbar}\, e^{-i(m_\mathrm{e}/2)v^2 t/\hbar}. \tag{1.6}$$

Here, $\varphi_j$ and $\varphi_k$ are eigenfunctions of the target and the projectile Hamiltonian with eigenenergies $\epsilon_j$ and $\epsilon_k$, respectively. In Eq. (1.6), the last two exponentials constitute the translation factor in the laboratory system, which ensures the Galilean invariance of the moving projectile wave function [McC70, BrM92]. By inserting the expansion (1.4) into the Schrödinger equation (1.2), and by projecting upon the basis states $\psi_j$ and $\psi_k$, one obtains a set of linear first-order coupled differential equations for the time-dependent occupation amplitudes $a_j(t)$ and $a_k(t)$, which are subsequently solved numerically with the initial condition that for a given initial (target) state 0, the starting amplitude is $a_0(t \to -\infty) = 1$ while all other amplitudes at $t \to -\infty$ are zero. Once the coupled equations have been solved, the cross section for the particular transition $0 \to j, k$ is obtained by integrating the final impact-parameter dependent probability $|a_{j,k}(t \to \infty)|^2$ over the entire impact-parameter plane as

$$\sigma_{0\to j,k} = \int |a_{j,k}(t \to \infty)|^2\, d^2b, \tag{1.7}$$

where $d^2b = 2\pi b\, db$. Within the semiclassical approximation adopted, the only approximation consists in the unavoidable truncation of the basis sets in Eq. (1.4), dictated by computer limitations. If the final state is a bound or a continuum state $j$ of the target, Eq. (1.7) yields the excitation or ionization cross section, respectively. If the final state is a bound state $k$ of the projectile, the electron has made a transition from the target to the projectile, which means that $\sigma_{0\to k}$ is the cross section for charge transfer or capture. Owing to computer limitations, the method becomes impractical if the collision velocity exceeds considerably the velocity of the active electron.

For relativistic collisions, there are no structural changes in the method. The Schrödinger equation (1.2) is replaced by the Dirac equation, the wave functions by four-spinors, the moving Coulomb potential of the projectile by the appropriate Lorentz-transformed (i.e., Liénard-Wiechert) potential, and the requirement of Galilean invariance in Eq. (1.6) is replaced with the requirement of Lorentz invariance. From the technical point of view, all manipulations are a bit more complicated. A novel feature demanding attention is the existence of negative-energy electron states in a complete basis set.

*Perturbative treatment of excitation and ionization*

As mentioned above, for higher collision energy, the method of coupled channels becomes difficult to apply. On the other hand, owing to the shorter interaction time, the coupling among basis states becomes very weak. It is then justified to assume that the initial occupation amplitude remains $a_0(t) \simeq 1$ at all times and that all other amplitudes – except for the final state $j$ considered – remain identical to zero. The differential equation for $a_j(t)$ obtained by projecting on the states 0 and $j$ only can be solved exactly with the result

$$a_j(t \to \infty) = -\frac{i}{\hbar}\int_{-\infty}^{\infty} dt\, e^{-i(\epsilon_j-\epsilon_0)t/\hbar} \int \varphi_j^*(\mathbf{r}_\mathrm{T})\,\frac{Z_\mathrm{P} e^2}{r_\mathrm{P}}\,\varphi_0(\mathbf{r}_\mathrm{T})\, d^3r \qquad (1.8)$$

with $\mathbf{r}_\mathrm{P} = \mathbf{r}_\mathrm{T} - \mathbf{R}(t)$. Equation (1.8) represents the first-order Born approximation which may be derived independently of the coupled-channel approach [BrM92].

There are several ways to evaluate Eq. (1.8), the most common one employing a reformulation of the Coulomb interaction $Z_\mathrm{P}e^2/r_\mathrm{P}$ in terms of its Fourier transform [McC70, BrM92]. The cross section is again calculated by inserting Eq. (1.8) into Eq. (1.7). If $j$ describes a bound state, one obtains the excitation cross section, if $j$ is a continuum state, one gets the ionization cross section for the specified electron momentum $\mathbf{p}_j$.

The corresponding relativistic formulation is quite straightforward. Besides replacing the wave functions with the relevant four-spinors, one has to substitute the Liénard-Wiechert (i.e., Lorentz-transformed Coulomb) interaction in the matrix element (1.8).

It is worth mentioning that a new process, occurring only in relativistic collisions, can formally be described by the same amplitude. Namely, if the initial state 0 is a negative-energy solution of the Dirac equation and the final state $j$ is a (positive-energy) bound or continuum state, the process can be interpreted as electron-positron pair production, either with a bound or with a free electron.

*Perturbative treatment of charge transfer*

In a similar way as for excitation, a simple perturbative expression for charge transfer can be derived from the coupled-channel formalism. The only difference is that here the initial state 0 and the final state $k$ belong to different centers. Hence the initial and final states are not mutually orthogonal, so that a finite overlap matrix element results from the projection procedure following the insertion of the expansion (1.4) into Eq. (1.2). Usually, the overlap matrix element is neglected as being small compared to unity. In this way, or directly from the Born approximation, one derives the amplitude for charge transfer as

$$\begin{aligned} a_{0\to k}(t\to\infty) &= -\frac{i}{\hbar}\int_{-\infty}^{\infty} dt\, e^{-i(\epsilon_0-\epsilon_k)t/\hbar}\, e^{i(m_e/2)v^2t/\hbar} \\ &\quad \times \int \varphi_k^*(\mathbf{r}_\mathrm{P})\, e^{-im_e\mathbf{v}\cdot\mathbf{r}_\mathrm{T}/\hbar}\, \frac{Z_\mathrm{P}e^2}{r_\mathrm{P}}\, \varphi_0(\mathbf{r}_\mathrm{T})\, d^3r. \qquad (1.9) \end{aligned}$$

This is the charge exchange amplitude of the Oppenheimer-Brinkman-Kramers (OBK) approximation.

In contrast to Eq. (1.8), one observes the occurrence of the translation factor needed to describe the final state as a moving orbital.

Strictly speaking, it is not justified to consider initial and final states to be undistorted as in Eqs. (1.8) and (1.9), because the Coulomb potentials in Eq. (1.3) have an infinite range and hence impose a distortion upon the states (1.5) and (1.6), even at asymptotic separations. While this effect is innocuous for excitation and ionization, it is very important for charge transfer, nonrelativistically as well as relativistically (see Sec. 8.3).

The relativistic version of Eq. (1.9) is derived with the usual replacements for the wave functions and potentials and, in addition, by replacing the translation factor occurring in Eqs. (1.6) and (1.9) with the corresponding phase factor generating a Lorentz transformation instead of a Galilean transformation of the final state from the projectile to the target inertial frame.

## 1.3 Parametric dependencies of the cross sections

The preceding discussion points out the way in which a generalization of known nonrelativistic results leads to cross sections for relativistic atomic collisions. The details are the subject of this book, but we give here the rough dependence of reaction cross sections on the main parameters entering the picture, namely the collision energy $E$ (in MeV/u or GeV/u), and the projectile and target charge numbers $Z_{\mathrm{P}}$ and $Z_{\mathrm{T}}$, respectively.

In general, the parametric dependence is very complicated and can be obtained only by systematic numerical studies. Nevertheless, it is possible to extract the dominant behavior, in particular for the extreme relativistic energy regime with, say, $E > 10$ GeV/u.

It is often convenient to express the collision energy in terms of the Lorentz factor $\gamma = (1 - v^2/c^2)^{-\frac{1}{2}}$, which is the ratio of the total energy of the projectile to its rest energy $M_{\mathrm{P}}c^2$. Since the nucleon rest energy is about 0.93 GeV/u, the Lorentz factor $\gamma$ roughly equals the collision energy measured in GeV/u. In the following, we give the dominant dependencies on the nuclear charges and the Lorentz factor $\gamma$ for the main elementary processes. For the particular case of $\mathrm{U}^{92+}$ + U collisions the rough magnitude of these cross sections is displayed in Fig. 1.1.

*Ionization*

For the ionization of the target by the projectile, Eq. (1.8) provides the perturbative proportionality to the projectile charge as $Z_{\mathrm{P}}^2$ while a $Z_{\mathrm{T}}^{-2}$ target charge dependence arises from the scaling of the target dipole moment. A more detailed consideration (see Chap. 6) yields the $\gamma$ dependence, and one finds that the ionization cross section is composed of two parts. In the first part, the angular momentum projection $\mu_j$ does not change (or changes by more than one unit), in the second part $\mu_j$ changes by just one unit. They have the asymptotic behavior

$$\sigma_{\mathrm{fi}}^{\mathrm{ion}}(\mu_{j_i} \to \mu_{j_i}) \propto Z_{\mathrm{P}}^2 Z_{\mathrm{T}}^{-2} \tag{1.10}$$

and

$$\sigma_{\mathrm{fi}}^{\mathrm{ion}}(\mu_{j_i} \to \mu_{j_i} \pm 1) \propto Z_{\mathrm{P}}^2 Z_{\mathrm{T}}^{-2}\,(2\ln\gamma - 1), \tag{1.11}$$

where $i$ and $f$ identify the initial and final electron states, respectively. The expression (1.10) is mainly due to the Coulomb interaction while (1.11) is caused exclusively by the vector potential. The asymptotic increase of the cross section (1.11) as $(\ln\gamma)$ can be understood by the extension in the transverse direction of the Liénard-Wiechert potential shown in Fig. 3.3

exchange, a limited body of experimental data is also available, so that a comparison with theoretical predictions is possible.

Capture processes with the simultaneous emission of a photon are the subject of Chap. 9. After discussing the basic qualitative features in the case of nonrelativistic collisions, it is argued that presently the best description of radiative electron capture is provided by treating it essentially as the inverse of the photoelectric effect. Therefore, several ways, both approximate and rigorous, of obtaining photoelectric cross sections are discussed and subsequently used to derive the desired REC cross sections. Comparisons with experimental results are given.

Chapter 10 discusses the production of electron-positron pairs. This is a process that becomes particularly important at very high collision energies and has no nonrelativistic counterpart. Creation of free pairs is treated within perturbation theory and by adopting the equivalent-photon or Weizsäcker-Williams method. Bound-electron – free positron pair production is described in first-order perturbation theory and within the coupled-channel method. In some cases, a comparison with first experimental results is possible. A brief outlook is given on the problem of multiple-pair production which may become appreciable at the highest projected collision energies and the highest target and projectile charges. The creation of heavier leptons and exotic particles like magnetic monopoles, W-pairs, b-quarks, and the Higgs particle lie outside the scope of this book.

In Chaps. 11 – 13, the basic experimental methods are described. These chapters, respectively, present the three main components of any measurement with relativistic ion beams, namely the preparation of the projectile, arrangements of the target, and the use of detectors to determine yields or cross sections. The preparation of the projectile typically requires an appropriate stripper. Targets must be chosen according to the nature of the accelerator producing the projectiles and the desired outcome of the experiment. The detectors must be chosen taking into account the spatial and temporal structure of the beam and the parameters to be detected: charge states of the projectiles and target recoil ions, and the radiations (photons, electrons, positrons) from the colliding partners.

## 1.5 Comment on units

Different fields of physics have their own sets of practical units designed to avoid unessential complications in the formulation. In atomic physics, it is an almost universal custom to use atomic units, in which Planck's constant, the electron mass, and the electron charge are set equal to unity, i.e., $\hbar = m_{\mathrm{e}} = e = 1$. In relativistic physics, one uses relativistic or "natural units," in which Planck's constant, the electron mass, and the velocity of light are set equal to unity, i.e., $\hbar = m_{\mathrm{e}} = c = 1$.

Since this book treats atomic as well as relativistic physics, we adopt the following policy. Full units, in which all natural constants are retained, are kept as far as possible, in particular in the general outline of a theory and in structural discussions. This helps to identify various terms in a formulation and allows for a transition to any other set of units by equating the appropriate elementary constants to unity. However, in some explicit formulations when detailed analytical developments are needed, it would be awfully clumsy to keep all the constants. In such cases, we locally adopt atomic or relativistic units, always in accordance with the common usage in the particular context. In each case, we carefully point out which set of units is used. In an Appendix, we give the connections between the various units for the quantities of physical interest.

# Part I

# Theoretical Methods

# Chapter 2

# Relativistic Kinematics

It is the aim of this chapter to give a brief introduction to relativistic kinematics, which may serve both for later reference and for the planning and evaluation of experiments. We start by introducing Lorentz transformations and by discussing their properties to the extent that is needed for the present purpose. As an application, we derive relativistic one-body kinematics and then proceed to two-body systems. Here it turns out to be useful to express, as far as possible, measurable quantities like energies and angles in terms of Lorentz invariants. We give explicit formulas for the transformation of energies and momenta between two inertial frames, for example, the center-of-momentum and the laboratory system. Finally, the transformation of the differential cross sections for the emission of particles or photons is discussed. In each case, the general procedure is illustrated by examples.

## 2.1 Lorentz transformation

The fundamental experimental fact that stands at the beginning of the Theory of Special Relativity is the observation by Michelson and Morley that the speed of light $c$ is independent of the frame of reference, that is

$$c^2t^2 - \mathbf{x}^2 = c^2t'^2 - \mathbf{x}'^2 = 0. \tag{2.1}$$

Here, $(\mathbf{x}, t)$ and $(\mathbf{x}', t')$ are the space and time coordinates of a spherical shell of radiation observed in two different coordinate frames that move with a constant velocity $v$ with respect to each other and coincide at $t = t' = 0$. The transition between the two frames is mediated by a "Lorentz transformation." The simultaneous transformation of space and time coordinates

$\mathbf{x} = (x, y, z)$ and $t$, subject to Eq. (2.1), suggests that one combine time and space coordinates into a four-vector

$$\mathsf{x} = x^\mu = (x^0, x^1, x^2, x^3) = (ct, x, y, z) = (ct, \mathbf{x}). \tag{2.2}$$

Before discussing the transformation properties, let us establish the notation. In order to account for the minus sign in the invariant quadratic form (2.1), it is customary [Jac75, Sch61, Mes62, BjD64, BeL82] to introduce two different forms of a given four-vector $\mathsf{a}$, namely the *contravariant* form $a^\mu = (a^0, a^1, a^2, a^3)$ and the *covariant* form $a_\mu = (a_0, a_1, a_2, a_3)$. Both are related by $a_0 = a^0$, $a_\mu = -a^\mu$ for $\mu = 1, 2, 3$, so that

$$\begin{aligned} a^\mu &= (a^0, \mathbf{a}) \qquad \text{contravariant vector,} \\ a_\mu &= (a^0, -\mathbf{a}) \qquad \text{covariant vector.} \end{aligned} \tag{2.3}$$

One can rewrite this definition with the aid of a metric tensor $g_{\mu\nu}$ in the form

$$a_\mu = \sum_{\nu=0}^{3} g_{\mu\nu} a^\nu \equiv g_{\mu\nu} a^\nu, \tag{2.4}$$

where the quantities $g_{\mu\nu}$ are elements of the matrix

$$g_{\mu\nu} = g^{\mu\nu} = \begin{pmatrix} 1 & 0 & 0 & 0 \\ 0 & -1 & 0 & 0 \\ 0 & 0 & -1 & 0 \\ 0 & 0 & 0 & -1 \end{pmatrix}. \tag{2.5}$$

In Eq. (2.4) and in the following, we use the summation convention which defines that the occurrence of repeated Greek-letter indices implies a summation over the labels 0,1,2,3.

The metric of Eq. (2.5) is introduced so that the invariant quadratic form of Eq. (2.1) can be interpreted as a scalar product $\mathsf{x}\cdot\mathsf{x}$. More generally, invariant quantities can be obtained as scalar products of two four-vectors $\mathsf{a}$ and $\mathsf{b}$ defined as

$$\mathsf{a} \cdot \mathsf{b} = a_\mu b^\mu = a^\mu b_\mu = a^0 b^0 - \mathbf{a} \cdot \mathbf{b}. \tag{2.6}$$

Another convention [Gol81, Sak67, Eic90] to secure the negative sign in Eq. (2.6) is to use a formally Euclidean metric $g'_{\mu\nu} = \delta_{\mu\nu}$ with an imaginary fourth component $a_4 = ia_0$. It avoids distinction between covariant and contravariant four-vectors, so that $a_\mu = a^\mu = (\mathbf{a}, ia_0)$. While this convention is easier to handle for Special Relativity and is probably more transparent, it is regarded as old-fashioned, and we do not use it in this book.

For $\beta_{\rm cm}$ and $\gamma_{\rm cm}$, we may equally well substitute any values corresponding to an arbitrary constant velocity in the direction of the positive $z$-axis. The primed quantities then refer to this Lorentz system.

### 2.3.2 Transformation of the differential cross section

Total cross sections are invariant quantities, $\sigma' = \sigma$, because they can be obtained by a mere counting of events, an operation that must be the same in all Lorentz frames.

*Single-differential cross sections*

Differential cross sections are transformed from one Lorentz frame to another by requiring that the number of particles passing through a differential solid angle in one frame be equal to the number of particles passing through the *corresponding* differential solid angle in the other frame. Particle conservation then demands

$$\frac{d\sigma'(\theta',\varphi')}{d\Omega'}\,|d\Omega'| = \frac{d\sigma(\theta,\varphi)}{d\Omega}\,|d\Omega|, \tag{2.43}$$

where $d\sigma'(\theta',\varphi')/d\Omega'$ and $d\sigma(\theta,\varphi)/d\Omega$ are differential cross sections in the center-of-mass frame and in the laboratory system, respectively. Since the azimuthal angles $\varphi' = \varphi$ are unaffected by the Lorentz transformation, we have $|d\Omega'/d\Omega| = |d\cos\theta'/d\cos\theta|$ which can be calculated [Ded62] from Eq. (2.41) as

$$\left|\frac{d\Omega'}{d\Omega}\right| = \gamma_{\rm cm}^2\,\frac{[(\beta_{\rm cm}/\beta' + \cos\theta')^2 + \gamma_{\rm cm}^{-2}\sin^2\theta']^{3/2}}{|1 + \beta_{\rm cm}/\beta'\;\cos\theta'|} \tag{2.44}$$

so that

$$\frac{d\sigma(\theta,\varphi)}{d\Omega} = \left|\frac{d\Omega'}{d\Omega}\right|\,\frac{d\sigma'(\theta',\varphi')}{d\Omega'}. \tag{2.45}$$

In writing down these equations we have assumed that the quantities $\beta', \gamma'$, and $\theta'$ in the center-of-mass system (or any other natural coordinate frame) are independent variables. This is a convenient choice in many cases because most relations are simpler in the cm system than in the laboratory system and, furthermore, to each cm angle there corresponds only one laboratory angle which may be obtained from Eq. (2.41).

Thus, for calculating a differential cross section in the laboratory system, we first choose a fixed reduced velocity $\beta'$ of the particle in the cm system. Next we cover the angular range of $\theta'$ in the cm system by a sufficiently dense mesh of points, and for each point evaluate Eq. (2.44) and the known cm differential cross section. The resulting value for Eq. (2.45) is subsequently assigned to the laboratory angle obtained from Eq. (2.41). This transformation is also valid for any two Lorentz frames with a relative velocity of the primed with respect to the unprimed system in the direction of the positive $z$-axis.

*Double-differential cross sections*

Occasionally, one is interested in the double-differential cross section. In that case, the energy $E'$ is not taken as fixed in the cm system but instead, the energy dependence in the *laboratory system* is measured along with the angle. If $N(E, \cos\theta)$ is the number of particles in the energy interval between $E$ and $E+dE$ and in the angle interval $d\Omega = |d\varphi d\cos\theta|$ around $\theta$ and $\varphi$, and with corresponding quantities defined in the cm system, the conservation of particle number in a transformation between the two coordinate systems requires

$$N'(E', \cos\theta')\,|dE'\,d\Omega'| = N(E, \cos\theta)|\,dE\,d\Omega\,| \tag{2.46}$$

where energies and angles are related through the Lorentz transformation Eqs. (2.28) and (2.29). The energy-angle intervals are related by

$$dE'\,d\Omega' = \frac{\partial(E', \cos\theta')}{\partial(E, \cos\theta)}\,dE\,d\Omega, \tag{2.47}$$

where the Jacobian is given by

$$\begin{aligned}\frac{\partial(E', \cos\theta')}{\partial(E, \cos\theta)} &= \begin{vmatrix} \dfrac{\partial E'}{\partial E} & \dfrac{\partial \cos\theta'}{\partial E} \\[2ex] \dfrac{\partial E'}{\partial \cos\theta} & \dfrac{\partial \cos\theta'}{\partial \cos\theta} \end{vmatrix} \\ &= \frac{p}{p'} = \frac{\sin\theta'}{\sin\theta}\,. \end{aligned} \tag{2.48}$$

In the Jacobian, we use the usual convention that the variables $\cos\theta$ or $E$, with respect to which a derivative is *not* taken explicitly, are kept fixed. For the last equality, Eq. (2.28) has been used. Correspondingly, with $\Omega = (\theta, \varphi)$ and $\Omega' = (\theta', \varphi')$, the double-differential cross section is transformed as

$$\frac{1}{p}\,\frac{d^2\sigma(E, \Omega)}{dE\,d\Omega} = \frac{1}{p'}\,\frac{d^2\sigma'(E', \Omega')}{dE'\,d\Omega'} \tag{2.49}$$

and similarly,

$$\frac{E}{p^2}\frac{d^2\sigma(p,\Omega)}{dp\,d\Omega}=\frac{E'}{p'^2}\frac{d^2\sigma'(p',\Omega')}{dp'\,d\Omega'}\ . \tag{2.50}$$

Equations (2.49) and (2.50) are valid only if there is a one-to-one relationship between the variable pairs $(E,\cos\theta)$ and $(E',\cos\theta')$. If several pairs $(E',\cos\theta')$ contribute to a given set $(E,\cos\theta)$ then one has to add the contributions of these points on the left-hand side of Eq. (2.46). It can be shown [Ded62] that a multivalued relationship between $(E,\cos\theta)$ and $(E',\cos\theta')$ occurs if the Jacobian (2.48) either vanishes or is infinite. These points are called critical points.

The actual evaluation of Eqs. (2.49) and (2.50) requires the calculation of $\sigma'$ on a grid of values $E',\Omega'$ or $p',\Omega'$ and the subsequent assignment through Eq. (2.40) of values $E,\Omega$ or $p,\Omega$ to each pair of values in the primed system.

The equations given here are valid not only for the transformation between the cm system and the laboratory frame, but quite generally apply to a transformation between any two Lorentz frames. In this case, the parameters $\beta_{\rm cm}$ and $\gamma_{\rm cm}$ are replaced by the appropriate values relating the two inertial frames.

For example, if it is assumed in a relativistic ion-atom collision that electron-positron pairs are produced in the projectile system with a known cross section $\sigma'(E',\Omega')$, the resulting laboratory cross section may be directly obtained from Eqs. (2.49) and (2.50). The cm parameters $\beta_{\rm cm}$ and $\gamma_{\rm cm}$ are replaced with the corresponding parameters $\beta_P$ and $\gamma_P$ of the projectile.

## 2.4 Emission of radiation

We now consider the emission of radiation from a moving system characterized by the velocity $\mathbf{v}$, with the reduced velocity $\beta$ and the associated Lorentz factor $\gamma$. Since electrodynamics is Lorentz invariant, see Sec. 3.1, the phase $\omega t-\mathbf{k}\cdot\mathbf{r}$ of an electromagnetic wave is an invariant quantity, the wave vector $\mathbf{k}$ and the frequency $\omega$ form a four-vector $\mathsf{k}=(\omega/c,\mathbf{k})$ which transforms according to Eq. (2.13) and is characterized by the invariant scalar product $\mathsf{k}\cdot\mathsf{k}=0$.

### 2.4.1 Transformation of angles and frequencies

Let us again denote quantities referring to the cm system by primed symbols and quantities relating to the laboratory system by unprimed symbols. In analogy to Eq. (2.29) we get then the Doppler shift formulas

$$\begin{aligned} k' \sin\theta' &= k \sin\theta \\ k' \cos\theta' &= \gamma k(\cos\theta - \beta) \\ \omega' &= \gamma\omega(1 - \beta\cos\theta) \end{aligned} \tag{2.51}$$

where $\theta$ and $\theta'$ are the angles of $\mathbf{k}$ and $\mathbf{k}'$ with respect to the direction of the velocity $\mathbf{v}$. The angles are related by

$$\tan\theta' = \frac{\sin\theta}{\gamma(\cos\theta - \beta)} \tag{2.52}$$

as in Eq. (2.41) or, alternatively,

$$\cos\theta' = \frac{\cos\theta - \beta}{1 - \beta\cos\theta}. \tag{2.53}$$

Since for photons the velocity does not depend on the energy, the transformation of angles is independent of the transformation of frequencies. This is different from the corresponding relation (2.41).

The inverse equations are obtained by interchanging primed and unprimed quantities and reversing the sign of $\beta$. The elements of the solid angle entering into the differential cross section are related by

$$\frac{d\Omega'}{d\Omega} = \frac{1}{\gamma^2(1 - \beta\cos\theta)^2}. \tag{2.54}$$

The last equation in Eqs. (2.51) shows that in relativistic kinematics there is a *transverse* Doppler shift which, owing to the presence of the factor $\gamma$, leads to a frequency change even when $\theta = \pi/2$.

### 2.4.2 Transformation of the differential cross section

Suppose we know the differential cross section $d\sigma'(\theta', \varphi')/d\Omega'$ in a moving (primed) system at a *fixed frequency* $\omega'$. From Eqs. (2.45) and (2.54) we then obtain the cross section in the (unprimed) laboratory system in the form

$$\frac{d\sigma(\theta, \varphi)}{d\Omega} = \frac{1}{\gamma^2(1 - \beta\cos\theta)^2} \frac{d\sigma'(\theta', \varphi')}{d\Omega'}, \tag{2.55}$$

where $\varphi' = \varphi$ and $\theta'$ is expressed by $\theta$ using Eq. (2.53).

If we are interested in the transformation of the double differential cross section $d^2\sigma'(\omega', \Omega')/(d\omega'\, d\Omega')$ with $\Omega' = (\theta', \varphi')$, we have to evaluate the Jacobian for the transformation $\omega', \Omega' \to \omega, \Omega$. Owing to Eq. (2.53),

$d\cos\theta'/d\omega = 0$, the Jacobian corresponding to (2.48) factorizes and simply reads

$$\frac{\partial(\omega', \cos\theta')}{\partial(\omega, \cos\theta)} = \frac{d\omega'}{d\omega}\,\frac{d\cos\theta'}{d\cos\theta}\;. \tag{2.56}$$

With the aid of the last equation (2.51) and with Eq. (2.54), one derives the transformation formula

$$\frac{d^2\sigma(\omega, \Omega)}{d\omega\; d\Omega} = \frac{1}{\gamma(1-\beta\cos\theta)}\;\frac{d^2\sigma'(\omega', \Omega')}{d\omega'\; d\Omega'}\;. \tag{2.57}$$

An example is provided by the emission of primary bremsstrahlung in ion-atom collisions [ScD74, JaK75, YaI81]. In this process, a target electron is ionized by the projectile with the simultaneous emission of a photon. Viewed from the rest system of a relativistic projectile, a target electron moves with the velocity $-\mathbf{v}$ (aside from a velocity spreading due to the momentum distribution in the target bound state) towards the projectile and scatters off the projectile Coulomb field. At the same time it emits a photon with frequency $\omega'$ into the angular element $d\Omega'$. This is interpreted as the emission of bremsstrahlung in the projectile system. The cross section $\sigma'(\omega', \Omega')$ can be calculated by folding the momentum distribution of the bound target electron into the well-known bremsstrahlung cross section for the free electron (see, e.g., Heitler [Hei54]), where the energy relationship is contained in Eq. (2.51). The cross section for primary or quasi-free electron bremsstrahlung in the laboratory system then is directly calculated from Eq. (2.57).

# Chapter 3

# Fields of Moving Charges

In a relativistic atomic collision, both the projectile motion with respect to the target and the orbital motion of inner-shell electrons around a high-$Z$ target or projectile nucleus proceed with velocities that are comparable to the speed of light. To a good approximation, the nuclear motion can be described by classical mechanics while the relativistic motion of inner-shell electrons in high-$Z$ atoms or ions is described by the Dirac theory.

In this chapter, we confine ourselves to the treatment of relativistically moving classical charges. While in Chapter 2 we discuss the consequences of Lorentz transformations on energies and momenta, we here consider the collision in space and time and study the electromagnetic fields produced by the collision partners.

After briefly summarizing the basic equations of electrodynamics, we discuss the classical motion of nuclei under the mutual influence of their electromagnetic fields. We show that it is appropriate for relativistic atomic processes to assume that the projectile moves with a constant velocity along a straight-line trajectory. Coordinate systems attached to the target and to the projectile nucleus then are inertial frames, and the Lorentz transformation as well as the Lorentz-transformed Coulomb fields are expressed in a simple form. For large values of the Lorentz factor $\gamma$ [Eq. (2.14)], the field generated in the laboratory system by a moving charge is quite similar to a pulse of electromagnetic radiation. The replacement of these time-dependent fields by electromagnetic waves constitutes the basis of the equivalent-photon method, also known as Weizsäcker-Williams method, which is discussed first in a space-time and subsequently in a momentum-space description.

## 3.1 Basic equations of electrodynamics

For later reference and for establishing the notation, we summarize here some of the basic equations of electrodynamics adopting Gaussian units [Jac75]. In the Lorentz family of gauges, the wave equations for the vector potential $\mathbf{A}$ and the scalar potential $\Phi$ are

$$\begin{aligned} \frac{1}{c^2}\frac{\partial^2 \mathbf{A}}{\partial t^2} - \nabla^2 \mathbf{A} &= \frac{4\pi}{c}\mathbf{j} \\ \frac{1}{c^2}\frac{\partial^2 \Phi}{\partial t^2} - \nabla^2 \Phi &= 4\pi\rho, \end{aligned} \tag{3.1}$$

where the current density $\mathbf{j}(\mathbf{r},t)$ and the charge density $\rho(\mathbf{r},t)$ satisfy the continuity equation

$$\frac{\partial \rho}{\partial t} + \nabla \cdot \mathbf{j} = 0 \tag{3.2}$$

and the potentials are subject to the Lorentz condition

$$\frac{1}{c}\frac{\partial \Phi}{\partial t} + \nabla \cdot \mathbf{A} = 0\,. \tag{3.3}$$

The electric field $\mathbf{E}$ and the magnetic field $\mathbf{B}$ are expressed in terms of the potentials $\mathbf{A}$ and $\Phi$ as

$$\begin{aligned} \mathbf{E} &= -\frac{1}{c}\frac{\partial \mathbf{A}}{\partial t} - \nabla\Phi \\ \mathbf{B} &= \nabla \times \mathbf{A}\,. \end{aligned} \tag{3.4}$$

If we now introduce the four-potential $A^\mu = (A^0, \mathbf{A}) = (\Phi, \mathbf{A})$, the four-current $j^\mu = (c\rho, \mathbf{j})$, and the four-derivative $\partial^\mu \equiv \partial/\partial x_\mu = (\partial/\partial(ct), -\nabla)$, see Eq. (2.9), we can write Eqs. (3.1) in the covariant form

$$\Box\, A^\mu = \frac{4\pi}{c}\, j^\mu, \tag{3.5}$$

where the d'Alembertian $\Box = \partial_\mu\partial^\mu \equiv \sum_\mu \partial_\mu\partial^\mu$. Furthermore, the continuity equation, Eq. (3.2), is rewritten as

$$\partial_\mu j^\mu = 0\,, \tag{3.6}$$

where we have again adopted the convention to sum over subscripts occurring twice. Similarly, the Lorentz condition, Eq. (3.3), becomes

$$\partial_\mu A^\mu = 0\,. \tag{3.7}$$

The electric and magnetic fields (3.4) are obtained as the components of the second-rank, antisymmetric field-strength tensor

$$F^{\mu\nu} = \partial^\mu A^\nu - \partial^\nu A^\mu = \begin{pmatrix} 0 & -E_x & -E_y & -E_z \\ E_x & 0 & -B_z & B_y \\ E_y & B_z & 0 & -B_x \\ E_z & -B_y & B_x & 0 \end{pmatrix} . \tag{3.8}$$

In terms of the field-strength tensor, the Maxwell equations can be written in covariant form as

$$\partial_\nu F^{\nu\mu} = \frac{4\pi}{c} j^\mu \tag{3.9}$$

and

$$\partial^\lambda F^{\mu\nu} + \partial^\mu F^{\nu\lambda} + \partial^\nu F^{\lambda\mu} = 0, \tag{3.10}$$

where $\lambda, \mu, \nu$ are any three of the integers 0, 1, 2, 3.

## 3.2 Motion of interacting point charges

The prototype system for an atomic collision is composed of three bodies: two nuclei and one electron. Since the electron is so much lighter than the nuclei, its influence on the nuclear motion can usually be ignored. As a first step, we therefore consider two point charges, the target nucleus with charge number $Z_\mathrm{T}$ and mass $M_\mathrm{T}$ and the projectile nucleus with charge number $Z_\mathrm{P}$ and mass $M_\mathrm{P}$. In Sec. 3.2.1 we assume that the motion of the nuclei can be described by classical mechanics and electrodynamics, and in Sec. 3.2.2 we discuss the limitations imposed by quantal effects. The connection with a full quantum treatment is presented in Sec. 5.2.

### 3.2.1 Particle scattering in classical mechanics

The scattering of classical particles by one another is described in terms of classical trajectories. We have in mind two particles in their center-of-mass coordinate system with an interaction force between them defined by a potential $V$. It should be remarked that relativistic potential scattering in general is a rather artificial construction, because static potentials are not Lorentz invariant. However, since electrodynamics is a relativistically invariant theory, it is possible to treat electromagnetic forces in a physically consistent manner. This holds precisely for ion-atom collisions in which we are interested.

For describing a classical collision of a particle with a fixed target, it is useful to introduce the concept of the impact parameter and of the deflection angle. During the collision, the projectile generally follows a curved

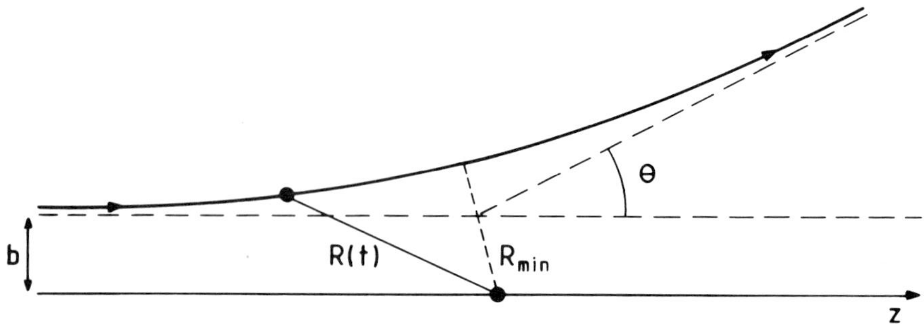


**Figure 3.1.** Classical trajectory of a particle in the laboratory system.

trajectory with an incoming and an outgoing branch. The asymptote to the incoming branch is parallel to the beam direction while the asymptote to the outgoing branch defines the deflection angle $\Theta$ with respect to the beam direction. The distance of closest approach of the incoming asymptote from the scattering center is denoted as the impact parameter $b$. Figure 3.1 illustrates a classical trajectory for a repulsive Coulomb potential.

For a general potential, with attractive and repulsive parts, the deflection angle $\Theta$ (also denoted as "deflection function") may have any value between $-\infty$ and $\pi$, where the negative values occur when the trajectory is bent around the scattering center. On the other hand, the observed "scattering angle" $\theta$, by definition, lies always between zero and $\pi$. The two angles are connected by

$$\theta = \pm(\Theta + 2\pi m), \tag{3.11}$$

where $m$ is a positive integer or zero. For a given deflection angle $\Theta$, the values are so chosen that the scattering angle $\theta$ lies between zero and $\pi$.

For a given collision energy $E$ and a given impact parameter $b$, the single-valued function $\Theta(b)$ is calculated from the classical equation of motion, see, e.g., [McC70, New66]. If one has a beam of particles (of equal energy) uniformly distributed over all impact parameters and all azimuthal angles, the differential cross section may be calculated as

$$\frac{d\sigma}{d\Omega} = \frac{b}{\sin\theta\,|d\Theta/db|}, \tag{3.12}$$

where $b(\Theta)$ is obtained by inverting the originally calculated function $\Theta(b)$.

*Collisions between two bare nuclei*

The results obtained for potential scattering can be applied to the relative motion of two particles interacting via a potential. For the purely repulsive interaction between two positive point charges in which we are interested, deflection angle $\Theta$ and scattering angle $\theta$ are identical and, within classical mechanics, there is a one-to-one relationship between the impact parameter and the scattering angle. Small impact parameters correspond to large scattering angles and vice versa. However, as is known from Rutherford's classic scattering experiment for projectile energies as low as about 1 MeV/u, large deflection angles of the order of $\pi/2$ correspond to impact parameters of the order a few fm ($= 10^{-13}$cm). On the other hand, the length scale for atomic processes is given by the atomic K-shell radius $a_{\mathrm{K}} = Z^{-1}a_0 = Z^{-1} \times 0.53 \times 10^{-8}$cm where Z is the nuclear charge number involved. Since the contribution to the cross section of an impact parameter ranges between $b$ and $b + db$ is weighted with $2\pi\, b\, db$, we expect that impact parameters of the order of a few fm which are associated with large deflection angles give a negligible contribution. Moreover, for atomic collision studies, these small impact parameters have to be avoided because the background from nuclear reactions would tend to mask atomic processes. Therefore, aside from deviations discussed in Sec.3.4.2, it is perfectly justified for most purposes to disregard the finite nuclear size and to substitute a positive point charge for the nucleus.

For the impact parameters relevant to most atomic processes, this means that the target nucleus effectively stays at rest in the laboratory system throughout the collision, thereby producing a static Coulomb potential. It is then convenient to express all quantities in the laboratory (target) system. Following Sommerfeld's classic treatment [Som31] of the relativistic Kepler orbit, one may derive [MaS87, New66] the deflection of a projectile of mass $M_{\mathrm{P}}$ and kinetic energy $T_{\mathrm{P}}$ by a space-fixed charge $Z_{\mathrm{P}}e$ (i.e., with an infinite target mass) as a function of the impact parameter $b$. Alternatively, one may consider two particles in their center-of-mass system. The symbol $M_{\mathrm{P}}$ then has the meaning of the reduced mass of the two particles, and the deflection angle refers to the cm frame.

It is often convenient to express the Lorentz factor $\gamma$ directly by the kinetic energy of the projectile as

$$\gamma = \frac{1}{\sqrt{1-\beta^2}} = 1 + \frac{T_{\mathrm{P}}\,[\mathrm{MeV/u}]}{931.494} \tag{3.13}$$

and to write the deflection angle as a function of the ratio

$$x = \frac{Z_{\mathrm{P}} Z_{\mathrm{T}} e^2/b}{M_{\mathrm{P}} c^2} = \frac{e^2}{\hbar c}\, Z_{\mathrm{P}} Z_{\mathrm{T}}\, \frac{\lambda\!\!\!^{-}_{\mathrm{c}}}{b} \tag{3.14}$$

between the Coulomb repulsion energy of the nuclei at the separation $b$ and the rest energy of the projectile. In the second equation, $\lambda\!\!\!^{-}_{\rm c} = \hbar/M_{\rm P}c$ is the Compton wavelength of the projectile (it is more convenient this expression, rather than $h/M_{\rm P}c$). Using the abbreviation

$$\eta = \sqrt{1 - \frac{x^2}{\gamma^2 - 1}}, \tag{3.15}$$

one obtains the deflection angle $\Theta$ as a function of the parameters $\gamma$ and $x$ as

$$\Theta = \theta = \pi - \frac{2}{\eta} \arctan \frac{(\gamma^2 - 1)\eta}{\gamma x}, \tag{3.16}$$

or for $x \ll 1$ and $\gamma \gg 1$

$$\theta \approx 2\,\frac{x}{\gamma}. \tag{3.17}$$

Let us consider U + U collisions as an example. The Compton wavelength (in the cm frame) for this system is $\lambda\!\!\!^{-}_{\rm c} = 1.78 \times 10^{-3}$ fm. From Eqs. (3.14) and (3.17) we then obtain the cm deflection angle

$$\theta = \frac{0.220}{\gamma\, b[\mathrm{fm}]} = \frac{3.83 \times 10^{-4}}{\gamma(b/a_{\rm K})}, \tag{3.18}$$

where in the first equation, $b$ is measured in fm, while in the second equation it is measured in units of the uranium K-shell radius $a_{\rm K}$. It is seen that down to impact parameters $b$ very small on the atomic scale, the Coulomb deflection is negligible for relativistic collisions. Even for $b = 15$ fm, when the uranium nuclei begin to touch, and for a kinetic energy of 1 GeV/u in the cm system, the more accurate estimate using Eq. (3.16) yields a deflection of only 9 mrad.

*Effects of retardation and of magnetic fields*

The estimates given above are based on potential scattering with a pure Coulomb repulsion, that is, effects of retardation and of magnetic fields are disregarded. Let us sketch here a more complete treatment [MaS87, MaF90]. Describing the collision in the laboratory system, the target nucleus experiences the Liénard-Wiechert potential produced by the projectile moving along its trajectory $\mathbf{x}_{\rm P} = \mathbf{x}_{\rm P}(t)$. At the position $\mathbf{x}_{\rm T}$ of the target nucleus, this potential has the form [Jac75]

$$\begin{aligned} \Phi_{\rm P}(\mathbf{x}_{\rm T}, t) &= \left(\frac{Z_{\rm P} e}{(1 - \boldsymbol{\beta}_{\rm P} \cdot \mathbf{n})R}\right)_{\rm ret} \\ \mathbf{A}_{\rm P}(\mathbf{x}_{\rm T}, t) &= \left(\frac{Z_{\rm P} e \boldsymbol{\beta}_{\rm P}}{(1 - \boldsymbol{\beta}_{\rm P} \cdot \mathbf{n})R}\right)_{\rm ret}. \end{aligned} \tag{3.19}$$

Here $\mathbf{R} = \mathbf{x}_\mathrm{T}(t) - \mathbf{x}_\mathrm{P}(t_\mathrm{ret})$ is the separation of the target nucleus from the projectile at the retarded time $t_\mathrm{ret} = t - R/c$, $\mathbf{n}$ is the unit vector in the direction of $\mathbf{R}$, and $\beta_\mathrm{P} = v_\mathrm{P}/c$ is the reduced projectile velocity at the retarded time. The electric and magnetic fields resulting from the potentials (3.19) determine the target motion. The target nucleus is initially (at $t = -\infty$) at rest and starts moving as the repulsion from the projectile becomes effective. Conversely, the recoiling target nucleus at the position $\mathbf{x}_\mathrm{T}$ is the source of Liénard-Wiechert potentials $\Phi_\mathrm{T}$ and $\mathbf{A}_\mathrm{T}$ acting on the moving projectile. In this way, the mutual projectile-target interaction leads to a set of eight coupled linear differential equations [MaS87, MaF90] for the nuclear motion which may be solved numerically. A rigorous treatment of this kind takes into account the effects of retardation and magnetic fields caused by the recoiling target.

As one might expect, it turns out that magnetic field corrections to the cross section for elastic scattering between nuclei are small (<3.5%) in a particular case considered (Xe + U at $\gamma \simeq 7$) even for a close collision in which the nuclei almost touch. This is so because the recoiling target nucleus does not acquire high enough speed to generate a sizable magnetic field and because the space-time interval in which deflection occurs is confined to such small internuclear separations that retardation effects on the projectile motion are negligible. It can also be shown that effects of the nuclear magnetic moments which might lead to spin-flip processes of the nuclei do not play any role because the nuclear magneton is by three orders of magnitude smaller than the Bohr magneton.

### 3.2.2 Quantal effects and the impact parameter picture

The conditions for the applicability of (nonrelativistic) classical physics to the scattering of a charged particle in a Coulomb field are analyzed by Niels Bohr in a classic paper [Boh48] and are further discussed in [AlW75, HaM70, Han90]. Limitations to the classical description are imposed by the Heisenberg uncertainty relation which restricts the one-to-one correspondence between a measured scattering angle $\theta$ and the impact parameter $b$. For relativistic Coulomb scattering, the argument is essentially the same. Starting from the relation (3.17) between $\theta$ and $b$, valid for small scattering angles, we have

$$\theta = \frac{2}{b}\frac{Z_\mathrm{P} Z_\mathrm{T} e^2}{\gamma M_\mathrm{P} c^2}, \tag{3.20}$$

which tells us that the relative uncertainties in scattering angle and impact parameter are connected by $|\Delta\theta/\theta| = |\Delta b/b|$. When $\Delta\theta$ is rewritten in terms of the uncertainty $\Delta p_\perp$ of the transverse momentum with respect to the longitudinal momentum $p_z$, the uncertainty relation $\Delta p_\perp \Delta b \geq \hbar$ yields

$$\Delta\theta = \frac{\Delta p_\perp}{p_z} \geq \frac{1}{p_z}\frac{\hbar}{\Delta b} = \frac{\hbar}{\gamma M_{\rm P} c\,\Delta b} \tag{3.21}$$

whence

$$\frac{\Delta b}{b} \geq \frac{1}{\Delta b}\frac{\hbar}{\gamma M_{\rm P} c\,\theta} .$$

Inserting the angle $\theta$ from Eq. (3.20) we have

$$\left(\frac{\Delta b}{b}\right)^2 \geq \frac{\hbar c}{2 Z_{\rm P} Z_{\rm T} e^2} . \tag{3.22}$$

In order to obtain sufficient precision in $b$, the right-hand side of Eq. (3.22) must be sufficiently small compared to unity. This requirement is conveniently expressed with the aid of the Sommerfeld parameter

$$\nu_c = \frac{Z_{\rm P} Z_{\rm T} e^2}{\hbar c} = \frac{Z_{\rm P} Z_{\rm T}}{137} \tag{3.23}$$

with the projectile velocity replaced by the speed of light. In addition to taking into account the limitations imposed by Eq. (3.22), we have to require that the energy loss $\Delta E$ of the particles during the ion-atom collision is negligible. In summary, the conditions for the validity of the concept of classical trajectories are

$$\begin{aligned} 2\nu_c &\gg 1 \\ \frac{\Delta E}{E} &\ll 1. \end{aligned} \tag{3.24}$$

Only if these conditions are satisfied is it possible to assign an impact parameter $b$ to a given scattering angle $\theta$. Hence the impact parameter in relativistic collisions cannot be determined unless the product of nuclear charge numbers $Z_{\rm P} Z_{\rm T} \gg 70$.

It should also be mentioned that within relativistic quantum theory, the least possible error in the measurement of the space coordinates of a point particle in its rest frame is given by the Compton wavelength $\lambda\!\!\!^{-}_c = \hbar/M_{\rm P}c$ of the particle. Since for a proton $\lambda\!\!\!^{-}_c = 2.1 \times 10^{-14}$ cm and $\lambda\!\!\!^{-}_c$ is still smaller for a heavy nucleus, this uncertainty does not place any serious limitation upon the definition of classical trajectories for heavy ions.

We see that in many relevant situations, the impact parameter is not directly accessible to measurement. However, this is not a reason to discard the concept of classical trajectories. It is known from nonrelativistic ion-atom collisions [McC70, Han90] that a semiclassical approximation, in which the projectile motion is treated classically but the electron motion quantum mechanically (see Secs. 1.1 and 5.2), is an extremely powerful method even when $\nu = Z_P Z_T e^2/\hbar v < 1$. The utility of this concept rests on the Born-Oppenheimer separation which allows one to separate consideration of electronic states from consideration of the nuclei of the atomic systems taking part in the atomic collision. Since the nuclear masses $M_P$ and $M_T$ are large compared to the electronic mass $m_e$, the energy transferred to the internal degrees of freedom of the atoms is much smaller than the initial kinetic energy of the relative motion of target and projectile. Consequently, the relative velocity practically does not change in the collision and scattering is essentially classical.

The separation of nuclear and electronic motion means that the electron may be regarded to move in a time-dependent potential, the time dependence arising from the motion of the nuclei (see Sec. 5.1). The method is generally called the "impact parameter treatment" or the "impact parameter picture" because the impact parameter **b** serves to specify the trajectory even though the latter may not be directly measurable. Impact parameter methods have a strong appeal in aiding the physical intuition. Usually, calculations based on this method are mathematically more tractable than fully quantum mechanical treatments. Both methods yield identical results for quite general cases of ion-atom collisions.

In the impact parameter picture, the classical one-to-one relation between scattering angle $\theta$ and impact parameter $b$ is replaced with an integral (see Sec. 5.4.2) over the relevant amplitudes in an atomic process. That is, the scattering amplitude $f_{fi}(\theta)$, Eq. (5.85), between initial and final states $i, f$ for a given angle $\theta$ receives contributions from the transition amplitudes $A_{fi}(b)$ for all impact parameters. Conversely, the impact parameter dependence of $A_{fi}(b)$ can be reconstructed, if the scattering amplitudes $f_{fi}(\theta)$, including the phases, are known for all angles.

Since the impact parameter method is intimately connected with the electronic motion, we defer a discussion to Sec. 5.1. Here, it may suffice to say that in conjunction with the impact parameter picture, a classical treatment of the nuclear motion is appropriate for relativistic atomic collisions up to a high level of accuracy.

## 3.3 Liénard-Wiechert potentials

As a consequence of the estimates given in Sec. 3.2, we assume in the following that the projectile moves along a classical rectilinear trajectory

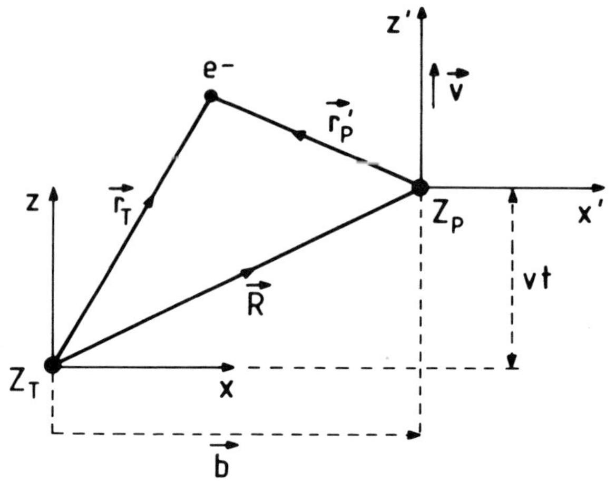


**Figure 3.2.** Coordinate systems for a relativistic collision between two atoms. The position of the target nucleus (charge $Z_{\mathrm{T}}$) is chosen as the origin of the coordinates $x, y, z$. The projectile nucleus (charge $Z_{\mathrm{P}}$) moves with the constant velocity $\mathbf{v}$ parallel to the $z$-axis along a trajectory that is displaced from the target by the impact parameter $\mathbf{b}$ along the $x$-axis. The projectile is located at the origin of the moving inertial frame with the coordinates $x', y', z'$. The electron $e^-$ has the coordinate $\mathbf{r}_{\mathrm{T}}$ with respect to the target frame and $\mathbf{r}'_{\mathrm{P}}$ with respect to the projectile frame. Owing to their definition in different inertial frames associated with different times, the vectors $\mathbf{R}$ (defined in the target frame) and $\mathbf{r}'_{\mathrm{P}}$ (defined in the projectile frame) *do not* add vectorially to $\mathbf{r}_{\mathrm{T}}$ as might be suggested by the figure.

which in the laboratory system is given by

$$\mathbf{R} = \mathbf{b} + \mathbf{v}t, \tag{3.25}$$

where $\mathbf{b}$ is the impact parameter vector and $\mathbf{v}$ is a constant velocity. This treatment of the projectile motion implies an important simplification for heavy-ion collisions with an energy exceeding a few MeV/u. When defining the coordinate systems, it is convenient to place the target nucleus at the origin of the laboratory system with the $x$ and $z$ axes taken in the directions of $\mathbf{b}$ and $\mathbf{v}$, respectively. Figure 3.2 illustrates the geometry. The inertial frame, in which the projectile nucleus is at rest, moves with the speed $v$ with respect to the laboratory frame, and its $x', y', z'$ axes are parallel to the $x, y, z$ laboratory axes, respectively. For the time coordinates, we choose the convention that the times $t_0, t'_0$ associated with the nuclear positions are both zero when the projectile reaches its closest approach to the target nucleus during the collision.

Let

$$\mathsf{x} = x^\mu = (x^0, x^1, x^2, x^3) = (ct, x, y, z) \quad \text{or} \quad (ct, \mathbf{r}) \tag{3.26}$$

and

$$\mathsf{x}' = x'^\mu = (x'^0, x'^1, x'^2, x'^3) = (ct', x', y', z') \quad \text{or} \quad (ct', \mathbf{r}') \tag{3.27}$$

denote the space-time coordinates of a point with respect to the two Lorentz frames. In analogy to Eq. (2.10), these coordinates are related by the inhomogeneous Lorentz transformation

$$x'^\mu = \Lambda^\mu{}_\nu \, x^\nu - \delta_{1\mu} \, b \,, \tag{3.28}$$

where we adopt the convention to sum over subscripts occurring twice and where the $\Lambda^\mu{}_\nu$ satisfy the orthogonality relation (2.11). Defining the unit vector $\hat{\mathbf{v}} = \mathbf{v}/v$, Eq. (3.28) can also be written in vector form as

$$\begin{aligned} \mathbf{r}' &= \mathbf{r} + (\gamma - 1)(\mathbf{r} \cdot \hat{\mathbf{v}})\hat{\mathbf{v}} - \gamma \mathbf{v} t - \mathbf{b} \\ t' &= \gamma(t - \mathbf{r} \cdot \mathbf{v}/c^2), \end{aligned} \tag{3.29}$$

or, for the inverse transformation,

$$\begin{aligned} \mathbf{r} &= \mathbf{r}' + (\gamma - 1)(\mathbf{r}' \cdot \hat{\mathbf{v}})\hat{\mathbf{v}} + \gamma \mathbf{v} t' + \mathbf{b} \\ t &= \gamma(t' + \mathbf{r}' \cdot \mathbf{v}/c^2) \,. \end{aligned} \tag{3.30}$$

Choosing the $z$-axis in the direction of $\mathbf{v}$, we have explicitly

$$\begin{pmatrix} ct' \\ x' \\ y' \\ z' \end{pmatrix} = \begin{pmatrix} \gamma & 0 & 0 & -\beta\gamma \\ 0 & 1 & 0 & 0 \\ 0 & 0 & 1 & 0 \\ -\beta\gamma & 0 & 0 & \gamma \end{pmatrix} \begin{pmatrix} ct \\ x \\ y \\ z \end{pmatrix} - \begin{pmatrix} 0 \\ b \\ 0 \\ 0 \end{pmatrix}, \tag{3.31}$$

and the distance of a point $(x, y, z, t)$ from the origin of the moving frame, measured in the moving frame, is

$$r' = \sqrt{(x-b)^2 + y^2 + \gamma^2 (z - vt)^2}. \tag{3.32}$$

The same transformations allow us to construct the four-potential generated by a uniformly moving charge.[1] In the projectile frame, the projectile charge $Z_{\mathrm{P}}e$ gives rise to the electrostatic potentials

$$\Phi'(\mathbf{r}', t') = \frac{Z_{\mathrm{P}} e}{r'}, \qquad \mathbf{A}'(\mathbf{r}', t') = 0. \tag{3.33}$$

[1] We always take the charge $e > 0$, in agreement with [Jac75, Ros61] and opposite to [BjD64] and [Sak67].

By the defining relation (2.7) of a four-vector, the corresponding four-potentials $\Phi$ and $\mathbf{A}$ in the target frame are obtained by the homogeneous part of the same Lorentz transformation (3.30) which carries the four-coordinates $t'$ and $\mathbf{r}'$ into $t$ and $\mathbf{r}$. By simple substitution, we derive the Liénard-Wiechert potentials produced by the projectile in the target frame in the forms

$$\begin{aligned} \Phi(\mathbf{r},t) &= \frac{\gamma Z_{\mathrm{P}} e}{\sqrt{(x-b)^2+y^2+\gamma^2(z-vt)^2}} \\ \mathbf{A}(\mathbf{r},t) &= \frac{\mathbf{v}}{c}\,\Phi(\mathbf{r},t) \end{aligned} \tag{3.34}$$

which satisfy the Lorentz gauge condition (3.3). In particular, at the origin of the target frame, the electric and magnetic fields due to the moving projectile are given by

$$\begin{aligned} E_x &= \frac{Z_{\mathrm{P}} e\,\gamma b}{(b^2+\gamma^2 v^2 t^2)^{3/2}} \\ E_y &= 0 \\ E_z &= -\frac{Z_{\mathrm{P}} e\,\gamma v t}{(b^2+\gamma^2 v^2 t^2)^{3/2}} \end{aligned} \tag{3.35}$$

and

$$\begin{aligned} B_x &= 0 \\ B_y &= \beta\, E_x \\ B_z &= 0, \end{aligned} \tag{3.36}$$

respectively. We see that the peak transverse electric field $E_x$ is increased by a factor of $\gamma$ with respect to its nonrelativistic value while the duration

$$\Delta t \simeq b/(\gamma v) \tag{3.37}$$

of appreciable field strength at the target nucleus is decreased by the same amount. We note that the ratio $E_x/E_z = -b/(vt)$ is just the tangent of the angle $\theta = \arccos(-\hat{\mathbf{R}}\cdot\hat{\mathbf{v}})$ formed between the vectors $-\mathbf{R}$ and the $z$-axis (Fig. 3.1). Hence the electric field produced by the charge $Z_{\mathrm{P}}e$ at the position of the target nucleus is directed radially from the projectile's present (i.e., not retarded) position to the observation point at the target nucleus. Writing $b = R\sin\theta$ and $vt = R\cos\theta$, we obtain the electric field

$$\mathbf{E} = \frac{-Z_{\mathrm{P}} e\,\mathbf{R}}{\gamma^2 R^3 (1-\beta^2\sin^2\theta)^{3/2}}. \tag{3.38}$$

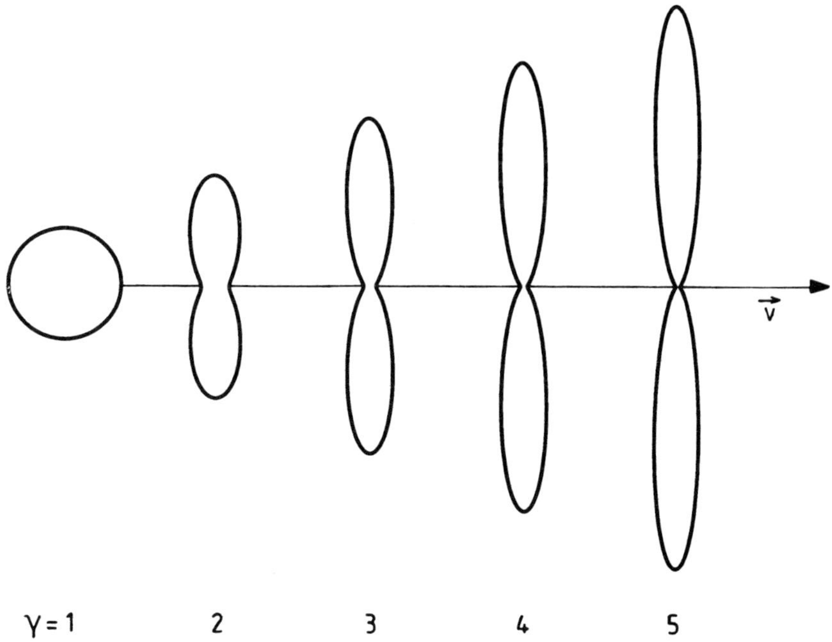


**Figure 3.3.** Polar diagrams for the angular dependence of the radial electric field strength produced by a point charge moving with the velocity **v** to the right. The numbers give the Lorentz factors $\gamma$.

The angular variation of the radial electric field strength is illustrated in Fig. 3.3 for various projectile velocities in terms of the Lorentz factor $\gamma$. Along the direction of motion, $\theta = 0$ or $\theta = \pi$, the field strength is decreased by a factor of $\gamma^{-2}$ as compared to a charge at rest. On the other hand, perpendicular to the trajectory, $\theta = \pi/2$, the field is increased by a factor of $\gamma$. The flattening of the surfaces into disk shapes is essentially an effect of the Lorentz contraction of the electromagnetic fields.

## 3.4 Equivalent-photon method

The electric field produced by a moving charge becomes almost transverse if $\gamma \gg 1$, see Eqs. (3.35) and (3.36), and is accompanied by a transverse magnetic field perpendicular to it and of almost equal strength. These electric and magnetic fields can be replaced approximately with the fields of a pulse of plane linearly polarized radiation propagating in the $z$-direction.

This replacement forms the basis of the equivalent-photon method which is originally due to Fermi [Fer24] and was developed by von Weizsäcker [Wei34] and Williams [Wil34, Wil35]. The method can be applied to a variety of electromagnetic processes induced by charged projectiles in nuclei and atoms, such as excitation of atoms and nuclei, ionization of atoms, emission of bremsstrahlung, and the creation of lepton pairs. These and other applications have been reviewed by Bertulani and Baur [BeB88]. The Weizsäcker-Williams or equivalent-photon method is valuable at extremely high energies in situations where rigorous methods are difficult to implement. It is then possible to use the known cross sections for photon-induced reactions to estimate the corresponding cross sections induced by moving charges, see Sec.10.2.3.

In Sec. 3.4.1 we follow the intuitively appealing space-time description of von Weizsäcker and Williams in order to explicitly exhibit the dependence of the fields and of the frequency spectra on the impact parameter $b$. In Sec. 3.4.2 we turn instead to the shorter but more formal derivation of the wave-number description which leads to essentially the same final result for the photon spectrum. The latter has the advantage that finite-size effects of the charge distribution can be accounted for by introducing a form factor in a natural way. The impact-parameter dependence of the frequency spectra can then be derived for any given form factor.

### 3.4.1 Formulation in the space representation

*Basic assumptions*

The main idea behind the equivalent-photon method is to replace the transient fields (3.35) and (3.36) by pulses of linearly polarized radiation. The field produced at a point $X$ by the moving particle $X'$ is indicated in Fig. 3.4. As has been mentioned above, the time interval for which the fields are appreciable are of the order $\Delta t \simeq b/\gamma v$. During this time, $|E_z| = |E_x|\,vt/b \simeq |E_x|/\gamma$.

Imagine now a pulse $P_1$ of plane radiation traveling parallel to $\mathbf{v}$ with an electric field $\mathbf{E}_x(t)$ and a magnetic field $\mathbf{B}_y(t)$ with $|B_y| = |E_x|$ at $X$ and another pulse $P_2$ traveling perpendicular to $\mathbf{v}$ and $\mathbf{E}_x(t)$ (that is in the positive or negative $y$-direction) with an electric field equal to $\mathbf{E}_z(t)$ and a magnetic field $\mathbf{B}_x(t)$ with $|B_x| = |E_z|$. These radiation pulses are constructed to accurately reproduce the electric field (3.35) originating from $X'$ at the point $X$, but they do not reproduce the magnetic field (3.36).

The magnetic fields of the pulses deviate from those of the moving charge by a component $|B_y^{\text{pulse}} - B_y^{\text{charge}}| = |E_x|(1-\beta)$ in the $y$-direction and by a

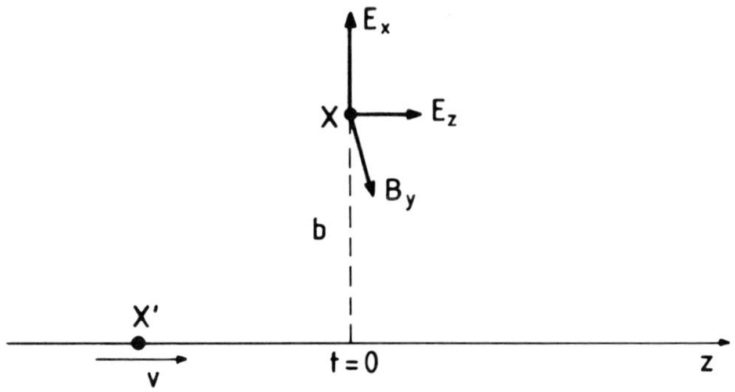


**Figure 3.4.** The electromagnetic fields (3.35) and (3.36) produced at the point $X$ by the charge at $X'$ moving with velocity $v$.

component $|B_x^{\text{pulse}} - B_x^{\text{charge}}| \simeq |E_z| \simeq |E_x|/\gamma$ in the $x$-direction. Since the ratio of magnetic to electric forces on a perturbed system (e.g., an atom) located at point $X$ is of the order $u/c$, where $u$ is a representative velocity within the perturbed system, it follows that the errors in the forces arising from the misinterpretation of the magnetic field of the moving charge by the pulses $P_1$ and $P_2$ are of the order $u/c\gamma$. These effects may be neglected if $u/c\gamma \ll 1$, a condition that is certainly satisfied for $\gamma \gg 1$.

Another error in the replacement is caused by the fact that the field produced by the moving charge varies in the transverse direction within the space region occupied by the perturbed system. Thus, if $\Delta b$ is its transverse extension, we have to require that $\Delta b \ll b$ in order to be able to approximate the particle field by the pulses $P_1$ and $P_2$.

*The pulses of radiation and their frequency spectra*

So far, we have discussed the assumptions needed to replace the transient electromagnetic field of a moving charge with pulses of electromagnetic radiation. In the next step, we assume these conditions to be satisfied and study the effect of the two pulses $P_1$ and $P_2$. In other words, we consider a swarm of real (on-shell) photons for which the relation $k = \omega/c$ between the wave number $k$ and the frequency $\omega$ holds. The task then consists in decomposing the classical fields into a photon spectrum and to calculate the number $N(\omega, b)$ of equivalent photons with a given frequency $\omega$ produced by the passage of a projectile at the impact parameter $b$.

In the laboratory system, the Poynting vector $\mathbf{S}$ describing the energy

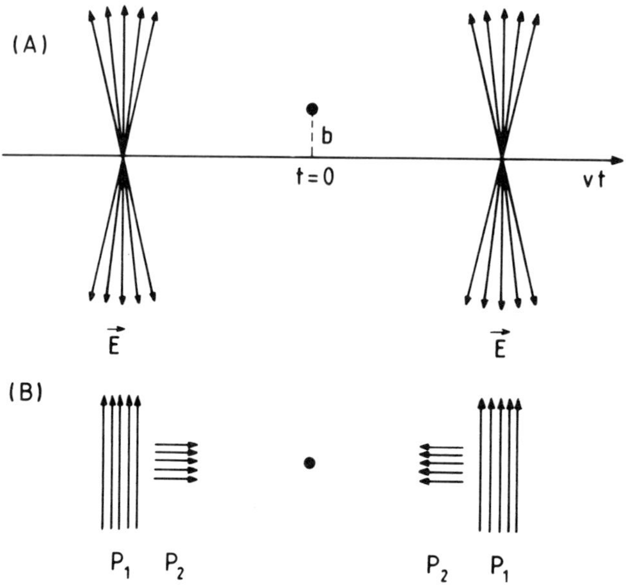


**Figure 3.5.** A projectile with velocity $v$ follows a trajectory shifted by the impact parameter $b$ with respect to the origin of the target system (indicated by a dot •). The arrows indicate the electric fields at positions before and after the projectile reaches the distance of closest approach. (A) Field strength accompanying the projectile, see Fig.3.3. (B) Equivalent-photon pulses $P_1$ and $P_2$ propagating in the $z$ and $y$ directions (perpendicular to the plane of the figure), respectively.

flux in the field of virtual photons is given by

$$\begin{aligned} \mathbf{S} &= \frac{c}{4\pi}(\mathbf{E}\times\mathbf{B}) \\ &= \frac{c}{4\pi}(\hat{\mathbf{e}}_z E_x^2 \pm \hat{\mathbf{e}}_y E_z^2) \end{aligned} \tag{3.39}$$

where $\hat{\mathbf{e}}_z$ is the unit vector in the $z$-direction and is associated with a packet of plane waves polarized in the $x$-direction and propagating in the $z$-direction. It represents the pulse $P_1$. The second term in Eq. (3.39) with the unit vector $\pm\hat{\mathbf{e}}_y$ corresponds to the propagation in the positive or negative $y$-direction and to a polarization in the $z$-direction. It represents the pulse $P_2$. It turns out that for high projectile velocities with $\gamma \gg 1$, the pulse $P_2$ is of minor importance and can be discarded, see Eq. (3.46), but for the time being, we retain the pulse $P_2$. Both pulses of radiation are schematically illustrated in Fig. 3.5.

In order to analyze the transient fields in terms of photon spectra, it is necessary to introduce the Fourier transform of the electric field as

$$\mathbf{E}(\omega) = \frac{1}{\sqrt{2\pi}} \int_{-\infty}^{\infty} \mathbf{E}(t) e^{i\omega t} dt \tag{3.40}$$

and its inverse

$$\mathbf{E}(t) = \frac{1}{\sqrt{2\pi}} \int_{-\infty}^{\infty} \mathbf{E}(\omega) e^{-i\omega t} d\omega \tag{3.41}$$

and corresponding relations for the magnetic field $\mathbf{B}$. We distinguish the function from its transform by explicitly displaying its argument(s) in question, in this case, $\omega$ and $t$. Since $\mathbf{E}(\mathbf{r}, t)$ is a real quantity, we may avoid reference to negative frequencies by noting that $|\mathbf{E}(-\omega)| = |\mathbf{E}^*(\omega)| = |\mathbf{E}(\omega)|$. By Fourier transforming the Poynting vector (3.39), we obtain the amount of energy per unit area and unit frequency interval incident on the target nucleus in the course of the collision as

$$\begin{aligned} I_1(\omega, b) &= \frac{c}{2\pi} |E_x(\omega, b)|^2 , \\ I_2(\omega, b) &= \frac{c}{2\pi} |E_z(\omega, b)|^2 , \end{aligned} \tag{3.42}$$

where $I_1$ and $I_2$ refer to the pulses $P_1$ and $P_2$, respectively, and a factor of two accounts for taking the frequencies to be positive. The dependence on the impact parameter $b$ and the charge number $Z$ (we drop the label $P$) is explicitly displayed in Eq. (3.35). The Fourier transform

$$E_x(\omega, b) = \frac{Ze\gamma b}{\sqrt{2\pi}} \int_{-\infty}^{\infty} \frac{e^{i\omega t}\, dt}{(b^2 + \gamma^2 v^2 t^2)^{3/2}} \tag{3.43}$$

can be performed conveniently by changing the integration variable to $\tau = \gamma vt/b$ and using the integral representation of the modified Bessel functions $K_0$ and $K_1$ of order 0 and 1 [Erd54, AbS65]. The result is [Jac75]

$$E_x(\omega, b) = \frac{Ze}{bv} \sqrt{\frac{2}{\pi}} \left[ \frac{\omega b}{\gamma v} K_1 \left( \frac{\omega b}{\gamma v} \right) \right] . \tag{3.44}$$

In a similar fashion, one derives

$$E_z(\omega, b) = -i \frac{Ze}{\gamma bv} \sqrt{\frac{2}{\pi}} \left[ \frac{\omega b}{\gamma v} K_0 \left( \frac{\omega b}{\gamma v} \right) \right] . \tag{3.45}$$

The field strength $E_z$ in Eq. (3.45) has an additional factor $1/\gamma$ compared to Eq. (3.44). Indeed, from the discussion following Eq. (3.37) and from

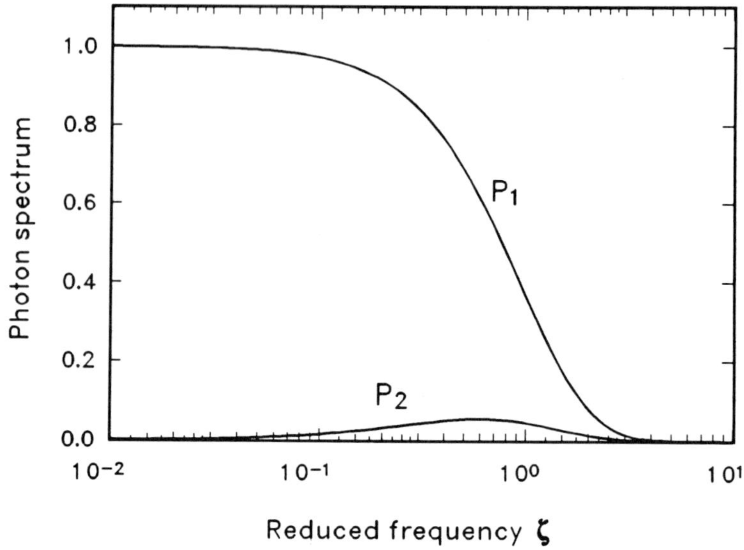


**Figure 3.6.** The shape of the equivalent-photon spectrum for the pulses $P_1$ and $P_2$ as a function of $\zeta = \omega b/\gamma v$ and assuming a Lorentz factor $\gamma = 2$. The cut-off frequency (3.47) corresponds to $\zeta \simeq 1$.

Fig. 3.3, one sees that the ratio $|E_z|/|E_x|$ decreases with $1/\gamma$. With the dimensionless quantity $\zeta = \omega b/\gamma v$, Eq. (3.42) yields the frequency spectra

$$\begin{aligned} I_1(\omega, b) &= \frac{1}{\pi^2}\frac{Z^2e^2}{cb^2}\,\beta^{-2}\zeta^2 K_1^2(\zeta)\,, \\ I_2(\omega, b) &= \frac{1}{\pi^2}\frac{Z^2e^2}{cb^2}\,\beta^{-2}\frac{1}{\gamma^2}\,\zeta^2 K_0^2(\zeta)\,. \end{aligned} \tag{3.46}$$

We see that the intensity of the pulse $P_2$ is reduced by a factor of $\gamma^{-2}$ compared to that of $P_1$, so that for extreme relativistic collisions, this pulse can be ignored. Figure 3.6 shows the frequency spectra of the pulses $P_1$ and $P_2$. The shape of the pulse $P_1$ reflects the fact that according to Eq. (3.35), the duration of the passage is given by $\Delta t \simeq b/\gamma v$ so that frequencies appreciably higher than

$$\omega_{\max} = \frac{\gamma c}{b} \tag{3.47}$$

cannot occur. In contrast, since $E_z$ passes through zero at $t = 0$, the pulse $P_2$ behaves approximately as $\sin^2(\gamma v t/b)$ near $t = 0$, and therefore the spectrum will be localized around $\omega = \gamma v/b$ or $\zeta \simeq 1$. Figure 3.6 confirms that the intensity of the pulse $P_2$ is considerably reduced with respect to that of $P_1$, even for a Lorentz factor as low as $\gamma = 2$. For larger values of $\gamma$, the contribution of $P_2$ becomes increasingly negligible.

*The cut-off frequency and the total radiated energy*

The cut-off frequency given by Eq. (3.47) depends on the impact parameter $b$. The maximum frequency in the pulse of radiation is determined by the minimum impact parameter $b_{\text{min}}$. This quantity is not always very well determined. For example, in the collision between two nuclei, one may take $b_{\text{min}} = R_0^{\text{T}} + R_0^{\text{P}}$ where $R_0^{\text{T}}$ and $R_0^{\text{P}}$ are the nuclear radii of target and projectile. Since the charge distribution of a nucleus is not precisely that of a uniformly charged sphere, nuclear form factors (see Sec. 3.4.2) may enter into the frequency cut-off. Moreover, the underlying assumption of a rectilinear projectile trajectory and of a constant field across the transverse extension of the system breaks down for such close encounters. The problem of the frequency cut-off often limits the usefulness of the Weizsäcker-Williams method. With these reservations in mind, we may integrate the intensities (3.42) over the impact parameter plane to get the total radiated energy within a given frequency range $d\omega$ as

$$I(\omega)\, d\omega = 2\pi \int_{b_{\text{min}}}^{\infty} [I_1(\omega, b) + I_2(\omega, b)]\, b\, db\, d\omega.$$

By using recursion formulas [GrR80] for the functions $K_0$, $K_1$, and their derivatives, one can show that

$$I(\omega) = \frac{2Z^2e^2}{\pi c \beta^2} \left\{ \zeta_{\text{min}} K_0(\zeta_{\text{min}}) K_1(\zeta_{\text{min}}) - \frac{\beta^2}{2} \zeta_{\text{min}}^2 [K_1^2(\zeta_{\text{min}}) - K_0^2(\zeta_{\text{min}})] \right\}, \tag{3.48}$$

where the contributions of $I_1$ and $I_2$ are no longer separated and where

$$\zeta_{\text{min}} = \frac{\omega b_{\text{min}}}{\gamma v}.$$

In the limiting cases of $\zeta_{\text{min}} \ll 1$ and $\zeta_{\text{min}} \gg 1$, approximate analytical expressions for the modified Bessel functions $K_0$ and $K_1$ are available [AbS65, GrR80] which lead to the estimates

$$I(\omega) \simeq \frac{2}{\pi} \frac{Z^2e^2}{c\beta^2} \left[ \ln\left( \frac{1.123\, \gamma v}{\omega b_{\text{min}}} \right) - \frac{\beta^2}{2} \right] \quad \text{for} \quad \omega \ll \frac{\gamma v}{b_{\text{min}}} \tag{3.49}$$

and

$$I(\omega) \simeq \frac{Z^2e^2}{c\beta^2}\left(1-\frac{\beta^2}{2}\right)e^{-2\omega b_{\min}/\gamma v} \qquad \text{for} \quad \omega \gg \frac{\gamma v}{b_{\min}}. \tag{3.50}$$

The energy spectrum[2] is dominated by the low-frequency part with a tail extending to about $\omega \sim 2\gamma v/b_{\min}$.

The number of virtual quanta per energy interval is obtained from the frequency spectrum by the relation $I(\omega)\,d\omega = \hbar\omega N(\hbar\omega)\,d(\hbar\omega)$, so that

$$N(\hbar\omega) = \frac{1}{\hbar^2\omega}I(\omega)\,, \tag{3.51}$$

and in the low-frequency limit

$$N(\hbar\omega) \simeq \frac{2}{\pi}\frac{e^2}{\hbar c}\frac{Z^2}{\beta^2\hbar\omega}\left[\ln\left(\frac{1.123\,\gamma v}{\omega b_{\min}}\right)-\frac{\beta^2}{2}\right]. \tag{3.52}$$

The minimum impact parameter $b_{\min}$ plays the role of a parameter which has to be chosen to suit the problem at hand. Depending on the system considered, $b_{\min}$ may be the sum of two nuclear radii, an atomic radius, or the Compton wavelength of an elementary particle created in the collision. In some cases, one may try to use a different approximation for impact parameters around or below $b_{\min}$. As was stated above, this is an unavoidable ambiguity in the application of the equivalent-photon method.

### 3.4.2 Formulation in the wave-number representation

In Sec. 3.4.1, we start from the Liénard-Wiechert potential with the fields (3.35) and (3.36), of a point charge $Ze$ moving along a straight-line trajectory. It is instructive to reconsider the process in a formulation [BeL82] in which (a) the Lorentz transformation of the electric fields produced by a general spherical charge distribution is explicitly carried out, and (b) from the very beginning, only the wave numbers and frequencies of the emitted virtual photons play a role. The impact-parameter dependence is recovered when form factors are introduced.

We proceed as follows. We write the projectile charge distribution in its rest system in a covariant form, both in the space representation and in the Fourier-transformed wave-number representation. The covariant form allows us to evaluate the corresponding four-current in a coordinate frame in which the projectile moves and the target is at rest. From the four-current

[2] The numerical factor 1.123 in Eq. (3.49) arises from the occurrence of Euler's constant in the series expansion of $K_0$ and $K_1$.

thus obtained, one can infer directly the electromagnetic four-potential in the wave-number representation which, in turn, yields the electric and magnetic fields and hence the Poynting vector and, finally, the number of equivalent photons.

*The four-current for an extended charge distribution*

As before, quantities in the moving projectile system are marked with a prime while quantities in the laboratory system are unprimed. To begin with, the density of the four-current of the charge $Ze$ in its rest system is

$$j'^{\mu}(\mathsf{x}') = \rho(|\mathbf{r}'|)\, u'^{\mu}, \tag{3.53}$$

where $\mathsf{x}' = x'^{\mu}$ denotes the four-vector of the space-time coordinates, $\rho(|\mathbf{r}'|) = \rho(r')$ is a time-independent spherical charge density, and $\mathsf{u}' = u'^{\mu} = (c, 0, 0, 0)$ represents the four-velocity, here in the rest system of the charge density. In order to derive the decomposition of $j'^{\mu}(\mathsf{x}')$ with respect to wave numbers $\mathbf{k}$ and frequencies $\omega$, which are combined as a four-vector $\mathsf{k} = (\omega/c, \mathbf{k})$, we take the four-dimensional Fourier transform

$$\begin{aligned} j'^{\mu}(\mathsf{k}') &= \frac{1}{(2\pi)^2} \int d^3x\, dt\, \rho(r')\, u'^{\mu}\, e^{-i(\mathbf{k}'\cdot\mathbf{r}' - \omega' t')} \\ &= \sqrt{2\pi}\, \delta(\omega')\, \rho(\mathbf{k}')\, u'^{\mu}. \end{aligned} \tag{3.54}$$

Here, $\rho(\mathbf{k}') = \rho(k')$ is the three-dimensional Fourier transform of $\rho(r')$. For extended spherical charge distributions, it is convenient to define a form factor $f(k'^2)$, which is again spherical, by writing

$$\rho(k') = (2\pi)^{-3/2} Ze\, f(k'^2), \tag{3.55}$$

or, explicitly,

$$f(k'^2) = \frac{1}{Ze} \int \rho(r')\, e^{-i\mathbf{k}'\cdot\mathbf{r}'}\, d^3r'\,. \tag{3.56}$$

In particular, for a point charge $\rho(r') = Ze\, \delta(\mathbf{r}')$ we have $f_0(k'^2) = 1$. A bare nucleus, however, has an extended charge distribution and, for the present purpose, may be represented by a homogeneously charged sphere (hcs) with radius $R_0$ and total charge $Ze$. By direct evaluation of Eq. (3.56) one obtains the nuclear form factor

$$f_{\text{hcs}}(k'^2) = \frac{3\, j_1(k' R_0)}{k' R_0}, \tag{3.57}$$

where $j_1$ is a spherical Bessel function. In the limit $R_0 \to 0$, the form factor $f_0(k'^2) = 1$ for a point charge is retrieved.

If one assumes a Gaussian charge distribution, the corresponding form factor in wave-number space is also Gaussian and of the form

$$f_{\text{Gauss}}(k'^2) = \exp\left(-\frac{k'^2}{Q_0^2}\right), \tag{3.58}$$

where $Q_0 = 60$ MeV (in energy units) for Pb–Pb [DrE89].

The frequency $\omega'$, which occurs in the delta function of Eq. (3.54), is the zeroth component of a four-vector. By formally writing it as a Lorentz scalar $\omega' = \mathsf{k}' \cdot \mathsf{u}'$, the evaluation in any coordinate system is greatly simplified. Similarly, since Eq. (3.54) requires $\omega' = 0$, we can write $k'^2 = -\mathsf{k}' \cdot \mathsf{k}' = -\mathsf{k}'^2$ as a Lorentz invariant quantity. This means that Eq. (3.54) can be written in a covariant form when (3.55) is inserted. In the laboratory system, the four-current reads

$$j^\mu(\mathsf{k}) = \frac{1}{2\pi}\, Ze\, \delta(\mathsf{k} \cdot \mathsf{u})\, f(-\mathsf{k}^2)\, u^\mu, \tag{3.59}$$

where the velocity four-vector $\mathsf{u} = u^\mu = \gamma(c, \mathbf{v})$ with $\mathsf{u} \cdot \mathsf{u} = c^2$ explicitly refers to the projectile velocity $v$.

*The electromagnetic four-potential*

The current density (3.59) allows us to calculate the vector potential from Eq. (3.5) which in wave-number space takes the form

$$\mathsf{k} \cdot \mathsf{k}\, A^\mu(\mathsf{k}) = -\frac{4\pi}{c}\, j^\mu(\mathsf{k}). \tag{3.60}$$

By inserting Eq. (3.59) into Eq. (3.60) we get the wave-number decomposition of the electromagnetic four-potential produced by a moving charge as

$$A^\mu(\mathsf{k}) = -\frac{2}{c}\, Ze\, \delta(\mathsf{k} \cdot \mathsf{u})\, \frac{f(-\mathsf{k}^2)}{\mathsf{k}^2}\, u^\mu. \tag{3.61}$$

From $\omega' = 0$, we know that the Lorentz invariant argument of the delta function $\mathsf{k} \cdot \mathsf{u} = \gamma(\omega - \mathbf{k} \cdot \mathbf{v}) = \omega' = 0$ and hence that

$$k_z = \omega / v. \tag{3.62}$$

Therefore,

$$-\mathsf{k}^2 = \mathbf{k}_\perp^2 + k_z^2 - \frac{\omega^2}{c^2} = k_\perp^2 + \frac{1}{v^2}\left(\frac{\omega}{\gamma}\right)^2, \tag{3.63}$$

where $k_z$ denotes the wave-number component in the direction of the moving charge while $\mathbf{k}_\perp$ is the component perpendicular to $\mathbf{v}$.

We have now calculated the electromagnetic four-potential in wave-number representation in the laboratory system without taking recourse to the explicit Liénard-Wiechert potentials (3.34). The longitudinal component $k_z$ is fixed because the charge distribution in its rest system is time-independent so that its oscillation frequency $\omega' = 0$ in Eq. (3.54). In the space representation of $A^\mu$, this means that the fields do not depend on the $z$-coordinate.

*The total photon flux*

The next task is to construct the Poynting vector and the photon flux from the electric and magnetic fields. The transverse electric field $\mathbf{E}_\perp = (E_x, E_y, 0)$ is obtained from Eq. (3.4) as

$$\mathbf{E}_\perp(\omega, \mathbf{k}) = i\mathbf{k}_\perp A^0(\omega, \mathbf{k})$$

in the wave-number representation, where use has been made of $\mathbf{A}_\perp = 0$ [which is a consequence of $\mathbf{v}_\perp = 0$ in (3.61)]. In a first step, we perform only the inverse Fourier transform in the $z$-direction to get

$$|\mathbf{E}_\perp(\omega, \mathbf{k}_\perp, z)|^2 = \frac{2}{\pi}\, Z^2 e^2\, \frac{\mathbf{k}_\perp^2\, f^2(-\mathsf{k}^2)}{v^2\, \mathsf{k}^4}, \tag{3.64}$$

where $\mathsf{k}^2$ and $\mathbf{k}_\perp^2$ are connected by Eq. (3.63). So far, no approximation has been introduced. When calculating the longitudinal electric field and the magnetic field from Eq. (3.5), we take the limit $v \to c$. Then $u^\mu = \gamma(c, 0, 0, c)$ in Eq. (3.61), so that $A^\mu = (A^0, 0, 0, A^0)$ and, because of $k_z = \omega/c$, the longitudinal field in Eq. (3.8) is

$$E_z = -F^{03} = -i\left(\frac{\omega}{c} A^3 - k_z A^0\right) = 0.$$

In this limit, there are only transverse fields with $|\mathbf{E}| = |\mathbf{B}|$, and the pulse $P_2$ discussed in Sec. 3.4.1 is discarded.

We have now everything needed to calculate the energy flux (3.39) and to interpret it as an equivalent flux of photons with positive frequencies $\omega$. In order to establish the identification, we integrate $\mathbf{S}$ over time and over the coordinate plane perpendicular to $\mathbf{v}$.

$$\begin{aligned}
\int dt \int dx\, dy\, \mathbf{S} &= \frac{c}{2\pi}\int_0^\infty d\omega \int dx\, dy\, |\mathbf{E}_\perp(\omega, \mathbf{r})|^2 \\
&= \frac{c}{2\pi}\int_0^\infty d\omega \int dk_x\, dk_y\, |\mathbf{E}_\perp(\omega, \mathbf{k}_\perp, z)|^2 \\
&= \int_0^\infty d(\hbar\omega)\, \hbar\omega\, N(\hbar\omega).
\end{aligned} \tag{3.65}$$

In the first equation, a factor of two arises from $|\mathbf{E}(-\omega)| = |\mathbf{E}(\omega)|$, similarly as in Eq. (3.42); in the second equation, the remaining Fourier transformation is performed in the $x, y$-plane, with the aid of the Parseval relation [Mes62], that is, the invariance of the scalar product under Fourier transformations. The last equation defines the number of equivalent photons per energy unit. It leads to the identification

$$N(\hbar\omega) = \frac{2}{\pi} \frac{e^2}{\hbar c} \frac{Z^2}{\hbar\omega} \int_0^{k_{\max}} \frac{k_\perp^3 \, f^2(-\mathsf{k}^2)}{[k_\perp^2 + (\omega/\gamma c)^2]^2} \, dk_\perp \,, \tag{3.66}$$

where we recall that $f(-\mathsf{k}^2) = 1$ for a moving point charge. For a constant form factor, the integral diverges logarithmically at the upper limit as $k_{\max} \to \infty$. It is therefore necessary to introduce the cut-off parameter $k_{\max} \equiv 1/r_0$ with $r_0$ having the dimension of a length. This cut-off precisely reflects the cut-off $b_{\min}$ that we had to introduce for the impact parameter in Eq. (3.48). For a point charge, the integral (3.66) can be carried out. Discarding terms that are negligible for large values of $k_{\max}$ we obtain

$$N(\hbar\omega) = \frac{2}{\pi} \frac{e^2}{\hbar c} \frac{Z^2}{\hbar\omega} \left( \ln \frac{\gamma c}{\omega r_0} - \frac{1}{2} \right) . \tag{3.67}$$

This result is identical to Eq. (3.52) in the limit $v \to c$ if the cut-off radius $r_0$ is identified with $b_{\min}$ and the numerical factor 1.123 of Eq. (3.52) is replaced with 1. These minor differences are all within the limits of accuracy inherent in the method.

If the form factor $f_{\text{hcs}}(-\mathsf{k}^2)$ of Eq. (3.57) or $f_{\text{Gauss}}(-\mathsf{k}^2)$ of Eq. (3.58) is inserted in (3.66), the integral converges at the upper limit, so that we can take $k_{\max} = \infty$. Nevertheless, very large wave numbers may correspond to impact parameters that are smaller than the radius $R_0$ of the projectile nucleus. In this case, the assumption of a rectilinear trajectory breaks down and, furthermore, a mutual penetration of the colliding nuclei will occur, so that nuclear reactions can proceed in addition to the electromagnetic processes.

*The impact parameter-dependent photon flux*

For some applications, it is useful to know the impact parameter dependence of the photon flux. Similarly as in Eq. (3.64), we find by an additional Fourier transformation in the impact parameter plane that

$$|\mathbf{E}_\perp(\omega, \mathbf{b}, z)|^2 = \frac{1}{2\pi^3} (Ze)^2 \left| \int d^2k_\perp \, \frac{\mathbf{k}_\perp \, f^2(-\mathsf{k}^2)}{v \, \mathsf{k}^2} \, e^{i\mathbf{k}_\perp \cdot \mathbf{b}} \right|^2 . \tag{3.68}$$

In the limit $v \to c$ and with the identification of Eq. (3.65) (for positive frequencies $\omega$), we get for the photon flux in the infinitesimal area of the impact parameter plane between $b$ and $b + db$ and between the azimuth $\varphi$ and $\varphi + d\varphi$ the result

$$\begin{aligned} N(\hbar\omega, \mathbf{b}) &= 4\frac{\alpha Z^2}{\hbar\omega}\left|\int \frac{d^2k_\perp}{(2\pi)^2}\frac{\mathbf{k}_\perp f(k_\perp^2 + \omega^2/\gamma^2 v^2)}{k_\perp^2 + \omega^2/\gamma^2 v^2} e^{i\mathbf{k}_\perp \cdot \mathbf{b}}\right|^2 \\ &= 4\frac{\alpha Z^2}{\hbar\omega}\left|\int_0^\infty \frac{k_\perp^2 dk_\perp}{2\pi}\frac{f(k_\perp^2 + \omega^2/\gamma^2 v^2)}{k_\perp^2 + \omega^2/\gamma^2 v^2} J_1(k_\perp b)\right|^2, \end{aligned} \tag{3.69}$$

where $J_1$ is the Bessel function of the first kind and order one. In the first integral of Eq. (3.69), only the component of $\mathbf{k}_\perp$ parallel to $b$ contributes to the integral over the azimuthal angle between $\mathbf{k}_\perp$ and $\mathbf{b}$. By integrating $N(\hbar\omega, b)$ over the impact parameter plane, we retrieve Eq. (3.66). The result (3.69) can, of course, also be obtained directly from Eq. (3.43); however, for introducing the form factor $f(-\mathsf{k}^2)$ and the extreme relativistic limit, the wave number representation applied here is more convenient.

For a point charge, Eq. (3.46) tells us that the corresponding photon flux can be written as

$$N(\hbar\omega, b) = \frac{1}{\pi^2}\frac{\alpha Z^2}{\hbar\omega}\left(\frac{\omega}{\gamma v}\right)^2 \beta^{-2}\left[K_1^2(\zeta) + \frac{1}{\gamma^2}K_0^2(\zeta)\right], \tag{3.70}$$

where $\zeta = \omega b/\gamma v$, while $K_1$ and $K_0$ are the modified Bessel functions of order one and zero, respectively. In Fig. 10.12 of Sec. 10.2, we compare the influences of various form factors on the impact parameter-dependent photon distributions $N(\hbar\omega, b)$ for a fixed Lorentz factor $\gamma = 3500$.

In this section, we have calculated the frequency spectrum of equivalent photons representing the transient electromagnetic field of a moving charge. In Chap. 10, we apply and extend these results to estimate the cross section for electron-positron pair production. By replacing the electromagnetic fields generated by both of the colliding nuclei by equivalent photons, one can calculate pair production cross sections from the known corresponding cross sections for photon-photon collisions.

In applying the equivalent-photon method, one has to be aware of the limitations arising from the assumption of real photons and from the assumption of a constant field across the transverse extension of the perturbed system. The latter restriction may invalidate the method for impact parameters smaller than the Compton wavelength of the electron.

# Chapter 4

# Relativistic Electron Motion

Chapter 3 discusses the time-dependent electromagnetic field produced by a projectile nucleus moving along a classical rectilinear trajectory. The next step towards a description of ion-atom collisions is to study the electron motion in the static Coulomb field of the target nucleus alone. This is done in the present chapter. Atomic processes are then obtained by subjecting the electron both to the static target and transient projectile potential. This is deferred to Chap. 5.

Although, in general, the target atom is composed of many electrons, we describe the basic process by confining our attention to a single bound electron, the active electron. The effect of additional passive electrons is usually described in an approximate manner, for example by introducing an effective nuclear charge or by explicitly introducing a screening function.

The present chapter provides a reference for the following chapters. In Sec. 4.1, we introduce the Dirac equation in a covariant and in a Hamiltonian form and derive its properties under Lorentz transformations. The simplest solutions, plane waves, are the subject of Sec. 4.2, while Secs. 4.3 and 4.4 treat bound and continuum states in a Coulomb potential, respectively, introducing exact as well as approximate solutions.

## 4.1 The Dirac equation

The Dirac equation is the relativistic wave equation for a spin-$\frac{1}{2}$ particle. The wave function is given by a four-component spinor in which the upper two components, with spin up and spin down, refer to positive-energy states while the lower two components refer to the negative-energy states to be discussed later. Here, we confine ourselves to introducing the notation and to summarizing those properties of the Dirac equation and its solutions which are needed in subsequent chapters. For more details, the reader is referred to standard textbooks [Mes62, BjD64, Sak67, BeL82].

### 4.1.1 The covariant form

For an electron with mass $m_e$ and charge $-e$ subject to an external electromagnetic vector potential $A^\mu = (\Phi, \mathbf{A})$, the covariant form of the Dirac equation at the space-time point $\mathsf{x} = (ct, \mathbf{x})$ is given by

$$\left(i\hbar\gamma^\mu \frac{\partial}{\partial x^\mu} + \frac{e}{c}\gamma^\mu A_\mu(\mathsf{x}) - m_e c\right)\psi(\mathsf{x}) = 0. \tag{4.1}$$

The $4 \times 4$ matrices $\gamma^\mu$ satisfy the anticommutation relations

$$\gamma^\mu\gamma^\nu + \gamma^\nu\gamma^\mu = 2g^{\mu\nu}1, \tag{4.2}$$

where $g^{\mu\nu}$ is defined in Eq. (2.5) and 1 denotes the $4 \times 4$ unit matrix.[1] In a standard representation, the $\gamma$-matrices can be composed from subblocks consisting of the $2 \times 2$ matrix 0, the $2 \times 2$ unit matrix $I$, and the Pauli matrices

$$\sigma_1 = \begin{pmatrix} 0 & 1 \\ 1 & 0 \end{pmatrix}, \quad \sigma_2 = \begin{pmatrix} 0 & -i \\ i & 0 \end{pmatrix}, \quad \sigma_3 = \begin{pmatrix} 1 & 0 \\ 0 & -1 \end{pmatrix}. \tag{4.3}$$

With these building blocks, one may write

$$\gamma^0 = \begin{pmatrix} I & 0 \\ 0 & -I \end{pmatrix}, \quad \gamma^i = \begin{pmatrix} 0 & \sigma_i \\ -\sigma_i & 0 \end{pmatrix} \quad i = 1, 2, 3. \tag{4.4}$$

It follows from this definition that $\gamma^0$ is hermitian and that the $\gamma^i$ are anti-hermitian with $(\gamma^i)^2 = -1$. With the decomposition $\gamma^\mu = (\gamma^0, \boldsymbol{\gamma}) = (\gamma^0, \gamma^1, \gamma^2, \gamma^3)$ and Eq. (2.9), we have

$$\gamma^\mu \frac{\partial}{\partial x^\mu} = \gamma^0 \frac{\partial}{\partial ct} + \boldsymbol{\gamma} \cdot \nabla \tag{4.5}$$

[1] In general, if in an equation a matrix appears on one side and a number on the other side, we always imply that this number is multiplied with the unit matrix of the corresponding rank.

and

$$\gamma^{\mu} A_{\mu} = \gamma^{0}\Phi - \boldsymbol{\gamma} \cdot \mathbf{A}. \tag{4.6}$$

In Eq. (4.1), the four-momentum[2] operator $p^{\mu} = i\hbar\partial/\partial x_{\mu}$ occurs in the same combination $p^{\mu} + (e/c)A^{\mu}$ which is used in classical relativistic mechanics to describe the gauge invariant interaction of a point charge $-e$ with an applied field. Indeed, similarly as the classical Lagrangian, the Dirac equation is gauge invariant. This means that the simultaneous transformation

$$\begin{aligned} A_{\mu}(\mathsf{x}) &\rightarrow \tilde{A}_{\mu}(\mathsf{x}) = A_{\mu}(\mathsf{x}) - \frac{\partial\chi(\mathsf{x})}{\partial x^{\mu}}, \\ \psi(\mathsf{x}) &\rightarrow \tilde{\psi}(\mathsf{x}) = \psi(\mathsf{x}) \exp[-i\frac{e}{\hbar c}\chi(\mathsf{x})] \end{aligned} \tag{4.7}$$

leave Eq. (4.1) unaltered. This freedom of choosing the gauge can be exploited to simplify the problem, see e.g. Sec. 5.5.1 or Eqs. (5.61) or (5.62).

One may get a more complete understanding of the quantity $\gamma^{\mu}A_{\mu}$ in Eq. (4.6) by relating it to the electron four-current $j^{\mu}(\mathsf{x})$. If we introduce the Dirac adjoint spinor as $\overline{\psi} = \psi^{\dagger}\gamma^{0}$, we find from Eq. (4.1) and its adjoint that the vector

$$j^{\mu}(\mathsf{x}) = -ec\overline{\psi}(\mathsf{x})\gamma^{\mu}\psi(\mathsf{x}) \tag{4.8}$$

indeed qualifies as a current because it satisfies the continuity equation

$$\frac{\partial}{\partial x^{\mu}}\, j^{\mu}(\mathsf{x}) = 0 \tag{4.9}$$

with a positive definite probability density $\overline{\psi}\gamma^{0}\psi = \psi^{\dagger}\psi$. We then see that $-e\gamma^{\mu}A_{\mu}$ is the operator form of the interaction $j^{\mu}A_{\mu}/c$ of an electromagnetic current with an external electromagnetic field.

### 4.1.2 The Hamiltonian form

In the covariant form (4.1) of the Dirac equation, the derivatives with respect to the coordinates appear linearly. Singling out the time coordinate, Eq. (4.1) may be written in the Hamiltonian form

$$i\hbar\frac{\partial}{\partial t}\psi(\mathsf{x}) = H\psi(\mathsf{x}), \tag{4.10}$$

which has originally been given by Dirac and which for many practical applications is more convenient and transparent. We first introduce the

[2]The four-momentum $p^{\mu}$ used here has the dimension of a momentum, while in Chap. 2 it is convenient to adopt energy units, i.e., momentum times the velocity of light.

hermitian Dirac matrices [3]

$$\boldsymbol{\alpha} = \gamma^0 \boldsymbol{\gamma} \quad \text{with} \quad \boldsymbol{\alpha} = \begin{pmatrix} 0 & \boldsymbol{\sigma} \\ \boldsymbol{\sigma} & 0 \end{pmatrix} \tag{4.11}$$

whose components, according to Eq. (4.2), satisfy the anticommutation relations

$$\begin{aligned} \alpha_i \alpha_k + \alpha_k \alpha_i &= 2\delta_{ik} \\ \alpha_i \gamma^0 + \gamma^0 \alpha_i &= 0 \\ \alpha_i^2 = (\gamma^0)^2 &= 1. \end{aligned} \tag{4.12}$$

By using this notation and rewriting Eq. (4.1) in the form (4.10), we can identify the Dirac Hamiltonian as

$$H = -i\hbar c \,\boldsymbol{\alpha} \cdot \nabla - e\Phi + e\boldsymbol{\alpha} \cdot \mathbf{A} + m_e c^2 \gamma^0. \tag{4.13}$$

This form has the virtue that individual terms can be readily interpreted. For example, the first term in Eq. (4.13) represents the operator for the kinetic energy with the matrix $c\boldsymbol{\alpha}$ appearing as the operator transcription of the velocity. The term $-e\Phi + e\boldsymbol{\alpha} \cdot \mathbf{A}$ replaces the classical expression

$$H'_{\text{classical}} = -e\left(\Phi - \frac{1}{c}\mathbf{v} \cdot \mathbf{A}\right) \tag{4.14}$$

for the interaction of a moving point charge $-e$ with the electromagnetic field, and $m_e c^2 \gamma^0$ represents the electron rest energy.

For the electromagnetic potentials $\Phi$ and $\mathbf{A}$ occurring in Eq. (4.13), the gauge transformation (4.7) is rewritten as

$$\begin{aligned} \Phi &\rightarrow \tilde{\Phi} = \Phi - \frac{1}{c}\frac{\partial \chi}{\partial t}, \\ \mathbf{A} &\rightarrow \tilde{\mathbf{A}} = \mathbf{A} + \nabla\chi, \\ \psi &\rightarrow \tilde{\psi} = \psi \exp[-i\frac{e}{\hbar c}\chi]. \end{aligned} \tag{4.15}$$

Similarly, the electron charge density and the electron current of Eq. (4.8) are explicitly reformulated as

$$\begin{aligned} \rho(\mathbf{r}, t) &= -e\,\psi^\dagger(\mathbf{r}, t)\psi(\mathbf{r}, t) \\ \mathbf{j}(\mathbf{r}, t) &= -ec\,\psi^\dagger(\mathbf{r}, t)\boldsymbol{\alpha}\psi(\mathbf{r}, t). \end{aligned} \tag{4.16}$$

---

[3] Conventionally, one sets $\gamma^0 = \gamma_0 = \beta$; however, we wish to reserve the Greek letter $\beta$ for $v/c$ and therefore retain the notation $\gamma^0$ for the fourth Dirac matrix.

They satisfy the continuity equation (4.9) in the form

$$\frac{\partial}{\partial t}\rho(\mathbf{r},t)+\nabla\cdot\mathbf{j}(\mathbf{r},t)=0. \tag{4.17}$$

### 4.1.3 Lorentz transformations and covariance

We now want to establish the Lorentz covariance of the Dirac equation

$$\left(i\hbar\gamma^{\mu}\frac{\partial}{\partial x^{\mu}}+\frac{e}{c}\gamma^{\mu}A_{\mu}(\mathsf{x})-m_{\mathrm{e}}c\right)\psi(\mathsf{x})=0 \tag{4.18}$$

given in Eq. (4.1). Invariance under homogeneous Lorentz transformations mediated by the matrix $\Lambda^{\mu}{}_{\nu}$ of Eqs. (2.10) and (2.13) requires that it must be possible to rewrite Eq. (4.18) in a moving (primed) coordinate system in the same form, namely as

$$\left(i\hbar\gamma^{\mu}\frac{\partial}{\partial x'^{\mu}}+\frac{e}{c}\gamma^{\mu}A'_{\mu}(\mathsf{x}')-m_{\mathrm{e}}c\right)\psi'(\mathsf{x}')=0. \tag{4.19}$$

Here, for an observer in the primed system, $\psi'(\mathsf{x}')$ describes the same physical state which is expressed as $\psi(\mathsf{x})$ for an observer in the unprimed system. While coordinates and wave functions are transformed, it can be shown [BjD64, Sak67] that the representation of the $\gamma$-matrices may be retained.

Since the Dirac equation as well as the Lorentz transformation of the coordinates are linear, we seek a linear transformation $S=S(\Lambda)$ between $\psi$ and $\psi'$ which achieves

$$\psi'(\mathsf{x}')\equiv\psi'(\Lambda\mathsf{x})=S(\Lambda)\psi(\mathsf{x}). \tag{4.20}$$

Here $S(\Lambda)$ is a $4\times4$ matrix operating on the four-component column vector $\psi$. Through $\Lambda$, it depends on the relative velocity $\mathbf{v}$ and the relative orientation of the primed and the unprimed coordinate systems. Since the role of the coordinate systems can be interchanged, $S$ must have an inverse $S^{-1}$ so that, conversely,

$$\psi(\mathsf{x})=S^{-1}(\Lambda)\psi'(\mathsf{x}')\equiv S^{-1}(\Lambda)\psi'(\Lambda\mathsf{x}). \tag{4.21}$$

Interchanging the coordinate systems, the space-time coordinate $\mathsf{x}$ is obtained from $\mathsf{x}'$ by the inverse Lorentz transformation $\Lambda^{-1}$, so that, with the aid of Eq. (4.20), we write

$$\psi(\mathsf{x})\equiv\psi(\Lambda^{-1}\mathsf{x}')=S(\Lambda^{-1})\psi'(\Lambda\mathsf{x}).$$

which, by comparison with Eq. (4.21), allows the identification

$$S(\Lambda^{-1}) = S^{-1}(\Lambda). \tag{4.22}$$

In the following, we assume that the argument of $S$ is always $\Lambda$ so that we can suppress it in the notation. The remaining problem is now the construction of $S$.

Starting from Eq. (4.18), we transform the derivatives according to

$$\frac{\partial}{\partial x^\mu} = \frac{\partial x'^\nu}{\partial x^\mu}\frac{\partial}{\partial x'^\nu} = \Lambda^\nu{}_\mu \frac{\partial}{\partial x'^\nu}. \tag{4.23}$$

By inserting $\psi(\mathsf{x})$ from Eq. (4.21), the Dirac equation (4.18) takes the form

$$\left(i\hbar\gamma^\mu\Lambda^\nu{}_\mu \frac{\partial}{\partial x'^\nu} + \frac{e}{c}\gamma^\mu A_\mu(\mathsf{x}) - m_\mathrm{e}c\right) S^{-1}\psi'(\mathsf{x}') = 0. \tag{4.24}$$

Multiplication from the left with the matrix $S$ yields the result

$$\left(i\hbar\, S\gamma^\mu S^{-1}\,\Lambda^\nu{}_\mu \frac{\partial}{\partial x'^\nu} + \frac{e}{c}\, S\gamma^\mu S^{-1}\, A_\mu(\mathsf{x}) - m_\mathrm{e}c\right)\psi'(\mathsf{x}') = 0. \tag{4.25}$$

This equation is identical with Eq. (4.19), provided two conditions are satisfied. First, it must be possible to find a matrix $S$ such that

$$S\gamma^\mu S^{-1}\,\Lambda^\nu{}_\mu = \gamma^\nu$$

or, equivalently,

$$S^{-1}\gamma^\mu S = \Lambda^\mu{}_\nu\gamma^\nu. \tag{4.26}$$

Second, the term $\frac{e}{c}\, S\gamma^\mu S^{-1}\, A_\mu(\mathsf{x})$ has to be identified as the transformed interaction, or

$$\gamma^\mu A'_\mu(\mathsf{x}') = S\gamma^\mu S^{-1}\, A_\mu(\mathsf{x}). \tag{4.27}$$

In order to complete the proof of covariance, we confine ourselves to those transformations in which we are primarily interested, namely Lorentz boosts in the $z$-direction. In this case, one finds that the spinor transformation

$$\psi'(\mathsf{x}') = S\psi(\mathsf{x}) \tag{4.28}$$

of Eq. (4.20) is mediated by the $4\times 4$ matrix

$$S = S^\dagger = \sqrt{\frac{1+\gamma}{2}}\,(1-\delta\alpha_z), \tag{4.29}$$

where the Dirac matrix $\alpha_z$ is defined in Eq. (4.11) and

$$\delta = \sqrt{\frac{\gamma - 1}{\gamma + 1}} \tag{4.30}$$

is a parameter measuring the magnitude of the relativistic corrections arising from the Lorentz transformation.

Using the property

$$\gamma^0 S \gamma^0 = S^{-1} \tag{4.31}$$

and the explicit form (2.13) of the transformation matrix $\Lambda^{\mu}{}_{\nu}$, we may verify Eq. (4.26) by direct calculation. With this step, we have established the Lorentz covariance in the particular case of a Lorentz boost. For a general transformation, the explicit form (4.29) is replaced with a matrix that also contains the other $\alpha$-matrices and the rotation angles.

If the transformation is given by Eq. (4.29), the matrix $S = S(v)$ depends only on the relative velocity of the two coordinate frames with respect to one another. For later reference, we also give the explicit form

$$S^2(v) = S^{-2}(-v) = \gamma(1 - \beta\alpha_z) \tag{4.32}$$

with $\beta = v/c$, which is obtained from Eqs. (4.29) and (4.31) by direct matrix multiplication. Equations (4.29) and (4.32) play an important role in the formulation of relativistic atomic collisions. While the matrix (4.29) transforms the relativistic wave function, (4.32) transforms the potentials.

Before discussing the interaction term (4.27) in more detail, we remark that the preceding development is valid not only for homogeneous Lorentz transformations but also for the inhomogeneous transformation (3.28) or (3.31) which involves a lateral shift by the impact parameter **b**, see Fig. 3.2. The reason is simply that the derivatives in Eq. (4.18) transform according to the *homogeneous* part of the Lorentz transformation (3.28). It is only the relation between $\mathsf{x}$ and $\mathsf{x}'$ that changes, without, however, affecting the construction of the matrix $S$. In the following, we employ the inhomogeneous Lorentz transformation (3.28).

*Transformation of the electromagnetic interaction*

We now turn to the electromagnetic interaction (4.27) and specialize $A'_{\mu}(\mathsf{x}')$ to the simple form of a Coulomb potential $A'_0 = Z_{\mathrm{P}}e/r'_{\mathrm{P}}$ in the moving (primed) coordinate frame where $Z_{\mathrm{P}}e$ is the charge of the projectile nucleus and $r'_{\mathrm{P}}$ is the distance of the electron from this nucleus measured in the moving frame (Fig. 3.2). The interaction operator then is

$$-e\gamma^{\mu} A'_{\mu}(\mathsf{x}') = -\gamma^0 \frac{Z_{\mathrm{P}}e^2}{r'_{\mathrm{P}}}. \tag{4.33}$$

Inserting this expression into the left-hand side of Eq. (4.27) and using Eq. (4.31), we get the interaction in the (unprimed) laboratory system as

$$-e\gamma^\mu A_\mu(\mathrm{x}) = -\gamma^0 S^2 \frac{Z_\mathrm{P} e^2}{r'_\mathrm{P}} = -\gamma^0 \gamma(1-\beta\alpha_z)\frac{Z_\mathrm{P} e^2}{r'_\mathrm{P}}, \tag{4.34}$$

where the explicit form (4.32) of $S^2$ has been inserted. With the aid of the operator transcription $\boldsymbol{\alpha} \to \mathbf{v}/c$, the last term is identified with the classical interaction (4.14) between the electron charge $-e$ and the Liénard-Wiechert potential (3.34) produced by a charge $Z_\mathrm{P} e$ moving in the $z$-direction. This demonstrates that Eq. (4.27) gives a physically reasonable transformation of the interaction term.

*Space inversion*

In a similar way as the proper Lorentz transformations, we may consider space inversion or the parity transformation. If we proceed as before and write

$$\psi'(-\mathbf{r},t) = S_\mathrm{p}\psi(\mathbf{r},t) \tag{4.35}$$

where $S_\mathrm{p}$ is a $4\times 4$ matrix independent of the space-time coordinates, we find that the Dirac equation is form-invariant, provided

$$S_\mathrm{p} = \gamma^0 .$$

This allows us to define a parity operator $\Pi$ by

$$\Pi\psi(\mathbf{r},t) = \gamma^0\psi(-\mathbf{r},t). \tag{4.36}$$

If $\Pi$ commutes with the Hamiltonian, the solutions can be chosen to be eigenfunctions of $\Pi$ with the eigenvalues $\pm 1$. Since according to the definition (4.4), the matrix $\gamma^0$ multiplies the upper two components with $+1$ and the lower two components with $-1$, the orbital parity of the upper two components is opposite to the parity of the lower two components, provided the four-spinor has a definite parity.

### 4.1.4 Lorentz transformation of the derivative terms and the orthogonality properties

For various purposes, it is convenient to have at one's disposal the Lorentz transform of the scalar product of $\gamma^\mu$ with the derivative four-vector. By comparing the expressions (4.23) and (4.26), one confirms the relation

$$\gamma^\mu \frac{\partial}{\partial x'^\mu} = S\gamma^\mu \frac{\partial}{\partial x^\mu} S^{-1}, \tag{4.37}$$

which, with the aid of Eq. (4.31), can be cast into the explicit form

$$\frac{\partial}{\partial(ct')} + \boldsymbol{\alpha}\cdot\nabla' = S^{-1}\left(\frac{\partial}{\partial(ct)} + \boldsymbol{\alpha}\cdot\nabla\right)S^{-1}. \tag{4.38}$$

As a simple application, let us consider the orthogonality properties of solutions

$$\psi_k(\mathbf{r},t) = \varphi_k(\mathbf{r})\exp(-iE_kt/\hbar) \tag{4.39}$$

of the single-center Dirac equation

$$\left(-i\hbar\frac{\partial}{\partial t} - i\hbar c\boldsymbol{\alpha}\cdot\nabla - \frac{Ze^2}{r} + m_{\mathrm{e}}c^2\gamma^0\right)\psi_k(\mathbf{r},t) = 0. \tag{4.40}$$

As one would expect, in the rest frame of the generating potential $-Ze^2/r$, the solutions $\psi_k$ are orthogonal and can be normalized to give

$$\int \psi_l^\dagger\psi_k\,d^3r = \delta_{lk}. \tag{4.41}$$

The question arises whether an observer in a moving frame will arrive at the same conclusion. If we start from Eq. (4.40), then write down the adjoint equation for $\psi_l$, multiply with $\psi_l^\dagger$ and $\psi_k$, respectively, and integrate over $d^3r'$ in the moving system, we can use Eq. (4.38) to introduce the transformed derivatives and to obtain

$$\int \psi_l^\dagger(\mathbf{r},t)S\left(-i\hbar\frac{\overleftrightarrow{\partial}}{\partial t'}\right)S\psi_k(\mathbf{r},t)\,d^3r' = 0. \tag{4.42}$$

Here, the double arrow denotes the difference between the derivatives taken of the right-hand and of the left-hand functions. The corresponding difference between the space derivatives can be written as a surface integral at infinity which vanishes for localized wave functions.

Owing to the Lorentz transformation $t = \gamma(t' + vz'/c^2)$, we get the identity

$$(E_k - E_l)\int (S\varphi_l(\mathbf{r}))^\dagger e^{-i\gamma(E_k-E_l)vz'/\hbar c^2}\,S\varphi_k(\mathbf{r})\,d^3r' = 0, \tag{4.43}$$

which is satisfied if

$$\int (S\psi_l(\mathbf{r},t))^\dagger\,S\psi_k(\mathbf{r},t)\,d^3r' = \delta_{lk}. \tag{4.44}$$

This shows that in the moving coordinate system, the orthogonality is retained. In order to prove the normalization, we use the hermitian property

(4.29) of the matrix $S$ and the explicit form (4.32) of $S^2$. Since the Dirac $\alpha_z$-matrix has vanishing matrix elements between identical states ($\alpha_z$ has odd parity), the operator $S^2$ in Eq. (4.44) reduces to the factor $\gamma$. Remembering that by Lorentz contraction $d^3r' = d^3r/\gamma$, Eq. (4.44) can be seen to be identical to Eq. (4.41).

*Loss of orthogonality for approximate solutions*

Equation (4.44) is valid only if $\psi_k$ and $\psi_l$ are *exact* solutions of the Dirac equation (4.40). Approximate solutions, constructed, for example, by diagonalization in a finite space of basis functions, still may satisfy Eq. (4.41), but not, in general, Eq. (4.44) [ToE89]. In other words, *states that are orthogonal in the rest frame are orthogonal in a moving Lorentz frame only if they are exact eigenstates of the Dirac Hamiltonian.* Similarly, if the Hamiltonian matrix is diagonal in the rest frame, it is diagonal in a moving Lorentz frame only for exact eigenstates. These observations are relevant for the use of pseudostates, and we return to this problem in Sec. 6.5.3.

Let us examine why the problem of losing the orthogonality property does not arise under Galilean transformations. Using again the notation $E_{k,l}$ for the relativistic energies including the rest mass, and $\epsilon_{k,l}$ for the nonrelativistic energies, we may expand the exponent describing the time oscillation in the moving frame as

$$\gamma E_k \left( t' + \frac{vz'}{c^2} \right) = m_e c^2 t' + \epsilon_k t' + \frac{1}{2} m_e v^2 t' + m_e \mathbf{v} \cdot \mathbf{r}' + \cdots . \qquad (4.45)$$

Aside from an immaterial term due to the electron rest mass and the nonrelativistic frequency term $\epsilon_k t'$, we recognize in the last two terms the time and space dependence entering into the translation factor, see Eq. (1.6). Since space- and time coordinates transform separately, the space term $m_e \mathbf{v} \cdot \mathbf{r}'$ remains state-independent and cancels in the overlap matrix element (4.43). This confirms that orthogonality is preserved in any Galilei-transformed coordinate frame.

## 4.2 Plane-wave solutions

In this section we want to study the solutions of the Dirac equation for free electrons, that is, in the absence of any interactions. This implies that in Eq. (4.1) or in Eq. (4.13), the four-potential $A^\mu = (\Phi, \mathbf{A})$ is set equal to zero. The equation to be solved then reads

$$i\hbar \frac{\partial}{\partial t} \psi = -i\hbar c\, \boldsymbol{\alpha} \cdot \nabla \psi + m_e c^2 \gamma^0 \psi. \qquad (4.46)$$

There are various ways to construct solutions of this equation. We choose a derivation that makes explicit use of the transformation (4.28).

### 4.2.1 Construction of eigenstates

The simplest case arises if the particle is at rest. Then the first term on the right-hand side of Eq. (4.46) is discarded and, with the representation (4.4) of $\gamma^0$, we can write for the independent solutions

$$\psi_1 = \begin{pmatrix} 1 \\ 0 \\ 0 \\ 0 \end{pmatrix} e^{-im_e c^2\, t/\hbar} \;, \qquad \psi_2 = \begin{pmatrix} 0 \\ 1 \\ 0 \\ 0 \end{pmatrix} e^{-im_e c^2\, t/\hbar} \qquad (4.47)$$

and

$$\psi_3 = \begin{pmatrix} 0 \\ 0 \\ 1 \\ 0 \end{pmatrix} e^{im_e c^2\, t/\hbar} \;, \qquad \psi_4 = \begin{pmatrix} 0 \\ 0 \\ 0 \\ 1 \end{pmatrix} e^{im_e c^2\, t/\hbar} . \qquad (4.48)$$

Here, the first two functions are "positive-energy" solutions while the last two are "negative-energy" solutions, since in this case the eigenvalues of the Hamiltonian operator (4.13) are $\pm m_e c^2$, depending on whether the eigenvalues of $\gamma^0$ are $\pm 1$. The existence of negative-energy solutions is intimately related to the property of the Dirac theory that it can accommodate a positron. This is further discussed in Sec. 4.2.2.

The spinors $\psi_1$, $\psi_2$ and $\psi_3$, $\psi_4$ are distinguished by the spin degree-of-freedom of a Schrödinger-Pauli electron residing in the two upper and the two lower components, respectively. In order to describe the spin of a Dirac electron, we introduce a $4 \times 4$ spin matrix $\boldsymbol{\Sigma}$ with the components

$$\Sigma_k = \frac{i}{2}(\gamma^i\gamma^j - \gamma^j\gamma^i) = \begin{pmatrix} \sigma_k & 0 \\ 0 & \sigma_k \end{pmatrix} , \qquad (ijk) \text{ cyclic}, \qquad (4.49)$$

where the relation

$$\sigma_1\sigma_2 = -\sigma_2\sigma_1 = i\sigma_3$$

for the Pauli matrices (4.3) has been used, and $(ijk)$ cyclic stands for $(ijk) = (1,2,3)$ $(2,3,1)$ or $(3,1,2)$.[4] We see that the spinors

[4] It can be shown [BjD64, Sak67] that $\frac{1}{2}\hbar\Sigma$ is the appropriate spin operator for four-component Dirac spinors. For example, in a central-force problem, the sum of $\mathbf{r}\times\mathbf{p}+\frac{1}{2}\hbar\Sigma$ is indeed a constant of the motion, to be identified with the total angular momentum, see Sec. 4.3.

(4.47) and (4.48) are eigenfunctions of $\Sigma_z = \Sigma_3$ with eigenvalues +1 and −1, corresponding to "spin up" and "spin down."

The Lorentz transformation (4.29) may be used to construct the free-particle solutions for an arbitrary speed of the electron. By transforming to a coordinate system with velocity $-\mathbf{v} = -v\hat{\mathbf{e}}_z$ relative to that of the solutions at rest, we obtain free-particle wave functions for an electron with the velocity $\mathbf{v} = v\hat{\mathbf{e}}_z$.

As an illustration, let us explicitly perform the transformation for $\psi_1$ of Eq. (4.47). Denoting the quantities in the moving frame with a prime, we have from Eq. (4.29) and from the Lorentz transformation (3.30) for the time coordinate (but with the velocity $-v$) the transformed spinor

$$S(-v)\psi_1 = \sqrt{\frac{1+\gamma}{2}}\,(1+\delta\alpha_z)\begin{pmatrix}1\\0\\0\\0\end{pmatrix} e^{-im_\mathrm{e}c^2\gamma(t'-vz'/c^2)/\hbar}.$$

We now use Eq. (2.18) to rewrite the exponent as $-i(E't'-p'z')/\hbar$ and to cast the normalization factor and the parameter $\delta$ in the form

$$\sqrt{\frac{1+\gamma}{2}} = \sqrt{\frac{E'+m_\mathrm{e}c^2}{2m_\mathrm{e}c^2}} \tag{4.50}$$

and

$$\delta = \sqrt{\frac{\gamma-1}{\gamma+1}} = \frac{p'c}{E'+m_\mathrm{e}c^2}\,, \tag{4.51}$$

so that

$$S\psi_1 = \sqrt{\frac{E'+m_\mathrm{e}c^2}{2m_\mathrm{e}c^2}}\begin{pmatrix}1\\0\\p'c/(E'+m_\mathrm{e}c^2)\\0\end{pmatrix} e^{-i(E't'-p'z')/\hbar}.$$

The general case can be treated either in a similar way by a Lorentz transformation from the rest frame of the electron or, alternatively, by directly solving Eq. (4.46). Dropping the primes, we get the general free-electron wave functions for an arbitrary momentum $\mathbf{p}$ as

$$\psi_\rho(\mathbf{p}) = N_\pm\, u^{(\rho)}(\mathbf{p})\, e^{i(\mathbf{p}\cdot\mathbf{r}-Et)/\hbar}, \qquad (\rho = 1,2,3,4) \tag{4.52}$$

where $N_\pm$ are normalization factors for $E>0$ and $E<0$, respectively.

The positive-energy ($E > 0$) spinors are

$$u^{(1,2)}(\mathbf{p}) = \sqrt{\frac{E + m_e c^2}{2m_e c^2}} \begin{pmatrix} \chi^{(s)} \\ \frac{\boldsymbol{\sigma} \cdot \mathbf{p}\, c}{E + m_e c^2}\, \chi^{(s)}, \end{pmatrix} \tag{4.53}$$

where $\chi^{(s)} = \chi^{(s)}_{1/2}$ is a Pauli spinor and $s = \pm 1/2$ denotes the spin projection. Explicitly, we have

$$u^{(1)}(\mathbf{p}) = \sqrt{\frac{E + m_e c^2}{2m_e c^2}} \begin{pmatrix} 1 \\ 0 \\ p_z c/(E + m_e c^2) \\ (p_x + i p_y) c/(E + m_e c^2) \end{pmatrix} \quad \text{for spin up} \tag{4.54}$$

and

$$u^{(2)}(\mathbf{p}) = \sqrt{\frac{E + m_e c^2}{2m_e c^2}} \begin{pmatrix} 0 \\ 1 \\ (p_x - i p_y) c/(E + m_e c^2) \\ -\, p_z c/(E + m_e c^2) \end{pmatrix} \quad \text{for spin down.} \tag{4.55}$$

The negative-energy ($E < 0$) spinors are

$$u^{(3,4)}(\mathbf{p}) = \sqrt{\frac{|E| + m_e c^2}{2m_e c^2}} \begin{pmatrix} -\,\frac{\boldsymbol{\sigma} \cdot \mathbf{p}\, c}{|E| + m_e c^2}\, \chi^{(s)} \\ \chi^{(s)} \end{pmatrix} \tag{4.56}$$

or explicitly

$$u^{(3)}(\mathbf{p}) = \sqrt{\frac{|E| + m_e c^2}{2m_e c^2}} \begin{pmatrix} -\, p_z c/(|E| + m_e c^2) \\ -\,(p_x + i p_y) c/(|E| + m_e c^2) \\ 1 \\ 0 \end{pmatrix} \quad \text{for spin up} \tag{4.57}$$

and

$$u^{(4)}(\mathbf{p}) = \sqrt{\frac{|E| + m_e c^2}{2m_e c^2}} \begin{pmatrix} -(p_x - ip_y)c/(|E| + m_e c^2) \\ p_z c/(|E| + m_e c^2) \\ 0 \\ 1 \end{pmatrix} \quad \text{for spin down.} \tag{4.58}$$

For nonrelativistic or moderately relativistic electron motion, we have $pc \ll m_e c^2$. In this case, the two lower components of $u^{(1)}$ and $u^{(2)}$ and the two upper components of $u^{(3)}$ and $u^{(4)}$ are sometimes referred to as the "small components."

The spinors $u^{(\rho)}$ are eigenfunctions of the "helicity operator" $\boldsymbol{\Sigma} \cdot \hat{\mathbf{p}}$ (where $\hat{\mathbf{p}} = \mathbf{p}/p$) with eigenvalues $\pm 1$, that is, the spin has sharp values only parallel or antiparallel to the direction of motion. Hence, in contrast to nonrelativistic Pauli theory, it is not possible to choose solutions in such a way that they are eigenvectors of $\boldsymbol{\Sigma} \cdot \hat{\mathbf{n}}$ with an arbitrary unit vector $\hat{\mathbf{n}}$.

We also observe that in all cases, the parity of the upper two components is opposite to the parity of the lower two components. This is a consequence of the requirement that the four-component wave function is an eigenstate of the parity operator (4.36).

By inserting the solutions (4.52) into the free-particle Dirac equation (4.46), it can be shown that the spinors $u^{(\rho)}(\mathbf{p})$ satisfy the equation

$$(\gamma^\mu p_\mu - m_e c) u^{(\rho)}(\mathbf{p}) = 0, \tag{4.59}$$

regardless of whether $E > 0$ or $E < 0$. One may also verify the orthogonality and normalization relation

$$u^{(\rho)\dagger}(\mathbf{p})\, u^{(\rho')}(\mathbf{p}) = \frac{|E|}{m_e c^2}\, \delta_{\rho\,\rho'}. \tag{4.60}$$

In order to have a normalized total wave function,

$$\int \psi^\dagger(\mathbf{p}) \psi(\mathbf{p}')\, d^3r = \delta(\mathbf{p} - \mathbf{p}'), \tag{4.61}$$

the normalization coefficients $N_\pm$ in Eq. (4.52) have to compensate the spinor normalization (4.60). Rewriting this explicitly, we have for $E = \sqrt{|\mathbf{p}|^2 c^2 + m_e^2 c^4} > 0$ with $\rho = 1, 2$

$$\psi_+ = (2\pi)^{-3/2} \sqrt{\frac{m_e c^2}{E}}\, u^{(\rho)}(\mathbf{p})\, e^{i(\mathbf{p}\cdot\mathbf{r} - Et)/\hbar} \tag{4.62}$$

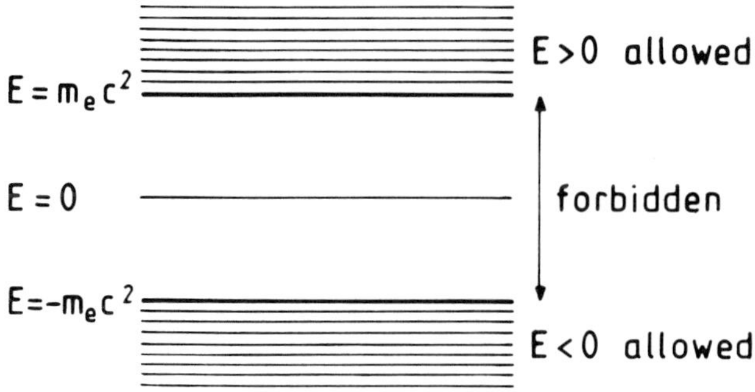


**Figure 4.1.** The allowed values for the energy $E$ of a free Dirac particle.

and for $E = -\sqrt{|\mathbf{p}|^2c^2 + m_e^2c^4} < 0$ with $\rho = 3, 4$

$$\psi_- = (2\pi)^{-3/2}\sqrt{\frac{m_e c^2}{|E|}}\, u^{(\rho)}(\mathbf{p})\, e^{i(\mathbf{p}\cdot\mathbf{r}+|E|t)/\hbar}. \tag{4.63}$$

The allowed values for $E$ form a continuous spectrum. For positive-energy free-particle solutions $m_e c^2 \leq E < \infty$ while for negative-energy solutions $-\infty < E \leq -m_e c^2$, as shown in Fig. 4.1.

### 4.2.2 Hole theory and field theoretic interpretation

*Difficulties with the negative-energy states*

At the present stage, the negative-energy solutions of the Dirac equation appear as an unavoidable consequence of the mathematical treatment. Negative-energy states to be discussed here are not confined to the case of free electrons, but also emerge in the presence of potentials like those treated in Secs. 4.3 and 4.4. These solutions have some disturbing features, for which we just give a few examples. (a) The ground state of a hydrogen atom is unstable when coupled to the radiation field: it may decay into a negative-energy state by the emission of a photon. (b) Negative-energy free-electron solutions are indispensable to build a well-localized positive-energy wave packet. Positive-energy plane waves alone forbid a localization to better than the Compton wavelength $\lambda\!\!\!^{-}_c = \hbar/m_e c$ [NeW49]. (c) Positive-energy bound-state wave functions, when expanded in terms of free-particle plane waves, contain some negative-energy components.

While (b) and (c) indicate the incompleteness of positive-energy states alone, the catastrophe of (a) requires the reinterpretation of the Dirac equation. Leaving the original ground of the one-particle equation, Dirac postulated that all the troublesome negative-energy states be filled up with noninteracting electrons, in accord with the Pauli principle, thus forming the "Dirac sea," The vacuum state is then one with all negative-energy levels occupied by noninteracting electrons and all positive-energy states empty. The stability of the hydrogen atom is assured because no additional electrons can be accommodated in the negative-energy sea on account of the Pauli principle.

*Hole theory*

Under suitable conditions, one of the negative-energy electrons in the Dirac sea with energy $E_- = -|E_-|$ can absorb a photon of energy $\hbar\omega > 2m_ec^2 + |E_-|$ and become a positive-energy electron with the energy $|E_+| = \hbar\omega - 2m_ec^2 - |E_-|$, as sketched in Fig. 4.2. As a result, a "hole" is created in the Dirac sea. The hole describes the absence of an electron (in the Dirac sea) of energy $-|E_-|$, momentum $\mathbf{p}$, and charge $-e$. Comparing this situation with the vacuum, an observer would interpret photoproduction of a hole as the simultaneous creation of an electron with energy $+|E_+|$ and charge $-e$ and a *positron* with energy $+|E_-|$ and charge $+e$. This is the basis of describing electron-positron pair production within the hole theory. Conversely, a hole in the Dirac sea may annihilate a positive-energy electron with the simultaneous emission of radiation. Physically, pair production or annihilation of free electrons and positrons is prohibited by energy-momentum conservation. It may, however, occur "off-shell" in the presence of a nucleus which takes up the excess energy and momentum.

With the hole theory, the single-particle interpretation of the Dirac equation has been sacrificed. Instead, we have a many-particle theory, describing particles of negative as well as of positive charges. While the hole theory assures the stability of the hydrogen atom, it still has disturbing features. It is certainly difficult to visualize a sea of noninteracting electrons in negative-energy states with the associated infinite mass and infinite charge. As a remedy, one usually defines a new vacuum by the requirement that it contains neither electrons nor positrons.

*Field-theoretic interpretation*

To this end, the Dirac equation is reinterpreted once again, namely as an equation to be satisfied by space-time-dependent *field operators* $\widehat{\psi}$. This is done within the formalism of field quantization (or "second quantization") permitting a description of systems for which the number of particles is

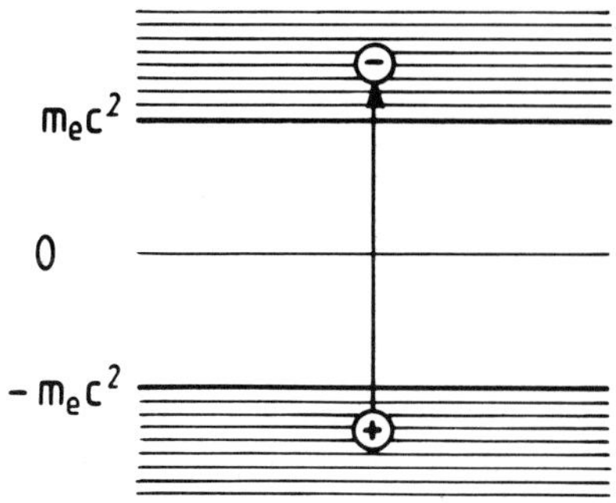


**Figure 4.2.** Schematic diagram of a pair creation process.

no longer a constant of the motion. For many-particle systems, a field theoretical description takes automatically into account the combinatorial aspects, that is, in the case of Dirac particles (fermions), the antisymmetry with respect to permutations of particles is satisfied, see Sec. 5.6.

If we refer all physical states to a "vacuum state," defined in Eq. (4.68) below, in which neither electrons nor positrons are present, space-time-dependent field operators $\widehat{\psi}$ or $\widehat{\psi}^\dagger$ (denoted by a "hat") are defined as creating or annihilating a particle at a given space-time point.

These operators are usually expanded as

$$\widehat{\psi}(\mathbf{x},t) = \sum_{k_+,s} \widehat{b}_{k_+}^{(s)} \psi_{k_+}^{(s)}(\mathbf{x},t) + \sum_{k_-,s} \widehat{d}_{k_-}^{(s)\dagger} \psi_{k_-}^{(s)}(\mathbf{x},t) \tag{4.64}$$

where, at each instant of time, the $\psi_{k_+}^{(s)}(\mathbf{x},t), \psi_{k_-}^{(s)}(\mathbf{x},t)$ form a complete set of orthonormal eigenstates of the Dirac equation, $\widehat{b}_{k_+}^{(s)}$ is the annihilation operator for an electron (a positive-energy state) with a discrete momentum $\mathbf{k}_+$ and a spin projection $s$, while $\widehat{d}_{k_-}^{(s)\dagger}$ is the creation operator for a positron (a negative-energy state) with momentum $\mathbf{k}_-$ and spin projection $s$. The operators $\widehat{b}, \widehat{b}^\dagger, \widehat{d}, \widehat{d}^\dagger$ obey fermion anticommutation relations [Sch61].

Defining anticommutators by braces as $\{A, B\} = AB + BA$, we have for the electrons

$$\begin{aligned} \{\widehat{b}_k^{(s)}, \widehat{b}_{k'}^{(s')\dagger}\} &= \delta_{ss'}\delta_{kk'} \\ \{\widehat{b}_k^{(s)}, \widehat{b}_{k'}^{(s')}\} &= \{\widehat{b}_k^{(s)\dagger}, \widehat{b}_{k'}^{(s')\dagger}\} = 0, \end{aligned} \tag{4.65}$$

and for positrons

$$\{\widehat{d}_k^{(s)}, \widehat{d}_{k'}^{(s')\dagger}\} = \delta_{ss'}\delta_{kk'}$$

$$\{\widehat{d}_k^{(s)}, \widehat{d}_{k'}^{(s')}\} = \{\widehat{d}_k^{(s)\dagger}, \widehat{d}_{k'}^{(s')\dagger}\} = 0, \tag{4.66}$$

and, finally, the mixed relations

$$\{\widehat{b}_k^{(s)}, \widehat{d}_{k'}^{(s')}\} = \{\widehat{b}_k^{(s)\dagger}, \widehat{d}_{k'}^{(s')\dagger}\} = \{\widehat{b}_k^{(s)\dagger}, \widehat{d}_{k'}^{(s')}\} = \{\widehat{b}_k^{(s)\dagger}, \widehat{d}_{k'}^{(s')\dagger}\} = 0. \tag{4.67}$$

These relations ensure that there cannot be more than one electron or positron in any given quantum state, so that the Pauli principle is always satisfied.

The functions $\psi_k^{(s)}$ are assumed to form an orthonormal and complete set of unperturbed eigenfunctions in the initial or, alternatively, in the final channel. One often uses plane-wave solutions; however, we may also take solutions discussed in Secs. 4.3 and 4.4.

Defining a vacuum state $|0\rangle$ by

$$\begin{aligned} \widehat{b}_{k_+}^{(s)}|0\rangle &= \widehat{d}_{k_-}^{(s)}|0\rangle = 0, \\ \langle 0|\,\widehat{b}_{k_+}^{(s)\dagger} &= \langle 0|\,\widehat{d}_{k_-}^{(s)\dagger} = 0, \end{aligned} \tag{4.68}$$

a free single-electron state is given by

$$|k_+\rangle = \widehat{b}_{k_+}^{(s)\dagger}|0\rangle \tag{4.69}$$

and a free electron-positron state by

$$|k_+, k_-\rangle = \widehat{b}_{k_+}^{(s)\dagger}\widehat{d}_{k_-}^{(s')\dagger}|0\rangle. \tag{4.70}$$

In this formalism, the Dirac wave functions serve as expansion coefficients while the Pauli principle is taken care of by the anticommutation relations satisfied by the creation and annihilation operators.

We now have to verify that the difficulties of hole theory with negative-energy states are cured in a ("occupation number") representation in which the vacuum is defined by Eq. (4.68). We first define the number operators

$$\begin{aligned} \widehat{N}_{k_+}^{(s)} &= \widehat{b}_{k_+}^{(s)\dagger}\,\widehat{b}_{k_+}^{(s)}, \\ \widehat{N}_{k_-}^{(s)} &= \widehat{d}_{k_-}^{(s)\dagger}\,\widehat{d}_{k_-}^{(s)} \end{aligned} \tag{4.71}$$

for electrons and positrons, respectively, with momentum $\mathbf{k}_\pm$ and spin projection $s$. These operators have the eigenvalue 0, if acting on a state in which there is no electron (or positron) with these quantum numbers, and the eigenvalue 1, if there is one electron (or positron) with these quantum numbers.

Since the operator for the total energy for free electrons can be expressed as

$$\begin{aligned} \widehat{H} &= \int \widehat{\psi}^\dagger(-i\hbar c\boldsymbol{\alpha}\cdot\nabla + m_{\rm e}c^2\gamma^0)\widehat{\psi}\, d^3x \\ &= \sum_{k_+,s} E_{k_+}\widehat{N}^{(s)}_{k_+} + \sum_{k_-,s} E_{k_-}\widehat{N}^{(s)}_{k_-}, \end{aligned} \tag{4.72}$$

where $E_{k_\pm} = \sqrt{k_\pm^2 c^2 + m_{\rm e}^2 c^4}$, and since the operator for the total charge is

$$\begin{aligned} \widehat{Q} &= -e\int \widehat{\psi}^\dagger\widehat{\psi}\, d^3x \\ &= -e\sum_{k_+,s}\widehat{N}^{(s)}_{k_+} + e\sum_{k_-,s}\widehat{N}^{(s)}_{k_-}, \end{aligned} \tag{4.73}$$

we see that the vacuum defined by Eq. (4.68) has zero energy and zero charge. For a state with several electrons and/or positrons, the energy eigenvalue obtained from (4.72) is necessarily positive definite and the total charge is $e$ times the number of positrons minus $e$ times the number of electrons. Thus we find that a reinterpretation of the Dirac equation in the formalism of field theory formally removes the difficulties of the hole theory.

We return to the field theoretical description in Sec. 5.3, when we discuss transition amplitudes, and in Chap. 10 in connection with single and multiple pair production.

## 4.3 Bound states in a Coulomb field

Before and after an ion-atom collision, the electrons in initial or final states are subject to the Coulomb potential produced by a single nucleus, either the target or the projectile. In the present and following sections, we discuss the bound and continuous eigenstates, respectively, of an electron moving in the Coulomb potential originating from a nuclear charge $Ze$. Writing the time-dependent wave function in the form

$$\psi(\mathbf{r},t) = \varphi(\mathbf{r})\, e^{-iEt/\hbar}, \tag{4.74}$$

hydrogenic states are defined as the solutions of the stationary Dirac equation

$$H\varphi(\mathbf{r}) = \left(-i\hbar c\boldsymbol{\alpha}\cdot\nabla - \frac{Ze^2}{r} + m_{\rm e}c^2\gamma^0\right)\varphi(\mathbf{r}) = E\varphi(\mathbf{r}), \tag{4.75}$$

where $\mathbf{r}$ is the coordinate of the electron with respect to the nucleus and E is the eigenenergy. Before summarizing the exact and approximate solutions of Eq. (4.75), we start by classifying the states in a spherical potential.

### 4.3.1 Classification of states in a spherical potential

The constants of the motion for a particle moving in a spherically symmetric potential $V(r)$ are independent of its particular form. In nonrelativistic quantum theory, the Hamiltonian commutes with the operators $\mathbf{L}^2, \mathbf{J}^2$, and $J_z$ where $\mathbf{L} = \mathbf{r} \times \mathbf{p}$ is the orbital angular momentum and $\mathbf{J} = \mathbf{L} + \frac{1}{2}\hbar\boldsymbol{\sigma}$ is the total angular momentum. The eigenstates may, therefore, be classified by the corresponding quantum numbers $l, j = l \pm \frac{1}{2}$, and $m_j$. In relativistic quantum theory, the operator $\mathbf{L}^2$ no longer commutes with the Hamiltonian $H$ defined by Eq. (4.75). On the other hand, the total angular momentum

$$\mathbf{J} = \mathbf{L} + \tfrac{1}{2}\hbar\boldsymbol{\Sigma}, \tag{4.76}$$

where $\boldsymbol{\Sigma}$ is defined in Eq. (4.49), commutes with $H$ and hence is a constant of the motion. While $\mathbf{L}$ and $\boldsymbol{\Sigma}$, taken separately, are not constants of the motion, it can be shown [BjD64, Sak67] that a spin-orbit operator $K$ defined by

$$K = \gamma^0(\boldsymbol{\Sigma} \cdot \mathbf{L} + \hbar) \tag{4.77}$$

does commute with $H$ and therefore is able to furnish the missing quantum number needed for a complete specification of spherical states, in analogy to nonrelativistic quantum mechanics. Indeed, since the operators $H, K, \mathbf{J}^2$, and $J_z$ all commute with one another, one can construct simultaneous eigenfunctions of $H, K, \mathbf{J}^2$, and $J_z$. Their corresponding eigenvalues are denoted by $E$, $-\hbar\kappa$, $j(j+1)\hbar^2$, and $m_j\hbar$. In order to derive an important relation between $\kappa$ and $j$, we compare

$$K^2 = \mathbf{L}^2 + \hbar\boldsymbol{\Sigma} \cdot \mathbf{L} + \hbar^2$$

with

$$\mathbf{J}^2 = \mathbf{L}^2 + \hbar\boldsymbol{\Sigma} \cdot \mathbf{L} + \tfrac{3}{4}\hbar^2$$

to get

$$K^2 = \mathbf{J}^2 + \tfrac{1}{4}\hbar^2. \tag{4.78}$$

Therefore, we must have

$$\kappa = \pm(j + \tfrac{1}{2}), \tag{4.79}$$

where $\kappa$ is a nonzero integer which can be positive or negative. It is now instructive to decompose the four-component spinor $\varphi$ into the upper components $\varphi_A$ and the lower components $\varphi_B$ so that

$$\varphi = \begin{pmatrix} \varphi_A \\ \varphi_B \end{pmatrix} \tag{4.80}$$

and correspondingly

$$K = \begin{pmatrix} \boldsymbol{\sigma} \cdot \mathbf{L} + \hbar & 0 \\ 0 & -\boldsymbol{\sigma} \cdot \mathbf{L} - \hbar \end{pmatrix}. \tag{4.81}$$

If indeed the energy eigenfunction $\varphi$ is a simultaneous eigenfunction of $K, \mathbf{J}^2$, and $J_z$, the upper and lower two-component spinors $\varphi_A$ and $\varphi_B$ obey the following set of eigenvalue equations

$$(\boldsymbol{\sigma} \cdot \mathbf{L} + \hbar)\varphi_A = -\kappa\, \hbar \varphi_A, \qquad (\boldsymbol{\sigma} \cdot \mathbf{L} + \hbar)\varphi_B = \kappa\, \hbar \varphi_B, \tag{4.82}$$

in addition to

$$\begin{aligned} \mathbf{J}^2 \varphi_{A,B} &= (\mathbf{L} + \tfrac{1}{2}\hbar\boldsymbol{\sigma})^2 \varphi_{A,B} = j(j+1)\hbar^2 \varphi_{A,B} \\ J_z \varphi_{A,B} &= (L_z + \tfrac{1}{2}\hbar\sigma_z)\varphi_{A,B} = m_j \hbar \varphi_{A,B}. \end{aligned} \tag{4.83}$$

We have used here the same notation for $\mathbf{J}$ and $\mathbf{L}$ in the two-component as in the four-component representation. In the former, the operators are connected by $\mathbf{L}^2 = \mathbf{J}^2 - \boldsymbol{\sigma} \cdot \mathbf{L}\hbar - \frac{3}{4}\hbar^2$. This means that any two-component eigenfunction $\varphi_A$ or $\varphi_B$ of $(\boldsymbol{\sigma} \cdot \mathbf{L} + \hbar)$ and $\mathbf{J}^2$ is automatically an eigenfunction of $\mathbf{L}^2$. Thus, whereas the four-component $\varphi$ is not an eigenfunction of $\mathbf{L}^2$, the two-component wave functions $\varphi_A$ and $\varphi_B$ separately are eigenfunctions of $\mathbf{L}^2$. We denote their eigenvalues with $l_A(l_A+1)\hbar^2$ and $l_B(l_B+1)\hbar^2$ where $l_{A,B} = j \pm \frac{1}{2}$. By comparing Eqs. (4.82) and (4.83) we obtain the relations

$$\begin{aligned} -\kappa &= j(j+1) - l_A(l_A+1) + \tfrac{1}{4} \\ \kappa &= j(j+1) - l_B(l_B+1) + \tfrac{1}{4}. \end{aligned} \tag{4.84}$$

For a given value of $\kappa$ we have

$$\begin{aligned} j &= |\kappa| - \tfrac{1}{2} \\ l_A &= \begin{cases} \kappa & \text{if } \kappa > 0 \\ |\kappa| - 1 & \text{if } \kappa < 0. \end{cases}, \\ l_B &= \begin{cases} \kappa - 1 & \text{if } \kappa > 0 \\ |\kappa| & \text{if } \kappa < 0. \end{cases} \end{aligned} \tag{4.85}$$

It is customary to denote relativistic wave functions by the total angular momentum and by the orbital angular momentum of the upper components. This establishes a connection to the nonrelativistic notation. For

**Table 4.1.** Relativistic quantum numbers and spectroscopic notation.

| $\kappa$ | $j$ | $l_A$ | $l_B$ | $l_j$ |
|---|---|---|---|---|
| $-1$ | 1/2 | 0 | 1 | $s_{1/2}$ |
| $+1$ | 1/2 | 1 | 0 | $p_{1/2}$ |
| $-2$ | 3/2 | 1 | 2 | $p_{3/2}$ |
| $+2$ | 3/2 | 2 | 1 | $d_{3/2}$ |
| $-3$ | 5/2 | 2 | 3 | $d_{5/2}$ |
| $+3$ | 5/2 | 3 | 2 | $f_{5/2}$ |
| $-4$ | 7/2 | 3 | 4 | $f_{7/2}$ |
| $+4$ | 7/2 | 4 | 3 | $g_{7/2}$ |

convenience, we give in Table 4.1 the quantum numbers $\kappa, j, l_A$, and $l_B$ of the lowest states together with the spectroscopic notation $l_j$. We see that the parities of $\varphi_A$ and $\varphi_B$ are necessarily opposite as required by the definition (4.36) of the parity operator $\Pi$ for a four-component wave function $\varphi$ with a definite parity.

### 4.3.2 The hydrogen atom

In the particular case of a Coulomb potential $-Ze^2/r$, we seek solutions of the stationary Dirac equation (4.75). Following the classification scheme of Sec. 4.3.1, we write the bound-state wave functions in the form

$$\varphi_{\kappa m_j}(\mathbf{r}) = \begin{pmatrix} g_\kappa(r)\chi_\kappa^{m_j}(\hat{\mathbf{r}}) \\ if_\kappa(r)\chi_{-\kappa}^{m_j}(\hat{\mathbf{r}}) \end{pmatrix}, \tag{4.86}$$

where the $\chi_\kappa^{m_j}$ are normalized spin-angular functions defined as eigenfunctions of $\mathbf{J}^2, \mathbf{L}^2$, and $J_z$, characterized by the angular momentum $j$ and the projection $m_j$

$$\chi_\kappa^{m_j}(\hat{\mathbf{r}}) = \sum_{m_l} \left( \begin{array}{cc|c} l & \frac{1}{2} & j \\ m_l & m_j - m_l & m_j \end{array} \right) Y_{lm_l}(\hat{\mathbf{r}})\, \chi_{\frac{1}{2}}^{m_j - m_l}. \tag{4.87}$$

Here $\hat{\mathbf{r}} = \mathbf{r}/r$ is a unit vector,

$$\left( \begin{array}{cc|c} j_1 & j_2 & j_3 \\ m_1 & m_2 & m_3 \end{array} \right) \equiv C(j_1 j_2 j_3; m_1 m_2) \equiv (j_1 m_1 j_2 m_2 | j_1 j_2 j_3 m_3) \tag{4.88}$$

is a Clebsch-Gordan coefficient [Ros57, Edm57], $Y_{lm}$ is a spherical harmonic, and $\chi_{\frac{1}{2}}^{\pm\frac{1}{2}}$ is a Pauli spinor.[5] Explicitly we have
for $j = l + \frac{1}{2}$

$$\chi_\kappa^{m_j} = \sqrt{\frac{l + m_j + \frac{1}{2}}{2l+1}}\, Y_{l,m_j-1/2} \begin{pmatrix} 1 \\ 0 \end{pmatrix} + \sqrt{\frac{l - m_j + \frac{1}{2}}{2l+1}}\, Y_{l,m_j+1/2} \begin{pmatrix} 0 \\ 1 \end{pmatrix} \tag{4.89}$$

and for $j = l - \frac{1}{2}$

$$\chi_\kappa^{m_j} = -\sqrt{\frac{l - m_j + \frac{1}{2}}{2l+1}}\, Y_{l,m_j-1/2} \begin{pmatrix} 1 \\ 0 \end{pmatrix} + \sqrt{\frac{l + m_j + \frac{1}{2}}{2l+1}}\, Y_{l,m_j+1/2} \begin{pmatrix} 0 \\ 1 \end{pmatrix}. \tag{4.90}$$

The radial functions depend on the quantum number $\kappa$. The factor $i$ has been inserted in Eq. (4.86) to make $f$ and $g$ real for bound-state problems.

The spin-angular functions satisfy the relation [Ros57]

$$(\boldsymbol{\sigma} \cdot \hat{\mathbf{r}})\, \chi_\kappa^{m_j} = -\chi_{-\kappa}^{m_j}. \tag{4.91}$$

When substituting (4.86) into the Dirac equation (4.75), decomposed in the form

$$\begin{aligned} c(\boldsymbol{\sigma} \cdot \mathbf{p})\, \varphi_B &= (E - V - m_e c^2)\varphi_A \\ c(\boldsymbol{\sigma} \cdot \mathbf{p})\, \varphi_A &= (E - V + m_e c^2)\varphi_B \end{aligned} \tag{4.92}$$

with $V = -Ze^2/r$, we need the relation

$$\begin{aligned} \boldsymbol{\sigma} \cdot \mathbf{p} &= \frac{\boldsymbol{\sigma} \cdot \mathbf{r}}{r^2} (\boldsymbol{\sigma} \cdot \mathbf{r})(\boldsymbol{\sigma} \cdot \mathbf{p}) \\ &= \frac{\boldsymbol{\sigma} \cdot \mathbf{r}}{r^2} \left( -i\hbar r \frac{\partial}{\partial r} + i \boldsymbol{\sigma} \cdot \mathbf{L} \right). \end{aligned}$$

[5] Throughout this book, we follow the phase convention of Condon and Shortley [CoS51, Ros57, Ros61, Mes62]. The phase convention by Bethe and Salpeter [BeS57] is different owing to an unconventional definition of the spherical harmonics.

**Table 4.2.** Relativistic quantum numbers of hydrogenic states and spectroscopic notation. Pairs of states degenerate according to Eq. (4.95) are grouped together.

| $n$ | $n' = n - \lvert\kappa\rvert$ | $\kappa = \pm(j + \frac{1}{2})$ | notation |
|---|---|---|---|
| 1 | 0 | $-1$ | $1s_{1/2}$ |
| 2 | 1 | $-1$ | $2s_{1/2}$ |
| 2 | 1 | $+1$ | $2p_{1/2}$ |
| 2 | 0 | $-2$ | $2p_{3/2}$ |
| 3 | 2 | $-1$ | $3s_{1/2}$ |
| 3 | 2 | $+1$ | $3p_{1/2}$ |
| 3 | 1 | $-2$ | $3p_{3/2}$ |
| 3 | 1 | $+2$ | $3d_{3/2}$ |
| 3 | 0 | $-3$ | $3d_{5/2}$ |

We then find that the spin-angular functions $\chi_\kappa^{m_j}$ appear as common factors in the coupled equations and hence can be discarded. With the substitutions $F(r) = r\,f(r)$, $G(r) = r\,g(r)$, we finally obtain the coupled radial equations

$$\begin{aligned}\hbar c\left(\frac{dF}{dr} - \frac{\kappa}{r}F\right) &= -(E - V - m_\mathrm{e}c^2)G \\ \hbar c\left(\frac{dG}{dr} + \frac{\kappa}{r}G\right) &= (E - V + m_\mathrm{e}c^2)F.\end{aligned} \tag{4.93}$$

When we apply the standard methods [Mes62] known from the treatment of the nonrelativistic hydrogen atom, we may obtain eigenenergies as well as eigenfunctions.

Introducing the fine-structure constant $\alpha = e^2/\hbar c = 1/137.036$, setting $\zeta = \alpha Z$ and

$$s = \sqrt{\kappa^2 - \zeta^2}, \tag{4.94}$$

we may write the Sommerfeld formula for the hydrogenic eigenenergies as

$$E_{n\kappa} = m_\mathrm{e}c^2\,W_{n\kappa}$$

$$W_{n\kappa} = \left[1 + \left(\frac{\zeta}{n' + s}\right)^2\right]^{-\frac{1}{2}}, \tag{4.95}$$

where $n' = 0, 1, 2, \cdots$ counts the nodes of the radial wave function and is related to the principal quantum number $n = 1, 2, 3, \cdots$ by $n = n' + |\kappa|$. Since $W_{n\kappa} < 1$, Eq. (4.95) can be expanded in terms of $\zeta = \alpha Z$ to give

$$W_{n\kappa} = 1 - \frac{1}{2}\frac{(\alpha Z)^2}{n^2} - \frac{1}{2}\frac{(\alpha Z)^4}{n^3}\left(\frac{1}{j+\frac{1}{2}} - \frac{3}{4n}\right) - \cdots, \tag{4.96}$$

where the second term, multiplied with $m_e c^2$, represents the nonrelativistic binding energy in a hydrogenic atom.

Now, if $\lambda\!\!\!^{-}_c = \hbar/m_e c$ is the electron Compton wavelength and

$$q = \frac{1}{\lambda\!\!\!^{-}_c}\sqrt{1 - W_{n\kappa}^2} = \frac{\zeta}{\lambda\!\!\!^{-}_c}\left[\zeta^2 + (n' + s)^2\right]^{-\frac{1}{2}} \tag{4.97}$$

is the bound-state wave number (suppressing the labels $n$ and $\kappa$ in $q$), the radial functions $g(r)$ and $f(r)$ can be expressed by confluent hypergeometric functions ${}_1F_1(a, c; x)$ as

$$\begin{aligned} g_\kappa(r) &= N_g\,(2qr)^{s-1}e^{-qr}[-n'\;{}_1F_1(-n'+1, 2s+1; 2qr) \\ &\quad -(\kappa - \frac{\zeta}{q\lambda\!\!\!^{-}_c})\;{}_1F_1(-n', 2s+1; 2qr)] \end{aligned}$$

$$\begin{aligned} f_\kappa(r) &= N_f\,(2qr)^{s-1}e^{-qr}[n'\;{}_1F_1(-n'+1, 2s+1; 2qr) \\ &\quad -(\kappa - \frac{\zeta}{q\lambda\!\!\!^{-}_c})\;{}_1F_1(-n', 2s+1; 2qr)] \end{aligned} \tag{4.98}$$

with

$$\begin{aligned} N_g &= \frac{\sqrt{2}\,q^{\frac{5}{2}}\lambda\!\!\!^{-}_c}{\Gamma(2s+1)}\left[\frac{\Gamma(2s+n'+1)(1+W_{n\kappa})}{n'!\,\zeta(\zeta - \kappa q\lambda\!\!\!^{-}_c)}\right]^{\frac{1}{2}}, \\ N_f &= -N_g\left(\frac{1 - W_{n\kappa}}{1 + W_{n\kappa}}\right)^{\frac{1}{2}}. \end{aligned} \tag{4.99}$$

For some purposes, it is sufficient to consider the hydrogenic ground state $1s_{1/2}$ with $\kappa = -1$. Here, $W_{n\kappa} = s$, $E = m_e c^2 W_{n\kappa}$, and with the Bohr radius $a_0 = \lambda\!\!\!^{-}_c/\alpha = \hbar^2/m_e e^2$, we have

$$\varphi^{\frac{1}{2}}_{1s_{1/2}}(\mathbf{r}) = (4\pi)^{-\frac{1}{2}}\begin{pmatrix} g(r) \\ 0 \\ -if(r)\cos\theta \\ -if(r)\sin\theta e^{i\phi} \end{pmatrix}$$

**Table 4.3.** Parameters defining the K- and L-shell radial wave functions for a Coulomb field [Ros61]. Here, $\zeta = \alpha Z$, $W = E/m_e c^2$, and $q$ is measured in units of the inverse electron Compton wavelength $\lambda\!\!\!^{-}_c{}^{-1}$.

| subshell | $s$ | $W$ | $q$ | $N$ |
|---|---|---|---|---|
| $1s_{1/2}: \kappa = -1$ | $\sqrt{1-\zeta^2}$ | $s$ | $\zeta$ | $\dfrac{(2\zeta)^{s+1/2}}{[2\Gamma(2s+1)]^{1/2}}$ |
| $2s_{1/2}: \kappa = -1$ | $\sqrt{1-\zeta^2}$ | $\sqrt{\dfrac{1+s}{2}}$ | $\dfrac{\zeta}{2W}$ | $\dfrac{(2\zeta)^{s+1/2}}{2(2W)^{s+1}} \times \left[\dfrac{2s+1}{\Gamma(2s+1)(2W+1)}\right]^{1/2}$ |
| $2p_{1/2}: \kappa = 1$ | $\sqrt{1-\zeta^2}$ | $\sqrt{\dfrac{1+s}{2}}$ | $\dfrac{\zeta}{2W}$ | $\dfrac{(2\zeta)^{s+1/2}}{2(2W)^{s+1}} \times \left[\dfrac{2s+1}{\Gamma(2s+1)(2W-1)}\right]^{1/2}$ |
| $2p_{3/2}: \kappa = -2$ | $\sqrt{4-\zeta^2}$ | $\frac{1}{2}s$ | $\frac{1}{2}\zeta$ | $\dfrac{\zeta^{s+1/2}}{[2\Gamma(2s+1)]^{1/2}}$ |

**Table 4.4.** Expansion coefficients for the K- and L-shell radial wave functions for a Coulomb field [Ros61].

| subshell | $b_0$ | $b_1$ | $c_0$ | $c_1$ |
|---|---|---|---|---|
| $1s_{1/2}: \kappa = -1$ | 1 | 0 | 1 | 0 |
| $2s_{1/2}: \kappa = -1$ | $2(W+1)$ | $-\dfrac{\zeta}{W}\dfrac{2W+1}{2s+1}$ | $2W$ | $-\dfrac{\zeta}{W}\dfrac{2W+1}{2s+1}$ |
| $2p_{1/2}: \kappa = 1$ | $2W$ | $-\dfrac{\zeta}{W}\dfrac{2W-1}{2s+1}$ | $2(W-1)$ | $-\dfrac{\zeta}{W}\dfrac{2W-1}{2s+1}$ |
| $2p_{3/2}: \kappa = -2$ | 1 | 0 | 1 | 0 |

$$\varphi_{1s_{1/2}}^{-\frac{1}{2}}(\mathbf{r}) \;=\; (4\pi)^{-\frac{1}{2}} \begin{pmatrix} 0 \\ g(r) \\ -if(r)\sin\theta e^{-i\phi} \\ if(r)\cos\theta \end{pmatrix} \tag{4.100}$$

with the radial wave functions

$$g(r) = N_g \left(\frac{r}{a_0}\right)^{s-1} e^{-Zr/a_0} \quad \text{and} \quad f(r) = N_f \left(\frac{r}{a_0}\right)^{s-1} e^{-Zr/a_0}, \tag{4.101}$$

where

$$\begin{aligned} N_g &= a_0^{-\frac{3}{2}} \frac{(2Z)^{s+\frac{1}{2}}}{[2\,\Gamma(2s+1)]^{\frac{1}{2}}} (1+s)^{\frac{1}{2}} \\ N_f &= -\left(\frac{1-s}{1+s}\right)^{\frac{1}{2}} N_g. \end{aligned} \tag{4.102}$$

Since for $|\kappa| = 1$ one has $s < 1$ according to Eq. (4.94), a mild singularity appears at the origin in the wave functions for $1s_{1/2}$ and, similarly, for $2p_{1/2}$ states.

When evaluating Eq. (4.95) for the energy of excited states, we find that within one principal shell $n$, the energy is the same for equal values of $j$ but for equal values of $l$ it is different. For a given value of $l$, the spin-orbit splitting between states with $j = l + \frac{1}{2}$ and $j = l - \frac{1}{2}$ gives rise to the fine structure in the spectrum of hydrogen-like atoms. Table 4.2 gives the lowest hydrogenic states together with their spectroscopic notation.

It may be useful to have an explicit form of the radial wave functions for the lowest few states. For the K- and L-shells, i.e., for the $1s_{1/2}$ and $2s_{1/2}$, $2p_{1/2}$, $2p_{3/2}$ states, the radial functions are written as

$$\begin{aligned} f &= -N\sqrt{1-W}\,r^{s-1}e^{-qr}(b_0 + b_1 r) \\ g &= +N\sqrt{1+W}\,r^{s-1}e^{-qr}(c_0 + c_1 r). \end{aligned} \tag{4.103}$$

Here, all lengths are expressed in units of the electron Compton wavelength $\lambda\!\!\!^{-}_c$, and the formulas for $s$, $W$, $q$, $N$ and $b_i$, $c_i$ are given in Tables 4.3 and 4.4, respectively.

Bound solutions of the Dirac equation for a Coulomb potential in *momentum space* are given in [Rub48] and [Lev51], a discussion of approximate forms can be found in [BeS57]. The two-center Coulomb wave functions for bound and continuum states are treated in [MuG76] and in [WiS87, RuG89], respectively.

### 4.3.3 Darwin wave functions for bound states

We are now in possession of exact relativistic wave functions for the bound Coulomb problem. Nevertheless, it is often more convenient, and for small values of $\zeta = \alpha Z$ also sufficient, to use approximate wave functions of a considerably simpler structure. In order to derive an approximate quasirelativistic bound-state wave function it is customary [Ros61, BeL82] to start from the exact second-order wave equation for the potential $V = -e^2 Z/r$. This equation is obtained from the time-independent Dirac equation

$$(E + i\hbar c\boldsymbol{\alpha} \cdot \nabla - m_{\rm e}c^2\gamma^0)\varphi = V\varphi \tag{4.104}$$

by acting on both sides with the operator $(E - i\hbar c\boldsymbol{\alpha} \cdot \nabla + m_{\rm e}c^2\gamma^0)$, so that

$$\begin{aligned}(E^2 + \hbar^2c^2\nabla^2 - m_{\rm e}^2c^4)\varphi &= (E - i\hbar c\boldsymbol{\alpha} \cdot \nabla + m_{\rm e}c^2\gamma^0)V\varphi \\ &= [-i\hbar c\boldsymbol{\alpha} \cdot (\nabla V) + V(2E - V)]\varphi. \end{aligned} \tag{4.105}$$

By reordering terms, we get

$$\left(-\frac{\hbar^2}{2m_{\rm e}}\nabla^2 - \frac{E^2}{2m_{\rm e}c^2} + \frac{EV}{m_{\rm e}c^2} + \frac{1}{2}m_{\rm e}c^2\right)\varphi = \frac{i\hbar}{2m_{\rm e}c}\boldsymbol{\alpha} \cdot (\nabla V)\varphi + \frac{V^2}{2m_{\rm e}c^2}\varphi. \tag{4.106}$$

Decomposing the relativistic total energy $E = m_{\rm e}c^2(1 + \epsilon/m_{\rm e}c^2)$ into the mass term and the total nonrelativistic energy $\epsilon$ and retaining only terms up to the order of $\alpha Z$, we rewrite Eq. (4.106) in the form

$$\left(-\frac{\hbar^2}{2m_{\rm e}}\nabla^2 + V - \epsilon\right)\varphi = \frac{i\hbar}{2m_{\rm e}c}\boldsymbol{\alpha} \cdot (\nabla V)\varphi. \tag{4.107}$$

Similarly, relativistic corrections in the wave function $\varphi = u^{(\pm)}(\varphi_0 + \varphi_1 + \cdots)$ are expanded in powers of $\alpha Z$, where $\varphi_0$ satisfies the Schrödinger equation

$$\left(-\frac{\hbar^2}{2m_{\rm e}}\nabla^2 + V - \epsilon\right)\varphi_0 = 0,$$

and $u^{(+)} = (1, 0, 0, 0)^\dagger$ and $u^{(-)} = (0, 1, 0, 0)^\dagger$ are the basic four-component spinors for a particle at rest with spin up and spin down, respectively. Up to order $\alpha Z$, Eq. (4.107) becomes

$$\left(-\frac{\hbar^2}{2m_{\rm e}}\nabla^2 + V - \epsilon\right)u^{(\pm)}\varphi_1 = \frac{i\hbar}{2m_{\rm e}c}\boldsymbol{\alpha} \cdot (\nabla V)\, u^{(\pm)}\varphi_0, \tag{4.108}$$

where the operator acting on the right-hand side can be produced by commuting the Schrödinger operator with $(-i\hbar/2m_ec)\boldsymbol{\alpha}\cdot\nabla$. The resulting Darwin wave function [Dar28]

$$\varphi^{(\pm)}(\mathbf{r}) = \left(1 - \frac{i\hbar}{2m_ec}\boldsymbol{\alpha}\cdot\nabla\right) u^{(\pm)}\varphi_0(\mathbf{r}) \tag{4.109}$$

is a quasirelativistic bound-state wave function accurate to first order in $\alpha Z$ in the relativistic correction (and normalized to the same order), with $\varphi_0$ being a nonrelativistic bound-state hydrogenic function. The method is also applicable to more general potentials.

## 4.4 Continuum states in a Coulomb field

In nonrelativistic quantum theory, the Schrödinger equation for a Coulomb potential is separable in spherical as well as in parabolic coordinates. This property allows one to expand continuum solutions into angular momentum eigenstates (partial waves) and also to write down closed-form solutions in parabolic coordinates. The latter form represents eigenstates corresponding to a definite asymptotic momentum.

The Dirac equation for a Coulomb potential is not separable in parabolic coordinates, so that exact closed-form continuum wave functions cannot be given. In Sec. 4.4.1, we therefore consider a partial-wave expansion of the exact wave function while an approximate closed-form representation is given in Sec. 4.4.2.

### 4.4.1 Partial-wave expansion of the exact solution

Similarly as in the case of bound-state wave functions, we may treat each partial wave, characterized by the quantum numbers $\kappa$ and $m_j$, by solving the corresponding radial equation (4.93) with the appropriate boundary conditions. While for the calculation of total cross sections, it is sufficient to know the individual partial waves [see Eq. (6.7)], the study of differential cross sections requires the construction of complete scattering solutions. We start with the partial waves and defer the scattering solutions to Eqs. (4.122) to (4.124).

If we express the energy of the electron in units of its rest energy by $W = E/m_ec^2$ (where now $|W| > 1$), we can introduce the electron (or positron) wave number

$$k = \frac{1}{\lambda\!\!\!^{-}_{c}}\sqrt{W^2-1} \tag{4.110}$$

and the relativistic generalization of the corresponding Sommerfeld parameter

$$\eta = \frac{\zeta W}{k\lambda\!\!\!^{-}_{c}}, \tag{4.111}$$

($\zeta = \alpha Z$) which is positive for an electron and negative for a positron in a nuclear Coulomb field. Furthermore, defining a phase factor $\delta_\kappa$ and a normalization factor $N_\kappa$ by

$$\begin{aligned} e^{2i\delta_\kappa} &= \frac{-\kappa + i\eta/W}{s + i\eta} \\ N_\kappa &= \frac{1}{(m_e c^2)^{\frac{1}{2}}} \frac{2\,k^{\frac{1}{2}}}{\pi^{\frac{1}{2}}\,\lambda\!\!\!^{-}_{c}}\, e^{\pi\eta/2}\, \frac{|\Gamma(s+i\eta)|}{\Gamma(2s+1)}, \end{aligned} \tag{4.112}$$

we can write the radial continuum wave functions entering in the spinor

$$\varphi_{E,\kappa m_j}(\mathbf{r}) = \begin{pmatrix} g_\kappa(r)\chi_\kappa^{m_j}(\hat{\mathbf{r}}) \\ if_\kappa(r)\chi_{-\kappa}^{m_j}(\hat{\mathbf{r}}) \end{pmatrix}, \tag{4.113}$$

for a given partial wave with $W > 1$ in the form

$$\begin{aligned} g_\kappa &= N_\kappa(W+1)^{\frac{1}{2}}(2kr)^{s-1}\,\mathrm{Re}\,[e^{-ikr}e^{i\delta_\kappa}(s+i\eta) \\ &\qquad \times {}_1F_1(s+1+i\eta, 2s+1; 2ikr)], \\ f_\kappa &= -N_\kappa(W-1)^{\frac{1}{2}}(2kr)^{s-1}\,\mathrm{Im}\,[e^{-ikr}e^{i\delta_\kappa}(s+i\eta) \\ &\qquad \times {}_1F_1(s+1+i\eta, 2s+1; 2ikr)]. \end{aligned} \tag{4.114}$$

Correspondingly, for negative energy $W < -1$, we have

$$\begin{aligned} g_\kappa &= N_\kappa(|W|-1)^{\frac{1}{2}}(2kr)^{s-1}\,\mathrm{Re}\,[e^{-ikr}e^{i\delta_\kappa}(s+i\eta) \\ &\qquad \times {}_1F_1(s+1+i\eta, 2s+1; 2ikr)], \\ f_\kappa &= -N_\kappa(|W|+1)^{\frac{1}{2}}(2kr)^{s-1}\,\mathrm{Im}\,[e^{-ikr}e^{i\delta_\kappa}(s+i\eta) \\ &\qquad \times {}_1F_1(s+1+i\eta, 2s+1; 2ikr)]. \end{aligned} \tag{4.115}$$

Since $\eta$ is negative for positrons, the phase factor $\delta_\kappa$ and the normalization factor $N_\kappa$ change correspondingly. In particular, for $r \to 0$, when we can set the confluent hypergeometric functions equal to unity, the quantity

$$e^{\pi\eta}|\Gamma(s+i\eta)|^2$$

in $N_\kappa^2$ determines the charge density of electrons and positrons near the nucleus. For small momenta, the ratio of the charge densities is given by

$$\rho_{\text{pos}}(0) = e^{-2\pi\eta}\,\rho_{\text{el}}(0). \tag{4.116}$$

In other words, as $k \to 0$, the probability of positrons to stay near the nucleus is very strongly suppressed as compared to the probability of slow electrons. This is what one would expect on account of the Coulomb repulsion and attraction, respectively.

The wave functions (4.114) and (4.115) are both normalized on the energy scale. This means that if $\varphi_E$ and $\varphi_{E'}$ are solutions with eigenenergies $E$ and $E'$ we have

$$\int \varphi^\dagger_{E',\kappa m_j}\varphi_{E,\kappa,m_j}\,d^3r = \delta(E-E'). \tag{4.117}$$

The asymptotic behavior for $W > 0$ is given by

$$\begin{aligned} g_\kappa &\simeq \frac{1}{(\hbar c)^{\frac{1}{2}} r}\left(\frac{W+1}{\pi k \lambda\!\!\!^{-}_c}\right)^{\frac{1}{2}} \cos(kr+\sigma_\kappa), \\ f_\kappa &\simeq -\frac{1}{(\hbar c)^{\frac{1}{2}} r}\left(\frac{W-1}{\pi k \lambda\!\!\!^{-}_c}\right)^{\frac{1}{2}} \sin(kr+\sigma_\kappa) \end{aligned} \tag{4.118}$$

and for $W < -1$ by

$$\begin{aligned} g_\kappa &\simeq \frac{1}{(\hbar c)^{\frac{1}{2}} r}\left(\frac{|W|+1}{\pi k \lambda\!\!\!^{-}_c}\right)^{\frac{1}{2}} \cos(kr+\sigma_\kappa), \\ f_\kappa &\simeq -\frac{1}{(\hbar c)^{\frac{1}{2}} r}\left(\frac{|W|-1}{\pi k \lambda\!\!\!^{-}_c}\right)^{\frac{1}{2}} \sin(kr+\sigma_\kappa). \end{aligned} \tag{4.119}$$

Since $\eta$ is negative for positrons, the phase factor $\sigma_\kappa$ changes correspondingly. Adopting the phase convention of Eqs. (4.118) and (4.119), the Coulomb phase shift is

$$\begin{aligned} \sigma_\kappa &= \Delta_\kappa + \eta\ln(2kr) \\ \Delta_\kappa &= \delta_\kappa - \arg\Gamma(s+i\eta) - \tfrac{1}{2}\pi s. \end{aligned} \tag{4.120}$$

Sometimes a phase convention for the radial wave functions is chosen which differs from the convention adopted in (4.118). Then the Coulomb phases (4.120) have to be modified accordingly. Effects of the finite nuclear size on the radial wave functions and the phases have been considered by Müller et al. [MuR73].

If it is the goal to calculate total cross sections for the emission of electrons or positrons, that is, if no information is needed regarding the direction of asymptotic propagation of the electron or the positron, it is sufficient to sum over all partial waves contributing to the cross section. If, however, the direction of propagation of an emitted electron is of interest, appropriate superpositions of partial waves must be used.

Momentum wave functions for a continuum electron in a Coulomb potential are given in [SoB94], and their use is illustrated in some applications. While space wave functions for the Coulomb continuum have an infinite extension, the corresponding momentum wave functions, by construction, are well localized in momentum space. The latter, however, have to be constructed numerically.

### 4.4.2 Exact eigenstates with well-defined asymptotic momenta

For the calculation of differential cross sections, it is necessary to specify the asymptotic momenta of the emitted electrons or positrons. We then have to transform from a partial-wave representation $\{k, \kappa, m_j\}$ to a representation $\{\mathbf{k}, m_s\}$ specifying $k$, the direction of emission and the spin projection $m_s$ with respect to a quantization axis. In addition, we have to impose boundary conditions which state whether the asymptotic wave function is composed of a plane wave (aside from the usual logarithmic phase factor) plus an *outgoing* or an *incoming* spherical wave [BeM54]. These cases are denoted by the superscript $\pm$, respectively.

If in an atomic process an *electron* is emitted, one has to choose *incoming* spherical waves as boundary conditions. If a *positron* is emitted, this corresponds to the absorption of a negative-energy electron, and hence one has to use *outgoing* spherical waves as boundary conditions.

Let us first suppose that the electron propagates in the positive $z$-direction. Then the solutions with outgoing and incoming spherical waves behave asymptotically for $|r - z| \to \infty$ as

$$\psi_{k_z}^{(\pm)m_s}(\mathbf{r}) \simeq u^{m_s}(\mathbf{k})e^{i[kz-\eta \ln k(r-z)]} + \sum_{m'_s} u^{m'_s}(\mathbf{k})\, M_{m'_s m_s}(\theta, \phi) \frac{e^{\pm i(kr+\eta \ln 2kr)}}{r}, \quad (4.121)$$

where $u^{m_s}(\mathbf{k})$ for $m_s = \pm\frac{1}{2}$ is defined in Eq. (4.53) and scattering matrix $M_{m'_s m_s}$ in the second term provides the possibility that the spin direction is changed by the interaction.

Introducing a partial-wave expansion into states characterized by $\kappa = (j, l)$ and $m_j$, the complete solution for electrons or positrons can be written as

$$\psi_{k_z}^{(\pm)m_s}(\mathbf{r}) = \sqrt{\frac{\pi}{2W\lambda\!\!\!^{-}_c k}} \sum_{\kappa} i^l \sqrt{4\pi(2l+1)} \left( \begin{array}{cc|c} l & \frac{1}{2} & j \\ 0 & m_s & m_s \end{array} \right) \times e^{\pm i\Delta_\kappa} \begin{pmatrix} g_\kappa \chi_\kappa^{m_s} \\ i f_\kappa \chi_{-\kappa}^{m_s} \end{pmatrix}, \tag{4.122}$$

where the spin-angular functions are defined in Eq. (4.87) and the normalization factor in front of the sum ensures the asymptotic behavior (4.118) or (4.119), and the coordinate-independent Coulomb phase shift $\Delta_\kappa$ is defined in Eq. (4.120). The radial wave functions are given by Eq. (4.114) for electrons and by Eq. (4.115) for positrons. In the latter case, $W$ as well as $\eta$ is negative.

If we are interested in electrons or positrons with a specified wave vector $\mathbf{k}$, we have to distinguish two cases.

*(a) Quantization of the electron spin in the z-direction*

We retain the axis of spin quantization in the direction of an arbitrary $z$-axis and just rotate the space part of the wave function. In this case the Wigner rotation matrices which rotate the $z$-direction into the $\mathbf{k}$-direction, $D^l_{m_l 0}(\hat{\mathbf{z}} \to \hat{\mathbf{k}}) = \sqrt{4\pi/(2l+1)}\, Y^*_{lm_l}(\hat{\mathbf{k}})$, reduce to spherical harmonics, so that

$$\psi_{\mathbf{k}}^{(\pm)m_s}(\mathbf{r}) = 4\pi \sqrt{\frac{\pi}{2W\lambda\!\!\!^{-}_c k}} \sum_{\kappa m_j} i^l \left( \begin{array}{cc|c} l & \frac{1}{2} & j \\ m_l & m_s & m_j \end{array} \right) \times e^{\pm i\Delta_\kappa}\, Y^*_{lm_l}(\hat{\mathbf{k}}) \begin{pmatrix} g_\kappa \chi_\kappa^{m_j} \\ i f_\kappa \chi_{-\kappa}^{m_j} \end{pmatrix}. \tag{4.123}$$

Since for relativistic electrons, the spin projection on any axis, except the direction of propagation, has no sharp values, this representation (which sometimes is easier to handle numerically) can be applied only if a summation over spin states is performed.

*(b) Helicity representation of the electron spin*

We use the generally valid helicity representation, i.e., we quantize the spin in the direction of the momentum. The wave functions then are eigenvectors of the helicity operator $\boldsymbol{\Sigma}\cdot\hat{\mathbf{k}}$ with the eigenvalues $2\sigma = \pm 1$ or with $\sigma = \pm\frac{1}{2}$. For an arbitrary direction of motion, we have

$$\begin{aligned}\psi_{\mathbf{k}}^{(\pm)\sigma}(\mathbf{r}) &= \sqrt{\frac{\pi}{2W\lambda_c k}} \sum_{\kappa m_j} i^l \sqrt{4\pi(2l+1)} \left( \begin{array}{cc|c} l & \frac{1}{2} & j \\ 0 & \sigma & \sigma \end{array} \right) \\ &\times e^{\pm i\Delta_\kappa} D^j_{m_j\sigma}(\hat{\mathbf{z}} \to \hat{\mathbf{k}}) \left( \begin{array}{c} g_\kappa \chi_\kappa^{m_j} \\ i f_\kappa \chi_{-\kappa}^{m_j} \end{array} \right). \end{aligned} \tag{4.124}$$

If the equations (4.122), (4.123), and (4.124) describe positrons, the proper radial wave functions (4.115) as well as the phases modified by the negative values of $\eta$ have to be used. In the case of electron/positron *emission* one has to impose boundary conditions corresponding to em incoming/ outgoing spherical waves and hence has to choose a phase factor $\exp(-i\Delta_\kappa)$ or $\exp(i\Delta_\kappa)$, respectively.

*Properties of the rotation matrices*

When working with the rotation matrices, the following properties are useful:

$$D^j_{m_1 m_2}(\hat{\mathbf{k}} \to \hat{\mathbf{z}}) = D^{j*}_{m_2 m_1}(\hat{\mathbf{z}} \to \hat{\mathbf{k}}) \tag{4.125}$$

$$D^{j*}_{m_1 m_2}(\hat{\mathbf{k}} \to \hat{\mathbf{z}}) = (-1)^{m_1 - m_2} D^j_{-m_1,-m_2}(\hat{\mathbf{k}} \to \hat{\mathbf{z}}). \tag{4.126}$$

The Clebsch-Gordan series states that the product of two matrix elements can be expressed by a sum over matrix elements:

$$D^j_{m_1 m_2} D^{j'}_{m_1' m_2'} = \sum_J \left( \begin{array}{cc|c} j & j' & J \\ m_1 & m_1' & M_1 \end{array} \right) \left( \begin{array}{cc|c} j & j' & J \\ m_2 & m_2' & M_2 \end{array} \right) D^J_{M_1 M_2}. \tag{4.127}$$

Here, the arguments of the rotation matrices are the same, and the quantum numbers $J, M_1, M_2$ assume all values compatible with the selection rules embodied in the Clebsch-Gordan coefficients. The group property states

$$\sum_m D^j_{m_1 m}(\hat{\mathbf{k}} \to \hat{\mathbf{z}})\, D^j_{m m_2}(\hat{\mathbf{z}} \to \hat{\mathbf{k}}') = D^j_{m_1 m_2}(\hat{\mathbf{k}} \to \hat{\mathbf{k}}'). \tag{4.128}$$

If at least one of the projection quantum numbers is zero, the elements of the rotation matrices take on particularly simple forms:

$$D^l_{m0}(\phi\theta\gamma) = \sqrt{\frac{4\pi}{2l+1}}\, Y^*_{lm}(\theta\phi) \tag{4.129}$$

$$D^l_{00}(\phi\theta\gamma) = P_l(\cos\theta), \tag{4.130}$$

where $\phi, \theta, \gamma$ are the Euler angles describing the rotation, and $Y_{lm}$ and $P_l$ are spherical harmonics and Legendre polynomials, respectively.

### 4.4.3 Sommerfeld-Maue wave functions for continuum states

In analogy to the bound-state case discussed in Sec. 4.3.3, one may also derive approximate relativistic continuum wave functions, accurate to the order $\alpha Z$ in the relativistic corrections. These functions, denoted as Sommerfeld-Maue [SoM35] or Furry [Fur34] wave functions, have been widely applied in the literature [BeM54]. As stated above, the nonrelativistic wave equation for continuum states in a Coulomb field is separable in parabolic coordinates, but the corresponding Dirac equation is not. Nevertheless, the Sommerfeld-Maue wave function, similarly as Eq. (4.109), is directly related to the solution of the nonrelativistic problem in parabolic coordinates. It is an advantage of the Sommerfeld-Maue approximation to render a decomposition of the continuum wave function into partial waves unnecessary.

The derivation is analogous to, albeit more complicated than, the derivation of the Darwin wave function given above [Ros61, BeL82]. If the wave function is normalized such that for a vanishing potential $V \to 0$ one gets the plane-wave solution $\varphi \to \exp(i\,\mathbf{k}\cdot\mathbf{r})\,u^{(\rho)}(\mathbf{k})$ where $u^{(\rho)}(\mathbf{k})$ is the four-component spinor for a free particle with energy $E$ and wave vector $\mathbf{k}$, defined in Eqs. (4.54) and (4.55), then the solution for *outgoing* spherical electron waves can be written as

$$
\begin{aligned}
\varphi^{(+)}_{E,k} &= e^{\frac{\pi}{2}\eta}\,\Gamma(1-i\eta)\,e^{i\,\mathbf{k}\cdot\mathbf{r}}\left(1-i\frac{\hbar c}{2E}\,\boldsymbol{\alpha}\cdot\nabla\right)\\
&\quad\times{}_1F_1[i\eta\,,1\,;i(kr-\mathbf{k}\cdot\mathbf{r})]u^{(\rho)}(\mathbf{k})
\end{aligned}
\tag{4.131}
$$

and for *incoming* spherical waves as

$$
\begin{aligned}
\varphi^{(-)}_{E,k} &= e^{\frac{\pi}{2}\eta}\,\Gamma(1+i\eta)\,e^{i\,\mathbf{k}\cdot\mathbf{r}}\left(1-i\frac{\hbar c}{2E}\,\boldsymbol{\alpha}\cdot\nabla\right)\\
&\quad\times{}_1F_1[-i\eta\,,1\,;-i(kr+\mathbf{k}\cdot\mathbf{r})]u^{(\rho)}(\mathbf{k}).
\end{aligned}
\tag{4.132}
$$

Here, the Sommerfeld parameter is $\eta = \alpha Z\,E/(\hbar ck)$ [Eq. (4.111)]. The solutions for negative-energy states are derived from Eqs. (4.131) and (4.132) by substituting $\mathbf{k} \to -\mathbf{k}$, $E \to -E$, and $Z \to -Z$, that is,
for *outgoing* spherical positron waves we have

$$
\begin{aligned}
\varphi^{(+)}_{|E|,k} &= e^{-\frac{\pi}{2}\eta}\,\Gamma(1+i\eta)\,e^{-i\,\mathbf{k}\cdot\mathbf{r}}\left(1+i\frac{\hbar c}{2|E|}\,\boldsymbol{\alpha}\cdot\nabla\right)\\
&\quad\times{}_1F_1[-i\eta\,,1\,;i(kr+\mathbf{k}\cdot\mathbf{r})]u^{(\rho)}(-\mathbf{k})
\end{aligned}
\tag{4.133}
$$

while for *incoming* spherical waves we write

$$\begin{aligned}\varphi^{(-)}_{|E|,k} &= e^{-\frac{\pi}{2}\eta}\,\Gamma(1-i\eta)\,e^{-i\,\mathbf{k}\cdot\mathbf{r}}\left(1+i\frac{\hbar c}{2|E|}\boldsymbol{\alpha}\cdot\nabla\right)\\ &\quad\times{}_1F_1[i\eta\,,1\,;-i(kr-\mathbf{k}\cdot\mathbf{r})]u^{(\rho)}(-\mathbf{k}).\end{aligned} \tag{4.134}$$

In all equations, $\eta$ is taken as a positive quantity, and in the last two relations the positron spinors are defined by Eqs. (4.57) and (4.58).

In the case of electron emission, one has to choose incoming spherical waves, in the case of positron emission, outgoing spherical waves [BeM54].

As has been shown by Bethe and Maximon [BeM54], the Sommerfeld-Maue wave functions are obtained by replacing $s=(\kappa^2-\alpha^2Z^2)^{1/2}$ with $|\kappa|$ and hence are good approximations for angular momenta $l \gg Z/137$. They can be used whenever the lowest partial waves $l \simeq 1$ are not important. This is the case, regardless of the nuclear charge $Z$, if total electron and positron energies are large compared to $m_ec^2$.

# Chapter 5

# Ion-Atom Collisions

In the preceding two chapters, we discussed the electromagnetic field produced by a point charge moving with relativistic velocity and, subsequently, the relativistic motion of a Dirac electron as a free particle or in a Coulomb potential. In the present chapter, we merge the results of Chaps. 3 and 4 by analyzing the electron motion in the combined potential of the space-fixed target and the moving projectile nucleus. Before discussing specific processes in Chaps. 6 – 10, we collect various results with general applicability.

We start out in Sec. 5.1 by writing down the coupled equations of motion which govern the interaction among the nuclear, electron, and photon fields. By adopting a field-theoretical formulation, the approximations leading to the time-dependent quantum description are clarified. In Sec. 5.2 we investigate under which assumptions a representation of the nuclear motion by classical trajectories can be derived from a quantum mechanical wave formulation. One arrives at a time-dependent Dirac equation whose properties are studied in Sec. 5.3. This establishes the basis for the developments of Sec. 5.4, where we discuss some general features of perturbative transition amplitudes for various processes, again starting from a field-theoretical formulation. One more remaining problem of some generality is the evaluation of transition matrix elements. To this end, in Sec. 5.5, it is shown how gauge transformations can be used to reformulate matrix elements of relativistic operators so that their evaluation becomes less sensitive to approximations. Finally, the general formulation of many-electron systems is discussed in Sec. 5.6.

## 5.1 Field-theoretic background

It is useful, as a starting point, to give a brief field-theoretic description of the coupled equations governing the electron field and the photon field and to discuss some of the consequences. This procedure allows us to address more clearly the approximations which are adopted in the rest of the book. The formalism described here is essential as a background for some of the more advanced treatments in the literature, but in this book, its full power is not needed.

### 5.1.1 Coupled equations of motion

For simplicity, we assume that the projectile nucleus moves along a classical rectilinear trajectory. This assumption is justified in Chap. 3 and is taken up in more detail in Sec. 5.2. Within a classical picture, the electromagnetic current produced by the moving nuclei can be regarded as an external current. The system then is represented: (i) by the nuclei which can emit and absorb photons via the coupling of their classical currents to the electromagnetic field, (ii) by the electron field which is also coupled to the photon field and which allows for the creation or annihilation of electron-positron pairs, and (iii) by the photons which are coupled both to the nuclei and to the electrons and which mediate the electromagnetic interaction between them.

The coupled equations [see Eqs. (4.1) and (3.9)] for the nuclear, electron, and photon fields are

$$\left( i\hbar\gamma_\mu \frac{\partial}{\partial x_\mu} + \frac{e}{c}\gamma_\mu \widehat{A}^\mu - m_e c \right) \widehat{\psi} = 0 \tag{5.1}$$

and

$$\frac{\partial}{\partial x^\nu} \widehat{F}^{\nu\mu} = \frac{4\pi}{c}\widehat{j}^\mu, \tag{5.2}$$

with the relation between the field-strength tensor $\widehat{F}^{\mu\nu}$ and the vector potential $\widehat{A}^\mu$ being given by Eq. (3.8). We use a hat symbol to characterize the field operators $\widehat{\psi}, \widehat{A}, \widehat{F}$, and $\widehat{j}$. Since the nuclei are treated classically, the operators for the external nuclear currents are replaced by their expectation values

$$\widehat{j}^\mu_{\rm ext} \longrightarrow \langle \widehat{j}^\mu_{\rm ext} \rangle = j^\mu_{\rm ext}, \tag{5.3}$$

so that the full current entering in Eq. (5.2) is given by

$$\widehat{j}^\mu = \widehat{j}^\mu_{\rm el} + \widehat{j}^\mu_{\rm ext} \longrightarrow -ec\widehat{\psi}^\dagger\gamma^0\gamma^\mu\widehat{\psi} + j^\mu_{\rm ext}. \tag{5.4}$$

Here, the definition (4.8) has been inserted. Likewise, the total radiation field is decomposed as

$$\widehat{A}^{\mu} = \widehat{A}^{\mu}_{\rm el} + \widehat{A}^{\mu}_{\rm ext} + \widehat{A}^{\mu}_{\rm free}. \tag{5.5}$$

Owing to Eq. (5.2), $\widehat{A}^{\mu}_{\rm el}$ is produced by the electron current $\widehat{j}^{\mu}_{\rm el}$, the field $\widehat{A}^{\mu}_{\rm ext}$ is produced by the nuclear current $j^{\mu}_{\rm ext}$, while $\widehat{A}^{\mu}_{\rm free}$ represents the free radiation field.

As the general coupled equations cannot be solved, one is forced to take recourse to a number of approximations: (a) The electron current is disregarded as a source of photons since its effects are small compared to those of the nuclear current for high-$Z$ nuclei and, by assumption, the electron current does not act back on the nuclear motion, which is described classically. (b) Correspondingly, the field $\widehat{A}^{\mu}_{\rm el}$ is neglected as compared to $\widehat{A}^{\mu}_{\rm ext}$. These two approximations are sufficient to decouple Eqs. (5.1) and (5.2). (c) Since the electromagnetic field of the nuclear motion produces a large number of photons, fluctuations can be disregarded and $\widehat{A}^{\mu}_{\rm ext} \rightarrow \langle \widehat{A}^{\mu}_{\rm ext} \rangle = A^{\mu}$, which is in accord with Eq. (5.3) and implies the corresponding relationship for the $\widehat{F}^{\nu\mu}$. (d) Radiative QED corrections, as for example vacuum polarization and self-energy corrections arising from the free photon field, are not taken into account here, hence $\widehat{A}^{\mu}_{\rm free} = 0$.

Having spelled out the various approximations, we return to the now decoupled equations

$$\left( i\hbar\gamma_{\mu}\frac{\partial}{\partial x_{\mu}} + \frac{e}{c}\gamma_{\mu}A^{\mu} - m_{\rm e}c \right) \widehat{\psi} = 0 \tag{5.6}$$

and

$$\frac{\partial}{\partial x^{\nu}} F^{\nu\mu} = \frac{4\pi}{c} j^{\mu}_{\rm ext}, \tag{5.7}$$

where Eq. (5.6) is the operator version of Eq. (4.1) for an electron field subject to an external electromagnetic potential, while in Eq. (5.7) the external nuclear current determines the field $F^{\mu\nu}$ and the potential $A^{\mu}$.

### 5.1.2 Expansion of the field operators

If the perturbation by a time-varying external potential is weak, it is possible, similarly as in Eq. (4.64), to expand the field operators

$$\widehat{\psi}(\mathbf{r}, t) = \sum_{j_+} \widehat{b}_{j_+} \psi_{j_+}(\mathbf{r}, t) + \sum_{j_-} \widehat{d}^{\dagger}_{j_-} \psi_{j_-}(\mathbf{r}, t), \tag{5.8}$$

in terms of time-independent (Schrödinger) creation and annihilation operators. The positive-energy wave functions $\psi_{j_+}$ and the negative-energy wave functions $\psi_{j_-}$ comprise a complete set of time-dependent stationary solutions of the Dirac equation (4.75), say in the target system $\mathbf{r} = \mathbf{r}_\mathrm{T}$, so that

$$i\hbar \frac{\partial}{\partial t} \psi_j(\mathbf{r}, t) = H_\mathrm{T} \psi_j(\mathbf{r}, t) \tag{5.9}$$

with

$$\psi_j(\mathbf{r}, t) = \varphi_j(\mathbf{r})\, e^{-iE_j t/\hbar} \tag{5.10}$$

and

$$H_\mathrm{T} = -i\hbar c \boldsymbol{\alpha} \cdot \nabla - \frac{Z_\mathrm{T} e^2}{r} + m_\mathrm{e} c^2 \gamma^0. \tag{5.11}$$

A corresponding set of equations can be written in the projectile system. The $\widehat{b}_{j_+}$ are destruction operators for electrons, and the $\widehat{d}^\dagger_{j_-}$ are creation operators for positrons, each characterized by the sets of quantum numbers $j_+$ and $j_-$, respectively. The commutation relations and the definition of the vacuum $|0\rangle$ are the same as in Eqs. (4.65) to (4.68) for free electrons.

### 5.1.3 Time evolution of the vacuum

In most parts of the book, it is assumed that perturbation theory is applicable, and hence the probability of producing electron-positron pairs remains negligible at all times. Then it is natural to consider the vacuum $|0\rangle$ to be unaltered during the collision so that the discussion of this and the following subsection is not needed.

If pair production occurs with large probabilities (Secs. 10.3.2 and 10.4), perturbation theory breaks down, and we have to examine the vacuum more carefully. In the Dirac hole theory, the vacuum represents the occupied Dirac sea composed of noninteracting negative-energy single-particle states. These single-particle states are affected by an external field and so is the vacuum. Consequently, for the case of stronger electromagnetic interactions, it is convenient to consider a modified vacuum which takes this process into account. Retaining the expansion in terms of stationary solutions (5.10) of the target (alternatively, the projectile) Dirac equation, this means that we have to replace the time-independent creation and annihilation operators by the corresponding time-dependent (Heisenberg) operators so that

$$\widehat{\Psi}(\mathbf{r}, t) = \sum_{j_+} \widehat{b}_{j_+}(t) \psi_{j_+}(\mathbf{r}, t) + \sum_{j_-} \widehat{d}^\dagger_{j_-}(t) \psi_{j_-}(\mathbf{r}, t) \tag{5.12}$$

instead of (5.8). Since for a sufficiently strong interaction during the collision, the vacuum $|0\rangle$ defined at $t \to -\infty$ will also evolve, we cannot require that at any given instant of time, the definitions $\widehat{b}_{j_+}|0\rangle = 0$ and $\widehat{d}_{j_-}|0\rangle = 0$

still hold. Rather, the "dressed" instantaneous vacuum $|\tilde{0}\rangle$ evolved from the initial vacuum $|0\rangle$ may contain pair and multiple pair amplitudes.

*Introducing a dressed vacuum*

It is convenient, therefore, to define a new set of time-independent creation operators $\widehat{\tilde{b}}^{\dagger}_{k_+}$, $\widehat{\tilde{d}}^{\dagger}_{k_-}$, so that in the expansion

$$\widehat{\Psi}(\mathbf{r},t) = \sum_{k_+} \widehat{\tilde{b}}_{k_+} \chi_{k_+}(\mathbf{r},t) + \sum_{k_-} \widehat{\tilde{d}}^{\dagger}_{k_-} \chi_{k_-}(\mathbf{r},t) \tag{5.13}$$

the time evolution is absorbed into the wave functions $\chi_{k_+}$ and $\chi_{k_-}$, where $k_+$ and $k_-$ denote sets of quantum numbers for positive- and negative-energy states, respectively. The new wave functions are solutions of the full time-dependent Dirac equation (5.6) which have evolved from the original stationary solutions (5.10). Therefore, the initial conditions are

$$\lim_{t\to-\infty} |\tilde{0}(t)\rangle = |0\rangle \tag{5.14}$$

and

$$\lim_{t\to-\infty} \chi_k(\mathbf{r},t) = \psi_k(\mathbf{r},t). \tag{5.15}$$

The time evolution of the dressed instantaneous vacuum $|\tilde{0}\rangle$ entails a canonical transformation of the associated annihilation and creation operators, which are defined as

$$\begin{aligned} \widehat{\tilde{b}}_{k_+}|\tilde{0}(t)\rangle &= 0, \qquad \widehat{\tilde{b}}^{\dagger}_{k_+}|\tilde{0}(t)\rangle = |\chi_{k_+}(t)\rangle \\ \widehat{\tilde{d}}_{k_-}|\tilde{0}(t)\rangle &= 0, \qquad \widehat{\tilde{d}}^{\dagger}_{k_-}|\tilde{0}(t)\rangle = |\chi_{k_-}(t)\rangle \end{aligned} \tag{5.16}$$

and preserve the commutation relations (4.65) to (4.67).

Since the representations (5.12) and (5.13) are equivalent, we have

$$\begin{aligned} &\sum_{j_+} \widehat{b}_{j_+}(t)\psi_{j_+}(t) + \sum_{j_-} \widehat{d}^{\dagger}_{j_-}(t)\psi_{j_-}(t) \\ &\quad = \sum_{k_+} \widehat{\tilde{b}}_{k_+}\chi_{k_+}(t) + \sum_{k_-} \widehat{\tilde{d}}^{\dagger}_{k_-}\chi_{k_-}(t), \end{aligned} \tag{5.17}$$

where the space dependence of the wave functions is suppressed for brevity. Using the orthogonality of the stationary wave functions $\psi_{j_+}, \psi_{j_-}$, one derives the transformations [WeO92]

$$\widehat{b}_{j_+}(t) = \sum_{k_+} \widehat{\tilde{b}}_{k_+}\langle\psi_{j_+}|\chi_{k_+}\rangle + \sum_{k_-} \widehat{\tilde{d}}^{\dagger}_{k_-}\langle\psi_{j_+}|\chi_{k_-}\rangle$$

$$\widehat{d}^{\dagger}_{j_-}(t) = \sum_{k_+} \widehat{\tilde{b}}_{k_+} \langle \psi_{j_-} | \chi_{k_+} \rangle + \sum_{k_-} \widehat{\tilde{d}}^{\dagger}_{k_-} \langle \psi_{j_-} | \chi_{k_-} \rangle, \tag{5.18}$$

in which, at $t \to -\infty$, the right-hand side of the equations yields the original time-independent operators $\widehat{b}_{j_+}, \widehat{d}^{\dagger}_{j_-}$. The relation (5.18) shows that for a time-independent basis set, the creation operators (and similarly the annihilation operators) evolve into a linear combination of creation and annihilation operators with respect to an instantaneous vacuum $|\tilde{0}\rangle$. Conversely, a creation or annihilation operator with respect to $|\tilde{0}\rangle$ is a linear combination of operators in the target basis. Only the latter ones can be directly associated with observable quantities.

*The number of electrons and positrons at intermediate times*

While the initial vacuum $|0\rangle$ is defined to contain no electrons nor positrons, we may wonder how many electrons (or positrons) are contained in the instantaneous vacuum $|\tilde{0}(t)\rangle$ at any given time. The expectation value for the number of observable electrons in the state $j_+$ with no regard to the associated positron state, i.e., the inclusive value, is given by [see Eq. (4.71)].

$$\langle \widehat{N}_{j_+}(t) \rangle = \langle \tilde{0}(t) | \widehat{b}^{\dagger}_{j_+}(t) \widehat{b}_{j_+}(t) | \tilde{0}(t) \rangle = \sum_{k_-} |\langle \phi_{j_+} | \chi_{k_-}(t) \rangle|^2, \tag{5.19}$$

where use has been made of the expansion (5.18) and of Eq. (5.10). This result is easiest to interpret in the Dirac picture: in the entrance channel, at time $t \to -\infty$, all electron states $k_+$ are empty while all negative-energy states $k_-$ are occupied. Equation (5.19) gives the sum of all probabilities that an initially occupied negative-energy state evolves into an occupied positive-energy state.

Likewise, the (inclusive) number of positrons in state $j_-$ is seen to be

$$\langle \widehat{N}_{j_-}(t) \rangle = \langle \tilde{0}(t) | \widehat{d}^{\dagger}_{j_-}(t) \widehat{d}_{j_-}(t) | \tilde{0}(t) \rangle = \sum_{k_+} |\langle \phi_{j_-} | \chi_{k_+}(t) \rangle|^2. \tag{5.20}$$

This is the sum of all probabilities that an initially vacant positive-energy state evolves into a vacant negative-energy state.

In a similar fashion, using the anticommutation relations (4.65) to (4.67) and (5.18), one may calculate the number of created electron-positron pairs as

$$\begin{aligned} \langle \widehat{N}_{j_+}(t) \widehat{N}_{j_-}(t) \rangle &= \langle \tilde{0}(t) | \widehat{b}^{\dagger}_{j_+}(t) \widehat{d}^{\dagger}_{j_-}(t) \widehat{d}_{j_-}(t) \widehat{b}_{j_+}(t) | \tilde{0}(t) \rangle \\ &= \langle \widehat{N}_{j_+}(t) \rangle \langle \widehat{N}_{j_-}(t) \rangle + \left| \sum_{k_-} \langle \phi_{j_+} | \chi_{k_-}(t) \rangle \langle \chi_{k_-}(t) | \phi_{j_-} \rangle \right|^2 . \end{aligned} \tag{5.21}$$

The first term is a product of independent probabilities while the second term describes coherent correlations.

Using the completeness relations for the sets of $|\phi_j\rangle$ and $|\chi_k\rangle$ (comprising positive as well as negative-energy states), one can show that the total number of electrons produced equals the number of positrons produced, as required by charge conservation, i.e.,

$$N_+(t) = \sum_{j_+} \langle \widehat{N}_{j_+}(t) \rangle = N_-(t) = \sum_{j_-} \langle \widehat{N}_{j_-}(t) \rangle. \tag{5.22}$$

Similarly, by summing Eq. (5.21) over all states $j_+, j_-$, one finds that one part of the correlation terms cancels the product term, so that the total number of pairs created equals the total number $N_+$ or $N_-$ of electrons or positrons produced.

### 5.1.4 Time evolution of single-electron states

According to Eq. (5.16), a single-electron state is given by

$$|\chi_{k_+}(t)\rangle = \widehat{\tilde{b}}^\dagger_{k_+} |\tilde{0}(t)\rangle. \tag{5.23}$$

If, initially, there is only one electron present, admixtures of two-electron-one-positron states and higher-order terms will develop during the collision. The expectation value for the number of observable electrons with quantum number $j_+$ contained in the single-particle state (5.23) is obtained as

$$\begin{aligned} \langle \widehat{N}_{j_+}(t) \rangle &= \langle \tilde{0}(t) | \widehat{\tilde{b}}_{k_+} \widehat{b}^\dagger_{j_+}(t) \widehat{b}_{j_+}(t) \widehat{\tilde{b}}^\dagger_{k_+} | \tilde{0}(t) \rangle \\ &= |\langle \phi_{j_+} | \chi_{k_+}(t) \rangle|^2 + \sum_{l_-} |\langle \phi_{j_+} | \chi_{l_-}(t) \rangle|^2. \end{aligned} \tag{5.24}$$

Here, $\langle \phi_{j_+} | \chi_{k_+}(t) \rangle$ is the occupation amplitude occurring in coupled-channel calculations discussed in Secs. 6.5, 8.7, and 10.3.2. The total number of electrons is obtained by summing over $j_+$ as

$$N_+(t) = 1 + \sum_{j_+, l_-} |\langle \phi_{j_+} | \chi_{l_-}(t) \rangle|^2. \tag{5.25}$$

In addition to the one electron originally present, there are electrons created during the collision from the vacuum. Correspondingly, the total number of positrons created is

$$N_-(t) = \sum_{j_-, l_+ \neq k_+} |\langle \phi_{j_-} | \chi_{l_+}(t) \rangle|^2. \tag{5.26}$$

Since the state $k_+$ is occupied in the entrance channel, it cannot develop into a hole in the Dirac sea and hence is excluded from the summation.

In most cases considered in this book (Chaps. 6 to 9), the perturbation by the projectile is sufficiently weak so that pair production does not occur with appreciable probability. In these cases, the vacuum remains the same throughout the collision, i.e., $|\tilde{0}(t)\rangle = |0\rangle$, and the number of electrons is conserved. A field-theoretic description is not needed in this case. While treating the collision with standard quantum mechanics in Chaps. 6 to 9, we return to the field-theoretic methods in Chap. 10. Still in Sec. 5.4, we use these methods to formulate the perturbative transition amplitudes, and in Sec. 5.6, we adopt similar techniques to describe many-electron systems.

## 5.2 Relativistic impact parameter treatment

In Chap. 3, we explain on an intuitive basis that a semiclassical treatment of the nuclear motion should provide a good approximation. By a semiclassical (or impact parameter) treatment one usually means a description in which the motion of the nuclei is assumed to follow a classical trajectory while the electronic motion obeys quantum mechanics. In this section, we show how under suitable assumptions a semiclassical treatment of the collision can be derived from a wave treatment. This derivation is of a more formal nature and is not essential for understanding the subsequent developments.

Within the realm of atomic collisions, the basic feature of the nuclear motion is the fact that the masses $M_{\mathrm{T}}$ and $M_{\mathrm{P}}$ of the target and the projectile nucleus, respectively, are very large compared to the electron mass $m_{\mathrm{e}}$. Assuming that the nuclei can be described as Dirac particles (other quantum descriptions would not lead to different results), we may safely ignore the negative-energy states of the nuclei. Moreover, the classical considerations of Chap. 3 show that in relativistic collisions (a) the target recoil is negligible under most conditions and (b) the projectile deflection does not play any role. We then can take the target nucleus as fixed in the laboratory frame ($M_{\mathrm{T}} \to \infty$) and use this coordinate system for describing the process. We follow the arguments given in [DeE94] for the nonrelativistic situation.

For a projectile moving in the positive $z$-direction, let the coordinates be $\mathbf{R} = (X, Y, Z)$ and the corresponding derivatives $\nabla_R = (\frac{\partial}{\partial X}, \frac{\partial}{\partial Y}, \frac{\partial}{\partial Z})$. Similarly, the electron coordinates and derivatives are $\mathbf{r} = (x, y, z)$ and $\nabla = (\frac{\partial}{\partial x}, \frac{\partial}{\partial y}, \frac{\partial}{\partial z})$, respectively.

In a time-independent wave description of the projectile as well as of the electron motion at a total energy $E$, the complete wave function for an incoming initial state $\Psi_{\mathrm{i}}^{+}(E)$ satisfies the time-independent Dirac equation

$$(H - E)\Psi_{\mathrm{i}}^{+}(E) = 0, \tag{5.27}$$

where the Dirac Hamiltonian is given by

$$\begin{aligned} H &= -i\hbar c\boldsymbol{\alpha}_R \cdot \nabla_R + \frac{Z_{\mathrm{T}} Z_{\mathrm{P}} e^2}{R} + M_{\mathrm{P}} c^2\, \gamma_R^0 \\ &\quad -i\hbar c\boldsymbol{\alpha} \cdot \nabla - \frac{Z_{\mathrm{T}} e^2}{r} + m_{\mathrm{e}} c^2 \gamma^0 + V_{\mathrm{P,e}}(\mathbf{R}, \mathbf{r}). \end{aligned} \tag{5.28}$$

Here, the Dirac matrices $\boldsymbol{\alpha}_R$, $\gamma_R^0$ act in the nuclear spinor space while $\boldsymbol{\alpha}$, $\gamma^0$ act in the electron spinor space, and the electron-projectile interaction $V_{\mathrm{P,e}}$ is yet to be determined.

In writing down the Hamiltonian, it is assumed that, owing to the heavy mass of the projectile, the projectile motion is independent of the electron motion. Then the solutions $\Psi_{\mathrm{i}}^{+}(E)$ of (5.27) can be written as direct spinor products for the (assumed spin $\frac{1}{2}$) nucleus and for the electron. Since it is shown in Sec. 3.2 that the deflection of the heavy nucleus is negligible, the nuclear wave functions obey the stationary free-particle Dirac equation, viz. (4.46),

$$(-i\hbar c\boldsymbol{\alpha}_R \cdot \nabla_R + M_{\mathrm{P}} c^2\, \gamma_R^0)\, \Phi_{E_{\mathrm{P}},\mathbf{K}} = E_{\mathrm{P}} \Phi_{E_{\mathrm{P}},\mathbf{K}}\,, \tag{5.29}$$

where

$$E_{\mathrm{P}}^2 = M_{\mathrm{P}}^2 c^4 + \hbar^2\, K^2 c^2$$

and

$$\Phi_{E_{\mathrm{P}},\mathbf{K}} = u_{\mathrm{P}}^{(\rho)}\, e^{i\mathbf{K}\cdot\mathbf{R}} \tag{5.30}$$

with a plane-wave solution for the wave vector $\mathbf{K} = K\hat{\mathbf{e}}_z$. The free-particle spinor with $\rho = 1, 2$ for the spin parallel and antiparallel to the $z$-axis is defined in Eqs. (4.54) and (4.55), respectively.

Because of the large ratio between their masses, electron and projectile are not treated on equal footing[1] in the Hamiltonian (5.28). We start from the interaction

$$j^\mu A_\mu / c = -e(\psi^{\dagger}\gamma^0\gamma^\mu\psi A_\mu) = -e(\psi^{\dagger}\psi\, A^0 - \psi^{\dagger}\boldsymbol{\alpha}\psi \cdot \mathbf{A}), \tag{5.31}$$

[1] A symmetric treatment, such as for the two electrons in helium, would lead to the approximate Breit interaction [BeS57].

see Sec. 4.4.1, between the electronic current density $j^\mu$ and the retarded vector potential $A_\mu$ generated by the time-independent relativistic projectile current. For the heavy projectile whose state remains unchanged during the collision, it is justified to replace the four-current by its expectation value

$$\langle \Phi_{E_\mathrm{P},\mathbf{K}} | Z_\mathrm{P} e c \gamma_R^0 \gamma_R^\mu | \Phi_{E_\mathrm{P},\mathbf{K}} \rangle = Z_\mathrm{P} e v^\mu = Z_\mathrm{P} e \, (\gamma c, \gamma \mathbf{v})$$

between the plane wave states (5.30), where $\mathbf{v} = \hbar \mathbf{K}/M_\mathrm{P}$ and $\gamma = (1 - \beta^2)^{-1/2}$ is the Lorentz factor. The Liénard-Wiechert potential calculated from this continuous charge flow is closely related to the potentials (3.34) produced by a single moving charge: at each position $\mathbf{R}$ of the current, the continuous charge flow provides charges $Z_\mathrm{P} e$ which, at the time $t = Z/v$, arrive at the $z$-coordinate $Z$. As a result, we obtain the effective electron-projectile interaction

$$V_{\mathrm{P,e}}(\mathbf{R}, \mathbf{r}) = -\gamma (1 - \beta \alpha_z) \frac{Z_\mathrm{P} e^2}{\sqrt{(x - X)^2 + (y - Y)^2 + \gamma^2 (z - Z)^2}} \,. \tag{5.32}$$

In order to solve Eq. (5.27), we insert the product ansatz

$$\Psi_\mathrm{i}^+(E) = f(\mathbf{r}, \mathbf{R}) \, \Phi_{E_\mathrm{P},\mathbf{K}}, \tag{5.33}$$

where $f(\mathbf{r}, \mathbf{R})$ is yet to be determined. When multiplying the resulting equation with the adjoint projectile spinor $u_\mathrm{P}^{(\rho')\dagger}$ from the left, we get a nonzero contribution only if $\rho = \rho'$ because $K_x/K_z$ and $K_y/K_z$ are zero in the initial state (and negligible in the final state even if a small Coulomb deflection occurs). The result is

$$\begin{aligned} u_\mathrm{P}^{(\rho')\dagger} \, (H - E) \Psi_\mathrm{i}^+(E) = & \\ & e^{i\mathbf{K}\cdot\mathbf{R}} \left( -i\hbar c \beta \frac{\partial}{\partial Z} + \frac{Z_\mathrm{T} Z_\mathrm{P} e^2}{R} - i\hbar c \boldsymbol{\alpha} \cdot \nabla \right. \\ & \left. - \frac{Z_\mathrm{T} e^2}{r} + V_{\mathrm{P,e}}(\mathbf{R}, \mathbf{r}) + m_\mathrm{e} c^2 \, \gamma^0 - \epsilon_\mathrm{i} \right) f(\mathbf{r}, \mathbf{R}) \\ = & \;\; 0, \end{aligned} \tag{5.34}$$

where $Z = \mathbf{R} \cdot \hat{\mathbf{e}}_z$ and $\epsilon_\mathrm{i} = E - E_\mathrm{P}$ is the difference between the total energy of the system and the total energy of the projectile alone. The vectors $\mathbf{R}$ and $\mathbf{r}$ are treated as independent variables, so that $\partial/\partial Z$ is performed with $\mathbf{r}$ kept fixed. If we decompose

$$\mathbf{R} = \mathbf{b} + Z \hat{\mathbf{e}}_z$$

into a transverse part $\mathbf{b}$ and a longitudinal part $Z\hat{\mathbf{e}}_z$, the vector $\mathbf{b}$ is not affected by any of the operators in Eq. (5.34) and remains constant within our approximation. We may now parametrize $Z$ as $Z = vt$ and write

$$f(\mathbf{r}, \mathbf{R}) = \psi_{\mathrm{i}}^{+}(\mathbf{r}, t)\, e^{i\epsilon_{\mathrm{i}} t/\hbar}. \qquad (5.35)$$

When this expression is inserted in Eq. (5.34), the differentiation of the phase factor with respect to $Z$ yields a term that cancels the energy $\epsilon_{\mathrm{i}}$ so that we obtain

$$\left(-i\hbar\frac{\partial}{\partial t} - i\hbar c\boldsymbol{\alpha}\cdot\nabla + \frac{Z_{\mathrm{T}}e^2}{r} + V_{\mathrm{P,e}}(t) + \frac{Z_{\mathrm{T}}Z_{\mathrm{P}}e^2}{R} + m_{\mathrm{e}}c^2\gamma^0\right)\psi_{\mathrm{i}}^{+}(\mathbf{r}, t) = 0 \qquad (5.36)$$

with

$$V_{\mathrm{P,e}}(t) = -\gamma(1-\beta\alpha_z)\frac{Z_{\mathrm{P}}e^2}{\sqrt{(x-b)^2 + y^2 + \gamma^2(z-vt)^2}}. \qquad (5.37)$$

Here, $\mathbf{r}$ and $t$ are independent variables. The $x$-axis has been chosen to point in the direction of $\mathbf{b}$, and the parameter $t$ is interpreted as the *electronic* time coordinate. In the limit $M_{\mathrm{T}}/m_{\mathrm{e}}$, $M_{\mathrm{P}}/m_{\mathrm{e}} \to \infty$, Eq. (5.33) reads

$$\Psi_{\mathrm{i}}^{+}(E) = u_{\mathrm{P}}^{(\rho)}\, e^{i\mathbf{K}\cdot\mathbf{R}}\psi_{\mathrm{i}}^{+}(\mathbf{r}, t)\, e^{i\epsilon_{\mathrm{i}} t/\hbar}. \qquad (5.38)$$

From this point on, Eq. (5.36) can be interpreted as a time-dependent Dirac equation for an electron initially bound by the target charge and in addition subject to the field produced by the charge of a projectile which follows a classical rectilinear trajectory. This constitutes the description of the collision in the impact parameter picture or in the semiclassical approximation referred to in Sec. 3.2.2.

We now have completed the development which starts from a time-independent wave formulation for a three-body system and ends with a time-dependent description of a single-body problem. In the following, we always adopt the time-dependent picture which aids the physical intuition and, from the outset, discards effects that would be difficult to retain but, fortunately, are negligible. Such effects include corrections of the order of the ratio between electron mass and nuclear mass, of nuclear negative-energy states, as well as of the nuclear spins.

The foregoing derivation is often referred to as the eikonal approximation [Gla59, WiW68, BrM92]. It exhibits the parallelism of the approximations made in the wave picture on the one hand and in the impact parameter picture on the other hand. Insofar, the result (5.36) is not surprising. Up to terms of the order of the ratio between the electron mass to the nuclear mass, the two descriptions are fully equivalent. It should be mentioned that in the corresponding nonrelativistic derivation [DeE94], a linearization of the kinetic-energy term is necessary (which is not needed in our relativistic treatment).

## 5.3 Moving Coulomb potentials

With Eq. (5.36) we have established what might have been written down from the outset, namely a time-dependent Dirac equation for an electron moving in electromagnetic potentials generated by two Coulomb sources which pass each other at a relativistic speed. This equation is the basis for the whole subsequent treatment.

### 5.3.1 Explicit form of the equations

Adopting the coordinates specified in Sec. 3.3 and in Fig. 3.2, we write down the equation (5.36) for an electron in the field of two moving point charges, both in the target frame and in the projectile frame. In the target frame, we have

$$i\hbar\frac{\partial}{\partial t}\Psi(\mathbf{r}_{\mathrm{T}},t)=H\Psi(\mathbf{r}_{\mathrm{T}},t) \tag{5.39}$$

with

$$H=-i\hbar c\,\boldsymbol{\alpha}\cdot\nabla_{\mathrm{T}}-\frac{Z_{\mathrm{T}}e^2}{r_{\mathrm{T}}}-S^2\,\frac{Z_{\mathrm{P}}e^2}{r'_{\mathrm{P}}}+m_{\mathrm{e}}c^2\gamma^0. \tag{5.40}$$

The Hamiltonian (5.40) describes an electron subject to the Coulomb potential $-Z_{\mathrm{T}}e^2/r_{\mathrm{T}}$ of the target nucleus and to the time-dependent projectile potential

$$-S^2\,\frac{Z_{\mathrm{P}}e^2}{r'_{\mathrm{P}}(t)}=-\gamma(1-\beta\alpha_z)\frac{Z_{\mathrm{P}}e^2}{r'_{\mathrm{P}}(t)}, \tag{5.41}$$

which is obtained by Lorentz transforming the potential $-Z_{\mathrm{P}}e^2/r'_{\mathrm{P}}(t)$ from the projectile to the target frame, see Eqs. (4.34) and (4.32).

Similarly, in the projectile frame, one has

$$i\hbar\frac{\partial}{\partial t'}\Psi'(\mathbf{r}'_{\mathrm{P}},t')=H'\Psi'(\mathbf{r}'_{\mathrm{P}},t') \tag{5.42}$$

with

$$H'=-i\hbar c\,\boldsymbol{\alpha}\cdot\nabla'_{\mathrm{P}}-S^{-2}\,\frac{Z_{\mathrm{T}}e^2}{r_{\mathrm{T}}}-\frac{Z_{\mathrm{P}}e^2}{r'_{\mathrm{P}}}+m_{e}c^2\gamma^0. \tag{5.43}$$

The Hamiltonian (5.43) describes an electron subject to the Coulomb potential $-Z_{\rm P}e^2/r_{\rm P}$ of the projectile nucleus and to the time-dependent target potential

$$- S^{-2} \frac{Z_{\rm T}e^2}{r_{\rm T}(t')} = -\gamma(1+\beta\alpha_z)\frac{Z_{\rm T}e^2}{r_{\rm T}(t')}, \tag{5.44}$$

which is obtained by Lorentz transforming the Coulomb potential $-Z_{\rm T}e^2/r_{\rm T}(t')$ from the target to the projectile frame, see Eqs. (4.34) and (4.32).

From the explicit form of the operators $S^2$ and $S^{-2}$ and from Eq. (4.34), one sees that the moving point charges $Z_{\rm P}e$ and $Z_{\rm T}e$, respectively, produce an electromagnetic interaction of the form $\rho\Phi - \mathbf{j}\cdot\mathbf{A}/c$ where $\Phi = A^0$ and $\mathbf{A}$ are the Liénard-Wiechert potentials given in Eqs. (3.34). The Lorentz invariance of these equations is demonstrated in Sec. 4.1.3.

### 5.3.2 A brief discussion of time

At this point, a brief discussion of time may be in order, because the role of the time variable may easily give rise to misunderstandings. In the target system (or laboratory system), the coordinates $\mathbf{r}_{\rm T}$ and $t$ are independent variables, that is, we may assign to each space point $\mathbf{r}_{\rm T}$ of the electron density distribution the same time coordinate $t$. The clock to measure the time $t$ may be provided, for example, by the motion of the projectile. In particular, the zero-point of the time axis can be chosen to coincide with the moment when the projectile is closest to the target nucleus. Then the projectile coordinate in the target frame is given by $\mathbf{R} = \mathbf{b}+\mathbf{v}t$, see Fig. 3.2. Nevertheless, $t$ is the time coordinate of the electron, not of the projectile.

The distinction between the time variables associated with the electron motion and associated with the projectile motion becomes clear when we perform a Lorentz transformation to the projectile inertial frame. From Eq. (3.31), the relevant coordinates transform as

$$\begin{aligned} z'_{\rm P} &= \gamma(z_{\rm T} - vt) \\ t' &= \gamma(t - vz_{\rm T}/c^2). \end{aligned} \tag{5.45}$$

One sees that the time coordinate $t'$ of the electron in the projectile frame depends on the position of the electron in the target frame. This shows that, even though the variable $t$ can be used to describe the trajectory, it is the *electronic* time associated with the spinor $\psi(\mathbf{r}, t)$.

If one wishes to assign a time coordinate $t_0$ to the projectile in the target frame, choosing again the closest approach to the target nucleus for the definition of the zero-point, then the $z$-coordinate $Z$ of the projectile in the target frame is also zero at this moment, and $\mathbf{R} = \mathbf{b} + \mathbf{v}t_0$. In its own rest system, instead of Eq. (5.45), the space-time coordinates of the projectile are

$$\begin{aligned} Z' &= 0 \\ t_0' &= t_0/\gamma. \end{aligned} \tag{5.46}$$

The last equation expresses the fact that the proper time is shorter than the time measured in a coordinate frame moving with respect to the rest frame. It is important to distinguish the two assignments of the time coordinate discussed here. Whereas in the former case, Eq. (5.45), for a quantum system, space and time coordinates are independent, in the latter case, Eq. (5.46), for a classical trajectory, the space coordinates are a function of time. For our considerations, we will always use the independent electronic times $t$ and $t'$.

### 5.3.3 Asymptotic solutions

The potentials appearing in the two-center time-dependent Dirac equations (5.39) and (5.42) are of long range so that conventional scattering theory is no longer applicable without special precautions. In general, it is convenient if not necessary to eliminate from the initial- and final-state electronic wave functions the asymptotic time dependence caused by the long-range Coulomb interaction. This can be done by imposing appropriate Coulomb boundary conditions, which is the same as applying an appropriate gauge transformation, so that the resulting transition amplitude contains a *short range* interaction. Here "short range" means that the perturbation falls off more rapidly than $1/r$ as a function of the electron-nucleus separation. The importance of such a procedure for the transition amplitudes has been demonstrated for nonrelativistic [DeE85, DeE86] and for relativistic collisions [Eic87a, ToE90b].

In order to establish well-defined channel wave functions that are devoid of the asymptotic time dependence caused by long-range interactions, we have to derive the asymptotic solutions of Eqs. (5.39) and (5.42) for $t \to \pm\infty$. We notice first that when the internuclear separation is large compared to the separation of the electron from the target or the projectile nucleus, respectively, we may replace

$$\begin{aligned} r_{\mathrm{T}} &\to R = \sqrt{b^2 + v^2 t^2} \\ r_{\mathrm{P}}' &\to R' = \sqrt{b^2 + v^2 t'^2}. \end{aligned} \tag{5.47}$$

Here, $t$ and $t'$ are the electronic time coordinates, referred to the target and projectile frame, respectively, see Secs. 5.2 and 5.3.2. Thus $R$ and

$R'$ measure the internuclear separations at times associated with the electronic positions $\mathbf{r}_T$ or $\mathbf{r}'_P$, respectively. These times are related according to Eq. (5.45) in contrast to (5.46), valid for the relative motion of the nuclei.

*Asymptotic equations*

In the limits expressed by Eqs. (5.47), the following asymptotic equations have to be satisfied [Eic87a]:

in the *target* frame, for an electron asymptotically bound to the *target* nucleus:

$$i\hbar\frac{\partial}{\partial t}\,\Phi_T^\infty(\mathbf{r}_T,t) = \left(-i\hbar c\boldsymbol{\alpha}\cdot\nabla_T - \frac{Z_T e^2}{r_T} - S^2\,\frac{Z_P e^2}{R'} + m_e c^2\gamma^0\right)\Phi_T^\infty(\mathbf{r}_T,t); \tag{5.48}$$

in the *target* frame, for an electron asymptotically bound to the *projectile* nucleus:

$$i\hbar\frac{\partial}{\partial t}\,\Phi_P^\infty(\mathbf{r}_T,t) = \left(-i\hbar c\boldsymbol{\alpha}\cdot\nabla_T - \frac{Z_T e^2}{R} - S^2\,\frac{Z_P e^2}{r'_P} + m_e c^2\gamma^0\right)\Phi_P^\infty(\mathbf{r}_T,t); \tag{5.49}$$

in the *projectile* frame, for an electron asymptotically bound to the *target* nucleus:

$$\begin{aligned} i\hbar\frac{\partial}{\partial t'}\,\Phi_T'^\infty(\mathbf{r}'_P,t') \;&=\; \Big(-i\hbar c\boldsymbol{\alpha}\cdot\nabla'_P \\ &\quad -S^{-2}\,\frac{Z_T e^2}{r_T} - \frac{Z_P e^2}{R'} + m_e c^2\gamma^0\Big)\,\Phi_T'^\infty(\mathbf{r}'_P,t'); \end{aligned} \tag{5.50}$$

in the *projectile* frame, for an electron asymptotically bound to the *projectile* nucleus:

$$\begin{aligned} i\hbar\frac{\partial}{\partial t'}\,\Phi_P'^\infty(\mathbf{r}'_P,t') \;&=\; \Big(-i\hbar c\boldsymbol{\alpha}\cdot\nabla'_P \\ &\quad -S^{-2}\,\frac{Z_T e^2}{R} - \frac{Z_P e^2}{r'_P} + m_e c^2\gamma^0\Big)\,\Phi_P'^\infty(\mathbf{r}'_P,t'). \end{aligned} \tag{5.51}$$

Since Eqs. (5.48) and (5.49) are completely analogous to Eqs. (5.50) and (5.51), we confine our discussion to the former expressions in the target system. For convenience, we introduce the dimensionless Sommerfeld parameters

$$\nu_T = \frac{Z_T e^2}{v\hbar} \quad \text{and} \quad \nu_P = \frac{Z_P e^2}{v\hbar}. \tag{5.52}$$

The same Eqs. (5.48) to (5.51) also apply if the electron is not *bound* to the target or projectile nucleus, but instead is represented by a continuum state referred to the target or to the projectile. This is so, because in a fast atomic collision, the asymptotic internuclear separation is large compared to the corresponding distance of the wave packet describing the emitted electron (or positron) from the emitting atom.

*Derivation of the asymptotic solutions*

We next show that the asymptotic solutions of Eqs. (5.48) and (5.49) factorize into a phase factor times an unperturbed eigenfunction satisfying the single-center Dirac equation in the target or projectile system, respectively, which are given by

$$i\hbar\frac{\partial}{\partial t}\,\psi_{\rm T}(\mathbf{r}_{\rm T},t) = \left(-i\hbar c\boldsymbol{\alpha}\cdot\nabla_{\rm T} - \frac{Z_{\rm T}e^2}{r_{\rm T}} + m_{\rm e}c^2\gamma^0\right)\,\psi_{\rm T}(\mathbf{r}_{\rm T},t) \qquad (5.53)$$

and

$$i\hbar\frac{\partial}{\partial t'}\,\psi'_{\rm P}(\mathbf{r}'_{\rm P},t') = \left(-i\hbar c\boldsymbol{\alpha}\cdot\nabla'_{\rm P} - \frac{Z_{\rm P}e^2}{r'_{\rm P}} + m_{\rm e}c^2\gamma^0\right)\,\psi'_{\rm P}(\mathbf{r}'_{\rm P},t'). \qquad (5.54)$$

The asymptotic solutions to be constructed are said to satisfy Coulomb boundary conditions because they take into account the asymptotic long-range potentials caused by the distant charges. Indeed, the parentheses occurring in Eqs. (5.48) and (5.53) differ only in the term $-S^2 Z_{\rm P}e^2/R'$, while those of Eqs. (5.51) and (5.54) in the projectile frame differ only in the term $-S^{-2}Z_{\rm T}e^2/R$. The removal of an electromagnetic four-potential may generally be achieved by a gauge transformation (4.7). We only have to rewrite the long-range potential term as the four-derivative of a gauge function $\chi(\mathsf{x})$ in order to be able to specify the corresponding phase factor of the wave function according to Eq. (4.7).

As an explicit illustration, we derive the asymptotic solution of Eq. (5.48). The simplest way to remove the long-range perturbing Liénard-Wiechert interaction

$$-\,S^2\frac{Z_{\rm P}e^2}{R'} = -\gamma(1-\beta\alpha_z)\frac{Z_{\rm P}e^2}{R'} \qquad (5.55)$$

is by transforming Eq. (5.48) into the projectile system thus obtaining Eq. (5.50). Here the perturbation has the operator-free Coulomb form $-Z_{\rm P}e^2/R'$. In order to cancel this term, we observe the identity

$$\frac{\partial}{\partial t'}\,e^{-i\nu_{\rm P}\ln(\sqrt{b^2+v^2t'^2}-vt')} = i\nu_{\rm P}v\,\frac{1}{\sqrt{b^2+v^2t'^2}}\,e^{-i\nu_{\rm P}\ln(\sqrt{b^2+v^2t'^2}-vt')}, \qquad (5.56)$$

with the definition (5.52) of $\nu_\mathrm{P}$. Using Eq. (5.47), this can be written as

$$i\hbar \frac{\partial}{\partial t'} e^{-i\nu_\mathrm{P} \ln(R'-vt')} = -\frac{Z_\mathrm{P} e^2}{R'} e^{-i\nu_\mathrm{P} \ln(R'-vt')}. \tag{5.57}$$

If we insert the ansatz[2]

$$\Phi'^{\infty}_\mathrm{T}(\mathbf{r}'_\mathrm{P}, t') = e^{-i\nu_\mathrm{P} \ln(R'-vt')} \psi'_\mathrm{T}(\mathbf{r}'_\mathrm{P}, t') \tag{5.58}$$

into Eq. (5.50), we find that the time differentiation produces a term on the left-hand side, which is identical to the long-range potential on the right-hand side, so that these terms cancel each other. The exponentials then appear as common factors and can be discarded, so that

$$i\hbar \frac{\partial}{\partial t'} \psi'_\mathrm{T}(\mathbf{r}'_\mathrm{P}, t') = \left( -i\hbar c \boldsymbol{\alpha} \cdot \nabla'_\mathrm{P} - S^{-2} \frac{Z_\mathrm{T} e^2}{r_\mathrm{T}} + m_\mathrm{e} c^2 \gamma^0 \right) \psi'_\mathrm{T}(\mathbf{r}'_\mathrm{P}, t'). \tag{5.59}$$

By Lorentz transforming the differential operators according to Eq. (4.38) we get

$$\begin{aligned} i\hbar S^{-1} \frac{\partial}{\partial t} S^{-1} \psi'_\mathrm{T}(\mathbf{r}'_\mathrm{P}, t') &= \Big( -i\hbar c S^{-1} \boldsymbol{\alpha} \cdot \nabla_\mathrm{T} S^{-1} \\ &\quad - S^{-1} \frac{Z_\mathrm{T} e^2}{r_\mathrm{T}} S^{-1} + m_\mathrm{e} c^2 \gamma^0 \Big) \psi'_\mathrm{T}(\mathbf{r}'_\mathrm{P}, t'). \end{aligned} \tag{5.60}$$

Upon multiplying Eq. (5.60) with $S$ from the left, identifying the inverse Lorentz transformation $S^{-1} \psi'_\mathrm{T}(\mathbf{r}'_\mathrm{P}, t') = \psi_\mathrm{T}(\mathbf{r}_\mathrm{T}, t)$, and realizing that according to Eq. (4.31) $S\gamma^0 = \gamma^0 S^{-1}$, we find that $\psi_\mathrm{T}(\mathbf{r}_\mathrm{T}, t)$ indeed satisfies the single-center Dirac equation (5.53). This means that Eq. (5.58) expresses the exact solution of the asymptotic equation (5.50) as a product of a phase factor and a single-center eigenfunction.

---

[2] A constant phase factor depending on the unit of length is dropped here and in the following phase factors.

*The phase-distorted asymptotic solutions*

Finally, the solution of Eq. (5.48) is obtained by transforming back to the laboratory system, so that

$$\Phi_{\mathrm{T}}^{\infty}(\mathbf{r}_{\mathrm{T}}, t) = e^{-i\nu_{\mathrm{P}} \ln(R' - vt')} \psi_{\mathrm{T}}(\mathbf{r}_{\mathrm{T}}, t). \tag{5.61}$$

The same result may also be derived by working in the target frame from the outset. In this case, because of the transformation $t' = \gamma(t - vz_{\mathrm{T}}/c^2)$, the time differentiation $\partial/\partial t$ leads to the cancellation of the scalar potential $-\gamma Z_{\mathrm{P}} e^2/R'$ contained in Eq. (5.48) through the identity (5.55), and the kinetic energy term $\boldsymbol{\alpha} \cdot \nabla_{\mathrm{T}}$ leads to the cancellation of the vector potential part $\gamma\beta\alpha_z Z_{\mathrm{P}} e^2/R'$ .

We still have to obtain the exact solution of the asymptotic equation (5.49). Since in this case, the long-range interaction is a simple Coulomb potential, there is no need for a transformation to the projectile system. In analogy to Eq. (5.61) we find

$$\Phi_{\mathrm{P}}^{\infty}(\mathbf{r}_{\mathrm{T}}, t) = e^{i\nu_{\mathrm{T}} \ln(R + vt)} \psi_{\mathrm{P}}(\mathbf{r}_{\mathrm{T}}, t). \tag{5.62}$$

The phase factor of Eq. (5.61), in contrast to its counterpart in Eq. (5.62), carries a dependence on the space coordinates in the laboratory system through the relation $t' = \gamma(t - vz_{\mathrm{T}}/c^2)$. While this space dependence is expected to give rise to small effects in most cases, it is essential for rigorously cancelling the magnetic part of the asymptotic interaction (5.55).

As has been said above, the elimination of the asymptotic scalar and vector potentials accompanied by an appropriate phase change of the wave function may be interpreted as a gauge transformation. Of course, gauge transformations may not only be applied to construct solutions satisfying asymptotic boundary conditions, but quite generally may be employed to add or to remove terms in the Hamiltonian in order to facilitate the problem at hand. Further applications are given in Sec. 5.5. Since these terms reappear in the concomitant phase factors, the physics remains unchanged.

*Influence of the internuclear interaction*

In the Hamiltonian (5.40), we have not incorporated the internuclear interaction, which is obtained from Eq. (3.34) as $V_{\mathrm{PT}} = Z_{\mathrm{P}} Z_{\mathrm{T}} e^2/(b^2 + \gamma^2 v^2 t^2)^{\frac{1}{2}}$. Since this term does not depend on the electronic space coordinates, it does not contribute to direct transitions between orthogonal states. However, in an eikonal treatment, see Sec. 5.2, it may be shown [BrM92] that the internuclear interaction enters into the differential cross section. For charge transfer, $V_{\mathrm{PT}}$ can be removed by a suitable gauge transformation, namely by including in Eqs. (5.61) and (5.62) another phase factor with a phase proportional to $\pm\ln[(b^2+\gamma^2v^2t^2)^{\frac{1}{2}}\mp\gamma vt]$, respectively. In the transition amplitude, these additional phase factors are combined to a term $(b/a_0)^{2i\nu}$ with $\nu = Z_\mathrm{P}Z_\mathrm{T}e^2/v\hbar$, which depends only on the impact parameter $b$. The internuclear interaction, therefore, does not affect the total cross section but only the differential cross section (5.84), see Sec. 5.4.2.

## 5.4 Transition amplitudes

In Sec. 5.3 we deal with the solutions of the two-center Dirac equation for an electron. Once we know the exact wave function, we are able to derive quantities of physical interest, for example, the probability amplitudes $A_\mathrm{fi}$ for a transition from an initial state $i$ to a final state $f$. Before we establish a perturbation description, we want to define the energy range in which such an approach is applicable.

### 5.4.1 Slow and fast collisions

For nonrelativistic collisions, it is convenient to distinguish low-energy and high-energy collisions, depending on whether the projectile velocity is smaller or larger than the velocity of the active electron. In the former case and in the transition region, the electron is able to follow the nuclear motion. This requires a theoretical description that pursues in detail the time evolution of the system during the collision. The most commonly used method, the coupled-channel (or close-coupling) approach, employs an expansion of the total time-dependent wave function in terms of appropriate stationary states and then considers the time evolution of the expansion coefficients, see Sec. 1.2.

If the projectile velocity is much larger than the electron velocity, the electron cannot follow the projectile motion. Consequently, the passage of the projectile acts as a small perturbation. This is the regime where perturbative methods are adequate.

The subdivision into two energy regimes implies that the term "slow" or "fast" has a meaning only for a specified charge number $Z$, since the nonrelativistic electron velocity in a Bohr orbital is $v_n = (\alpha Z/n)c$ where $\alpha = 1/137.036$ is the fine-structure constant and $n$ the principal quantum number. In other words, if K-shell electrons of high-$Z$ atoms are involved, much higher projectile energies are needed to render perturbation theory applicable than for low-$Z$ atoms. For excitation and ionization, the atomic charge number $Z$ may be identified with the target atomic number, while for electron transfer, the larger value $Z/n$ of target or projectile serves to assess the applicability of perturbation theory.

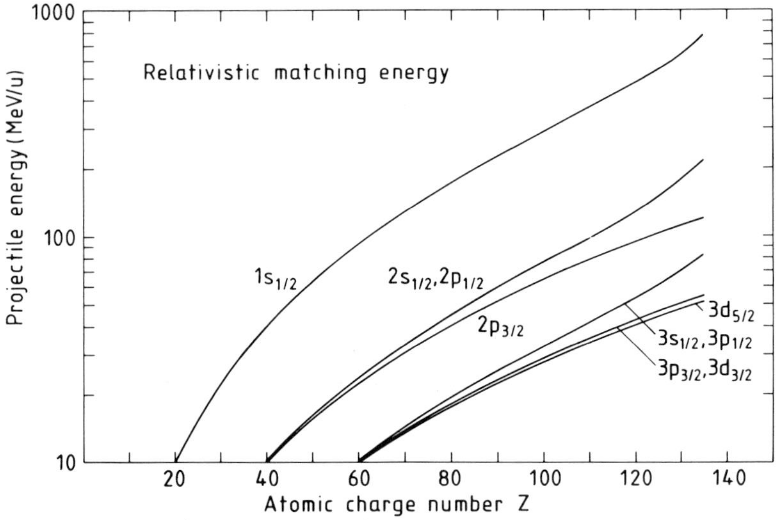


**Figure 5.1.** Collision energy in MeV/u for which energy matching according to Eq. (5.63) is obtained as a function of the atomic charge number for various electronic eigenstates.

The foregoing discussion refers to a nonrelativistic situation. For relativistic collisions, all velocities approach the speed of light so that a comparison of velocities is no longer meaningful. In this case, it is more appropriate to define a matching energy $T_m$ per atomic mass unit or, correspondingly, $\gamma_{\rm m} = 1 + T_{\rm m}[\text{MeV/u}]/931.5$, see Eq. (3.13), by demanding that the kinetic energy of a free electron traveling with the speed of the projectile, equals the binding energy $E_{\rm bind}$ of the electronic state, or

$$m_{\rm e}c^2(\gamma_{\rm m} - 1) = E_{\rm bind} = m_{\rm e}c^2[1 - \sqrt{1 - (\alpha Z)^2}], \tag{5.63}$$

where the second equation refers to atomic $1s_{1/2}$ states, see Table 4.3, which also contains the binding energies of other low-lying hydrogenic states. The definition (5.63) has the correct nonrelativistic limit $\beta = \alpha Z$. Figure 5.1 shows the matching energy $T_{\rm m}$ calculated from Eq. (5.63) for different atomic states as a function of the charge number $Z$. There are other possible prescriptions leading to similar estimates.

For example, for a collision involving $1s_{1/2}$ states of uranium, the matching energy of the projectile is $T_{\rm m}$=240 MeV/u, so that "fast" collisions start from, say, 1 GeV/u. Above this energy, we expect perturbative methods to be applicable.

### 5.4.2 Perturbative transition amplitudes and cross sections

In this section, we assume that the collision is sufficiently energetic and the interaction sufficiently weak to render perturbative methods valid. Since we wish to treat all possible transitions, including the production of electron-positron pairs, on the same footing, it is convenient in this subsection to use the language of field theory introduced in Sec. 4.2.2 and applied to ion-atom collisions in Sec. 5.1. In contrast to Sec. 5.1.3, however, we assume, within perturbation theory, that the probability for pair production remains small at all times and hence the vacuum state is unaltered during the collision. As before, we adopt the impact parameter representation.

In order to express measurable quantities, we first have to formulate the transition amplitudes $A_{\rm fi}$ that permit us to calculate the impact parameter-dependent transition probability

$$P_{\rm fi}(b) = |A_{\rm fi}(b)|^2 \tag{5.64}$$

and the total cross section

$$\sigma_{\rm fi} = \int P_{\rm fi}(b)\, d^2b \tag{5.65}$$

between any pair of specified initial and final states, $i$ and $f$, respectively. The $S$-matrix, from which the transition amplitudes may be derived, is generally given by

$$\begin{aligned} \widehat{S} &= \widehat{T} \exp\left\{-\frac{i}{\hbar}\int_{-\infty}^{\infty} dt\, \widehat{H}_{\rm I}(t)\right\} \\ &= 1 - \frac{i}{\hbar}\int_{-\infty}^{\infty} dt_1\, \widehat{H}_{\rm I}(t_1) \\ &\quad + \left(-\frac{i}{\hbar}\right)^2 \int_{-\infty}^{\infty} dt_1 \int_{-\infty}^{t_1} dt_2\, \widehat{H}_{\rm I}(t_1)\widehat{H}_{\rm I}(t_2) + \cdots \end{aligned} \tag{5.66}$$

Here $\widehat{T}$ is the Dyson time-ordering operator, and the interaction Hamiltonian

$$\widehat{H}_{\rm I}(t) = \int \widehat{\mathcal{H}}_{\rm int}\, d^3r \tag{5.67}$$

is expressed as a space integral over the basic Hamiltonian density

$$\widehat{\mathcal{H}}_{\rm int} = \widehat{j}^{\mu} A_{\mu}/c = -e\widehat{\overline{\psi}}\gamma^{\mu}\widehat{\psi} A_{\mu} \tag{5.68}$$

describing the interaction of an electron or positron current (4.8) with an electromagnetic field. The field may be external, that is classical, or may be quantized if photons are emitted or absorbed. The quantities $\widehat{\psi}$ and $\widehat{\overline{\psi}} = \widehat{\psi}^\dagger\gamma^0$ are time-dependent field operators. These operators are usually expanded according to Eq. (5.8), and by constructing states of one or more independent particles in analogy to Eqs. (4.69) and (4.70), we may calculate $S$-matrix elements from Eqs. (5.66).

*First-order transition amplitude in a covariant form*

In particular, the first-order transition amplitude appears in a manifestly covariant form

$$A_{\rm fi} = S_{\rm fi}^{(1)} = -\frac{i}{\hbar}\langle\Phi_{\rm f}| - e\int d^4x\,\widehat{\overline{\psi}}\gamma^\mu\widehat{\psi}A_\mu\,|\Phi_{\rm i}\rangle, \tag{5.69}$$

where $|\Phi_{\rm i}\rangle$ and $|\Phi_{\rm f}\rangle$ denote states in the occupation number representation introduced in Sec. 4.2.2. For electron excitation, ionization, or transfer, we use initial and final states of the type (4.69) and obtain

$$A_{\rm fi}^{\rm e,tr} = \frac{i}{\hbar}\int d^4x \sum_{k_+,l_+}\langle 0|\,\widehat{b}_{\rm f}\,\widehat{b}_{k_+}^\dagger\,\widehat{b}_{l_+}\,\widehat{b}_{\rm i}^\dagger\,|0\rangle\, e\psi_{k_+}^\dagger\gamma^0\gamma^\mu\psi_{l_+}A_\mu, \tag{5.70}$$

where $k_+, l_+$ are quantum numbers characterizing positive-energy states (while $k_-, l_-$ are used to denote negative-energy states). For excitation and ionization, we refer all states to an orthogonal basis of target states. After carrying out all field-theoretic manipulations, Eq. (5.70) simplifies to

$$A_{\rm fi}^{\rm e,tr} = \frac{i}{\hbar}\int d^4x\; e\overline{\psi}_{\rm f}\gamma^\mu\psi_{\rm i}\,A_\mu. \tag{5.71}$$

For electron transfer, we imagine that the wave functions centered around the projectile are expanded in terms of target functions, so that orthogonality is preserved and the vacuum is defined with respect to this basis set. At the end, one can return to projectile-centered states. We assume from the beginning that this has been done.

For pair creation we use $|0\rangle$ as the initial state and $\widehat{d}_{p_-}^\dagger\,\widehat{b}_{p_+}^\dagger|0\rangle$ as the final electron-positron state to obtain

$$A_{\rm fi}^{\rm pair} = \frac{i}{\hbar}\int d^4x \sum_{k_+,l_-}\langle 0|\,\widehat{b}_{p_+}\widehat{d}_{p_-}\,\widehat{d}_{l_-}^\dagger\,\widehat{b}_{k_+}^\dagger\,|0\rangle\, e\psi_{k_+}^\dagger\gamma^0\gamma^\mu\psi_{l_-}A_\mu \tag{5.72}$$

or

$$A_{\rm fi}^{\rm pair} = \frac{i}{\hbar}\int d^4x\; e\overline{\psi}_{p_+}\gamma^\mu\psi_{p_-}A_\mu. \tag{5.73}$$

The higher-order terms in Eq. (5.66) allow for excitation, ionization, or capture of more than one electron or for the creation of more than one electron-positron pair. In addition, within the space of single-electron states, these terms yield corrections to the unperturbed wave functions $\psi_\mathrm{i}$ or $\psi_\mathrm{f}$ appearing in Eqs. (5.71) and (5.73). A complete summation of the corresponding matrix elements of Eq. (5.66) replaces either $\psi_\mathrm{i}$ or $\psi_\mathrm{f}$ with the exact wave functions.

Before leaving the general discussion of the interaction Hamiltonian (5.67), let us recall that the gauge transformation (4.7) leaves the basic Dirac equation (5.6) to be satisfied by the field operators $\widehat{\psi}$ invariant. The invariance also holds for the interaction Hamiltonian

$$\widehat{\widetilde{H}}_\mathrm{I} = -e \int \widehat{\overline{\psi}} \gamma^\mu \widehat{\psi} \left( A_\mu - \frac{\partial \chi}{\partial x^\mu} \right) d^3x = \widehat{H}_\mathrm{I} \tag{5.74}$$

since integration by parts turns the term $\partial\chi/\partial x^\mu$ into a four-divergence $\partial \widehat{j}^\mu / \partial x^\mu$ of the four-current $\widehat{\overline{\psi}} \gamma^\mu \widehat{\psi}$. By construction, this four-divergence vanishes for any *exact solution* of the Dirac equation, and hence the interaction Hamiltonian is gauge invariant.

In this subsection so far, we have used the language of field theory in order to derive, from a single formula, the first-order amplitudes for all processes considered in this book. In the following, however, we will return to the language of wave functions.

*Explicit transition amplitudes for various processes*

It is useful here to rewrite the expressions for the transition amplitudes of the most relevant processes in some more detail. Since, for convenience, we work in the laboratory system, we have to express all quantities in Eqs. (5.71) and (5.73) in the target system. On the other hand, the projectile potential initially is best formulated in the projectile system, similarly the final-state wave function for charge transfer.

(a) *Excitation and ionization*

$$A^\mathrm{e}_\mathrm{fi} = -\frac{i}{\hbar} \int d^4x \, \overline{\psi}_\mathrm{f}(\mathbf{r}_\mathrm{T}, t) \left( e S^{-1} \gamma^\mu A'_\mu(\mathbf{r}'_\mathrm{P}, t') S \right) \psi_\mathrm{i}(\mathbf{r}_\mathrm{T}, t). \tag{5.75}$$

Since in the projectile system the interaction is simply Coulombic, we get from Eq. (4.34) the result

$$A^\mathrm{e}_\mathrm{fi} = -i\frac{e^2}{\hbar} \int dt \int d^3r_\mathrm{T} \, \psi_\mathrm{f}^\dagger(\mathbf{r}_\mathrm{T}, t) \, S^2 \left( -\frac{Z_\mathrm{P}}{r'_\mathrm{P}} \right) \psi_\mathrm{i}(\mathbf{r}_\mathrm{T}, t). \tag{5.76}$$

(b) *Charge transfer*

$$A^\mathrm{tr}_\mathrm{fi} = -\frac{i}{\hbar} \int d^4x \, \overline{(S^{-1}\psi'_\mathrm{f}(\mathbf{r}'_\mathrm{P}, t'))} \left( e S^{-1} \gamma^\mu A'_\mu(\mathbf{r}'_\mathrm{P}, t') S \right) \psi_\mathrm{i}(\mathbf{r}_\mathrm{T}, t) \tag{5.77}$$

where the final-state electronic wave function is transformed back to the target system (the bar indicates the Dirac adjoint $\overline{\psi} = \psi^{\dagger}\gamma^0$), hence

$$A_{\rm fi}^{\rm tr} = -i\frac{e^2}{\hbar}\int dt \int d^3r_{\rm T}\,\psi_{\rm f}'^{\dagger}(\mathbf{r}_{\rm P}', t')\,S\left(-\frac{Z_{\rm P}}{r_{\rm P}'}\right)\psi_{\rm i}(\mathbf{r}_{\rm T}, t). \tag{5.78}$$

(c) *Electron-positron pair creation*

$$A_{\rm fi}^{\rm pair} = -i\frac{e^2}{\hbar}\int dt \int d^3r_{\rm T}\,\psi_{p_+}^{\dagger}(\mathbf{r}_{\rm T}, t)\,S^2\left(-\frac{Z_{\rm P}}{r_{\rm P}'}\right)\psi_{p_-}(\mathbf{r}_{\rm T}, t), \tag{5.79}$$

similarly as for excitation, however with the initial wave function replaced by the positron wave function $\psi_{p_-}$ and the final wave function replaced by the electron wave function $\psi_{p_+}$. The transformation matrix $S$ and its square are given in Eqs. (4.29) and (4.32). The independent integration variables in all cases are $\mathbf{r}_{\rm T}$ and $t$, so that the Lorentz-transformed coordinates have to be inserted for $\mathbf{r}_{\rm P}'$ and $t'$ when the integrals are evaluated. Further approximations depend on the kind of wave function actually inserted for the four-spinors $\psi_{\rm i}$ and $\psi_{\rm f}$.

*Distorted-wave Born approximation. The post and the prior form*

For the discussion of scattering processes, it is often useful to start from the *exact* expressions for the transition amplitude. Let $\Psi_{\rm i}^{+}(t)$ be the exact outgoing-wave solution in the laboratory system evolving from the initial state $i$ and $\Phi_{\rm f}^{\infty}(t)$ be the exact asymptotic final state. This final state may be an asymptotic *target* eigenstate [e.g. (5.61)], occurring in a "direct reaction" like excitation, ionization, or pair production, or it may be an asymptotic *projectile* eigenstate [e.g. (5.62)], occurring for charge transfer. Then, by definition, the exact transition amplitude is given by

$$A_{\rm fi}^{(+)} = \lim_{t\to\infty}\langle\Phi_{\rm f}^{\infty}\,|\,\Psi_{\rm i}^{+}\rangle. \tag{5.80}$$

This version is called the *post* form of the transition amplitude. Alternatively, the *prior* form of the transition amplitude is obtained at $t \to -\infty$ as the projection of the exact incoming-wave solution evolving into the final state $f$ onto the exact asymptotic initial state (5.61), namely

$$A_{\rm fi}^{(-)} = \lim_{t\to-\infty}\langle\Psi_{\rm f}^{-}\,|\,\Phi_{\rm i}^{\infty}\rangle. \tag{5.81}$$

From the defining identical expressions, (5.80) and (5.81), one may derive the basic equation for the distorted-wave Born approximation.

Provided the overlap matrix elements within Eqs. (5.80) and (5.81) vanish for $t \to -\infty$ and $t \to +\infty$, respectively, one may write the *post* form as

$$\begin{aligned} A_{\rm fi}^{(+)} &= -\frac{i}{\hbar}\int_{-\infty}^{\infty} dt \left( \left\langle \Phi_{\rm f}^{\infty} \middle| i\hbar\frac{\partial}{\partial t}\Psi_{\rm i}^{+} \right\rangle - \left\langle i\hbar\frac{\partial}{\partial t}\Phi_{\rm f}^{\infty} \middle| \Psi_{\rm i}^{+} \right\rangle \right) \\ &= -\frac{i}{\hbar}\int_{-\infty}^{\infty} dt \left\langle \left( H - i\hbar\frac{\partial}{\partial t}\right)\Phi_{\rm f}^{\infty} \middle| \Psi_{\rm i}^{+} \right\rangle . \end{aligned} \tag{5.82}$$

and the *prior* form as

$$\begin{aligned} A_{\rm fi}^{(-)} &= \frac{i}{\hbar}\int_{-\infty}^{\infty} dt \left( \left\langle i\hbar\frac{\partial}{\partial t}\Psi_{\rm f}^{-} \middle| \Phi_{\rm i}^{\infty} \right\rangle - \left\langle \Psi_{\rm f}^{-} \middle| i\hbar\frac{\partial}{\partial t}\Phi_{\rm i}^{\infty} \right\rangle \right) \\ &= \frac{i}{\hbar}\int_{-\infty}^{\infty} dt \left\langle \Psi_{\rm f}^{-} \middle| H - i\hbar\frac{\partial}{\partial t} \middle| \Phi_{\rm i}^{\infty} \right\rangle . \end{aligned} \tag{5.83}$$

Being exact, both expressions are identical. If approximate wave functions are substituted for the exact ones, this is no longer true in all cases. This then is referred to as a *post-prior discrepancy.*

The physical meaning of the equations is the following. In the post form, the operator $(H - i\hbar\partial/\partial t)$ acting on $\Phi_{\rm f}^{\infty}$ yields the potential term, the *residual interaction*, not included in the definition of the asymptotic final state $\Phi_{\rm f}^{\infty}$ at $t \to +\infty$ (for charge transfer, this is the electron-target interaction). In the prior form, the operator $(H - i\hbar\partial/\partial t)$ acting on $\Phi_{\rm i}^{\infty}$ yields the potential term not included in the definition of the asymptotic initial state $\Phi_{\rm i}^{\infty}$ at $t \to -\infty$ (for charge transfer, this is the electron-projectile interaction).

The *Born approximation* is obtained by substituting unperturbed initial and final states. Examples of the corresponding operator $(H - i\hbar\partial/\partial t)$ are given in Eqs. (6.56) and (6.57). A variety of *distorted-wave Born approximations* can be constructed by inserting distorted states, in which a suitable part of the total Hamiltonian $H$ is included in the definition of the initial and final states. The operator $(H - i\hbar\partial/\partial t)$ then produces the perturbation associated with the assumed distortion.

### *Differential cross sections*

In principle, it is possible to perform measurements in which not only the atomic process, characterized by the transition amplitude $A_{\rm fi}(b)$, is identified but also the deflection of a heavy projectile is detected directly, or through target recoil (see Sec. 13.2.3). We furnish here the appropriate formulas. For the high collision energies considered and for small deflection angles, when the eikonal approximation discussed in Sec. 5.2 is valid, the dependence of the scattering amplitude on the transverse momentum transfer $\boldsymbol{\eta}$ and hence on the scattering angle $\theta$ may be obtained by Fourier transforming the impact parameter-dependent transition amplitude $A_{\mathrm{fi}}(b)$. Using the fact that the initial and final projectile momenta have very nearly the same magnitude $M_{\mathrm{P}}\gamma v$, one may write the differential cross section in the center-of-mass system as [BrM92, DeE94]

$$\frac{d\sigma_{\mathrm{fi}}}{d\Omega} = |f_{\mathrm{fi}}(\theta)|^2 \tag{5.84}$$

with[3]

$$f_{\mathrm{fi}}(\theta) = \frac{-iM_{\mathrm{PT}}\gamma v}{2\pi\hbar} \int A_{\mathrm{fi}}(b)\, b^{2i\nu} e^{-i\boldsymbol{\eta}\cdot\mathbf{b}}\, d^2b, \tag{5.85}$$

where $M_{\mathrm{PT}} = M_{\mathrm{P}}M_{\mathrm{T}}/(M_{\mathrm{P}} + M_{\mathrm{T}})$ is the reduced mass of target and projectile and the transverse momentum $\boldsymbol{\eta}$ with $\boldsymbol{\eta}\cdot\mathbf{v} = 0$ is related to the scattering angle $\theta$ in the center-of-mass system by

$$\eta = \frac{2M_{\mathrm{PT}}\gamma v}{\hbar} \sin\frac{\theta}{2}. \tag{5.86}$$

The phase factor $b^{2i\nu}$ with the Sommerfeld parameter $\nu = Z_{\mathrm{T}}Z_{\mathrm{P}}e^2/\hbar v$ results from the Coulomb deflection of the projectile by the target nucleus. Since the Coulomb repulsion term contributes mostly to larger scattering angles, where the cross section becomes very small, it is often neglected [Moi85, DeE91a].

For transitions of the electron between states with the magnetic quantum numbers $\mu_{\mathrm{i}}$ and $\mu_{\mathrm{f}}$, the integration over the azimuthal angle between $\boldsymbol{\eta}$ and $\mathbf{v}$ in Eq. (5.85) can be carried out, so that

$$\frac{d\sigma}{d\Omega} = \left|\frac{iM_{\mathrm{PT}}\gamma v}{\hbar} \int A_{\mathrm{fi}}(b)\, b^{2i\nu} J_{|\mu_{\mathrm{f}}-\mu_{\mathrm{i}}|}(\eta b)\, b\, db\right|^2, \tag{5.87}$$

where $J_{|\mu_{\mathrm{f}}-\mu_{\mathrm{i}}|}$ denotes the Bessel function of the first kind.

Equations (5.85) and (5.87) provide a relation between the impact parameter and the scattering angle which is not simply based on the classical one-to-one relation (3.20) between the two quantities. Instead, the whole range of impact parameters contributes to a given scattering angle, and by a Fourier transformation, it is seen that the opposite is also true. Even though finite deflections are taken into account by the integration over impact parameters, the straight-line approximation for the projectile trajectory can be used to calculate the atomic matrix elements.

[3] An irrelevant phase factor $(1/a_0)^{2i\nu}$ is discarded here and in the following phase factors.

## 5.5 Matrix elements and gauge transformations

In Sec. 5.4, we give perturbative transition amplitudes for various possible reaction channels. In all these cases, and actually in many others, the problem arises of calculating matrix elements of the Dirac operator $\boldsymbol{\alpha}$ or its components. As is known from nonrelativistic quantum theory, there are sometimes different ways to calculate a certain matrix element. A typical example is the "length form" and the "velocity form" of the electric dipole moment. Both forms are related by the operator identity

$$\mathbf{p} = i(m_\mathrm{e}/\hbar)\,[H, \mathbf{r}],$$

which is valid if the Hamiltonian $H$ does not contain velocity-dependent potentials (as, for example, the spin-orbit interaction). By taking matrix elements of both sides of the identity between states $|\mathrm{i}\rangle$ and $|\mathrm{f}\rangle$ with eigenenergies $E_\mathrm{i}$ and $E_\mathrm{f}$, we obtain the relation

$$\langle \mathrm{f}|\,\mathbf{p}\,|\mathrm{i}\rangle = i\,(m_\mathrm{e}/\hbar)(E_\mathrm{f} - E_\mathrm{i})\langle \mathrm{f}|\,\mathbf{r}\,|\mathrm{i}\rangle$$

between the dipole-velocity and the dipole-length forms of the matrix elements. This relation is rigorously valid as long as exact wave functions are employed in the evaluation; however, if approximate wave functions are used, both sides may differ. This relation can also be interpreted as the result of a gauge transformation, as is shown below.

In a similar way, we discuss general recipes for transforming matrix elements involving the Dirac $\boldsymbol{\alpha}$ operator. This is of particular importance if the Liénard-Wiechert interaction occurs in the transition amplitudes summarized in Sec. 5.4, because large cancellations appear for high projectile velocities, so that great care must be exerted in the evaluation. Following the general transformation, we discuss the representation and transformation of the Liénard-Wiechert interaction in wave-number space.

### 5.5.1 Reformulation of $\alpha$-matrix elements

Let us consider eigenstates $|\mathrm{i}\rangle$ and $|\mathrm{f}\rangle$ of the atomic Hamiltonian for the nuclear charge number $Z_\mathrm{T}$, that is

$$H_0 = -i\hbar c\boldsymbol{\alpha}\cdot\nabla - \frac{Z_\mathrm{T}e^2}{r} + m_\mathrm{e}c^2\gamma^0 \tag{5.88}$$

with eigenenergies $E_\mathrm{i}$ and $E_\mathrm{f}$. If $g(\mathbf{r})$ is an arbitrary differentiable function of $\mathbf{r}$, we have

$$[H_0, g(\mathbf{r})] = -i\hbar c\boldsymbol{\alpha}\cdot\nabla g(\mathbf{r}) \tag{5.89}$$

and the corresponding matrix element

$$\langle \mathrm{f}|\boldsymbol{\alpha}\cdot\nabla g(\mathbf{r})|\mathrm{i}\rangle = i\,\frac{E_{\mathrm{f}}-E_{\mathrm{i}}}{\hbar c}\,\langle \mathrm{f}|g(\mathbf{r})|\mathrm{i}\rangle. \tag{5.90}$$

In particular, for $g(\mathbf{r}) = x_j$, where $x_j$ is one of the space components of $\mathbf{r} = (x, y, z)$, we have the relation

$$\langle \mathrm{f}|\,\alpha_j\,|\mathrm{i}\rangle = i\,\frac{E_{\mathrm{f}}-E_{\mathrm{i}}}{\hbar c}\,\langle f|\,x_j\,|i\rangle \tag{5.91}$$

between the velocity and the length form of the relativistic dipole matrix element.

Whereas the right-hand sides of Eqs. (5.91) and (5.90) yield finite results if nonrelativistic Schrödinger wave functions are inserted, the left-hand sides would yield zero, because the Dirac $\boldsymbol{\alpha}$ matrix connects large and small components of the wave functions. We may therefore expect that the right-hand sides are less sensitive to approximations made in the evaluation.

*Connection with gauge transformations*

A more general and more powerful method for reformulating matrix elements is provided by gauge transformations. In analogy to the nonrelativistic case [SaA51,Eic74], the connection with gauge transformations is as follows: if the system described by Eq. (5.88) is subjected to an external vector potential $\mathbf{A}$, see Eq. (4.13), the resulting Hamiltonian is a functional

$$H\{\mathbf{A}\} = H_0 + H_1\{\mathbf{A}\} \tag{5.92}$$

with $H_1\{\mathbf{A}\} = e\boldsymbol{\alpha}\cdot\mathbf{A}$. The gauge invariance expressed by Eq. (4.15) can now be rewritten as

$$H\{\mathbf{A}\} = \exp(i\frac{e}{\hbar c}\chi)\,H\{\mathbf{A}+\nabla\chi\}\,\exp(-i\frac{e}{\hbar c}\chi), \tag{5.93}$$

or, in the absence of an external potential, $\mathbf{A} = 0$,

$$H\{\nabla\chi\} = \exp(-i\frac{e}{\hbar c}\chi)\,H_0\,\exp(i\frac{e}{\hbar c}\chi). \tag{5.94}$$

The right-hand side can be replaced by a nested commutator expansion whose first-order term $H_1$ for a time-independent gauge function $\chi(\mathbf{r})$ is written as

$$H_1\{\nabla\chi\} \equiv e\boldsymbol{\alpha}\cdot\nabla\chi = i\,\frac{e}{\hbar c}\,[H_0, \chi], \tag{5.95}$$

which is recognized as being identical to Eq. (5.90) if $g(\mathbf{r}) = \chi(\mathbf{r})$.

### 5.5.2 Liénard-Wiechert interaction in wave-number representation

In the transition amplitudes (5.76) for excitation and ionization and in Eq. (5.79) for pair production, we have to calculate matrix elements of the Liénard-Wiechert interaction $S^2(-Z_\mathrm{P}e/r'_\mathrm{P})$ between the relevant initial and final states $|\mathrm{i}\rangle$ and $|\mathrm{f}\rangle$. It is useful to consider the matrix elements generally before dealing with the specific cases.

Since the time dependence residing in the electron-projectile separation $r'_\mathrm{P}(t)$ is difficult to handle, one usually resorts to the wave-number representation (obtained by Fourier transformation), in which the complete time dependence is moved to an exponential. The time integration can then be carried out. Inserting the operator $S^2$ from Eq. (4.32) and adopting the bracket notation for the space integral in the target frame, we can write the generic matrix element for excitation, ionization, and pair production as

$$M_\mathrm{fi}(b) = \int_{-\infty}^{\infty} e^{i(E_\mathrm{f}-E_\mathrm{i})t/\hbar} \left\langle \mathrm{f} \left| (1-\beta\alpha_z)\frac{1}{r'_\mathrm{P}} \right| \mathrm{i} \right\rangle dt, \tag{5.96}$$

so that in Eqs. (5.76) and (5.79) the transition amplitude is expressed as $A_\mathrm{fi}^\mathrm{e,pair}(b) = i\gamma v\nu M_\mathrm{fi}(b)$. The coordinates are explained in Fig. 3.2 and

$$r'_\mathrm{P} = \sqrt{(x_\mathrm{T}-b)^2 + y_\mathrm{T}^2 + \gamma^2(z_\mathrm{T}-vt)^2} \tag{5.97}$$

is the electron-projectile distance measured in the projectile frame.

*The relativistic Bethe integral*

In order to construct the wave-number representation of the interaction, we derive the relativistic Bethe integral [AmA81, Moi85]. We start from the nonrelativistic Bethe integral

$$\int \frac{e^{i\mathbf{q}\cdot\mathbf{r}}}{r}\, d^3r = \frac{4\pi}{q^2} \tag{5.98}$$

in which a convergence-enforcing factor $\exp(-\mu r)$ with $\mu \to +0$ is assumed when performing the integration. In the following, we have to make use of the coordinate transformations (3.29) and (3.30). We also need the relation $\mathbf{r}_\mathrm{T} = \mathbf{r}_\mathrm{P} + \mathbf{R}$ in the target system and $\mathbf{r}'_\mathrm{P} = \mathbf{r}'_\mathrm{T} - \mathbf{R}'$ in the projectile system. Here, we decompose $\mathbf{R} = (\mathbf{b}, vt)$ and $\mathbf{R}' = (\mathbf{b}, vt')$ into the transverse and the longitudinal components. Similarly, we write $\mathbf{q} = (\mathbf{q}_b, q_z)$ and $\mathbf{r}_\mathrm{T} = (\mathbf{r}_{\mathrm{T}b}, z_\mathrm{T})$, where $\mathbf{r}_{\mathrm{T}b} = \mathbf{r}'_{\mathrm{P}b} + \mathbf{b}$.

We now consider the integral

$$\int \frac{1}{r'_{\mathrm{P}}} e^{i\mathbf{q}\cdot(\mathbf{r}_{\mathrm{T}}-\mathbf{R})}\, d^3 r'_{\mathrm{P}} = \int \frac{1}{r'_{\mathrm{P}}} e^{i[\mathbf{q}_b\cdot\mathbf{r}'_{\mathrm{P}b}+q_z(z_{\mathrm{T}}-vt)]}\, d^3 r'_{\mathrm{P}} = \int \frac{1}{r'_{\mathrm{P}}} e^{i(\mathbf{q}_b\cdot\mathbf{r}'_{\mathrm{P}b}+q_z z'_{\mathrm{P}}/\gamma)}\, d^3 r'_{\mathrm{P}}. \tag{5.99}$$

All coordinates now being expressed by $\mathbf{r}'_{\mathrm{P}}$, we can apply the Bethe integral (5.98) for the wave vector $(\mathbf{q}_b, q_z/\gamma)$, so that

$$\int \frac{1}{r'_{\mathrm{P}}} e^{i\mathbf{q}\cdot(\mathbf{r}_{\mathrm{T}}-\mathbf{R})}\, d^3 r'_{\mathrm{P}} = \frac{4\pi}{q_b^2 + q_z^2/\gamma^2} = \frac{4\pi}{q^2 - \beta^2 q_z^2}, \tag{5.100}$$

and the relativistic Bethe integral assumes the form

$$\int \frac{e^{i\mathbf{q}\cdot\mathbf{r}_{\mathrm{P}}}}{r'_{\mathrm{P}}}\, d^3 r'_{\mathrm{P}} = \frac{4\pi}{q^2 - \beta^2 q_z^2}. \tag{5.101}$$

We are now free to keep $\mathbf{r}'_{\mathrm{T}} = \mathbf{r}'_{\mathrm{P}} + \mathbf{R}'$, and hence $\mathbf{r}_{\mathrm{T}}$, fixed and integrate over $\mathbf{R}'$ instead of $\mathbf{r}'_{\mathrm{P}}$. This allows us to take $\exp(i\mathbf{q}\cdot\mathbf{r}_{\mathrm{T}})$ out of the integral and to write

$$\int \frac{1}{r'_{\mathrm{P}}} e^{-i\mathbf{q}\cdot\mathbf{R}}\, d^3 R' = 4\pi \frac{e^{-i\mathbf{q}\cdot\mathbf{r}_{\mathrm{T}}}}{q^2 - \beta^2 q_z^2}. \tag{5.102}$$

We may invert the Fourier transform on the left-hand side, taking into account $d^3 R' = d^2 b\, d(vt') = d^2 b\, d(\gamma vt) = \gamma\, d^3 R$. We then get the final result

$$\frac{1}{r'_{\mathrm{P}}} = \frac{1}{2\pi^2\gamma} \int \frac{e^{i\mathbf{q}\cdot(\mathbf{r}_{\mathrm{T}}-\mathbf{R})}}{q^2 - \beta^2 q_z^2}\, d^3 q. \tag{5.103}$$

*The Liénard-Wiechert interaction in momentum space*

With the aid of the Bethe integral (5.103), the Liénard-Wiechert interaction can be written in wave-number representation as

$$\left\langle \mathrm{f} \left| (1-\beta\alpha_z)\frac{1}{r'_{\mathrm{P}}} \right| \mathrm{i} \right\rangle = \frac{1}{2\pi^2\gamma} \int \frac{d^3 q}{q^2 - \beta^2 q_z^2}\, e^{-i\mathbf{q}\cdot\mathbf{R}(t)} \langle \mathrm{f}|(1-\beta\alpha_z) e^{i\mathbf{q}\cdot\mathbf{r}_{\mathrm{T}}}|\mathrm{i}\rangle. \tag{5.104}$$

By inserting this expression into Eq. (5.96), one may combine the time-dependent exponential of Eq. (5.96) with that of (5.104) arising from $\mathbf{q}\cdot\mathbf{R} =$

$\mathbf{q}_b \cdot \mathbf{b} + q_z vt$ and carry out the time integration. The resulting delta function $\delta(q_z - q_0)$ fixes the longitudinal wave number to the value

$$q_0 = \frac{E_\mathrm{f} - E_\mathrm{i}}{\hbar v}, \tag{5.105}$$

which is connected with the nonrelativistic minimum momentum transfer $\hbar q_0$ (1.1) and plays a central role in the wave picture, see Sec. 1.1. In the present impact parameter picture, in which neither energy nor momentum is formally conserved, $q_0$ merely acts as a convenient parameter.

The matrix element (5.96) then is

$$M_\mathrm{fi}(b) = \frac{1}{\pi\gamma v} \int \frac{d^2 q_b}{q_b^2 + (1-\beta^2) q_0^2} e^{-i\mathbf{q}_b \cdot \mathbf{b}} \langle \mathrm{f} | (1 - \beta\alpha_z) e^{i\mathbf{q}\cdot\mathbf{r}_\mathrm{T}} | \mathrm{i} \rangle. \tag{5.106}$$

For high collision velocities, where the upper and lower components of the spinor wave functions are almost the same, and for high projectile velocity where $\beta \simeq 1$, the two contributions to the matrix element may nearly cancel each other in some cases. This is related to the cancellation in the classical expression (4.14), when in Eq. (3.34) $v/c \simeq 1$. As a consequence, an evaluation of the matrix element with *approximate* wave functions may give rise to spurious contributions if the near cancellations are not appropriately taken care of. It has been pointed out by Amundsen and Aashamar [AmA81] that for ionization one may get an incorrect asymptotic behavior of the cross section at high energies if this point is not taken into account.

*Reformulation of the $\alpha$-matrix element*

In such a situation, it is mandatory to rewrite the $\alpha$-matrix element in order to avoid the combination of large and small spinor components in those terms in which cancellations occur. Applying Eq. (5.90) to $g(\mathbf{r}) = \exp(i\mathbf{q} \cdot \mathbf{r}_\mathrm{T})$, we get

$$\langle \mathrm{f} | \boldsymbol{\alpha} \cdot \mathbf{q}\, e^{i\mathbf{q}\cdot\mathbf{r}_\mathrm{T}} | \mathrm{i} \rangle = \frac{E_\mathrm{f} - E_\mathrm{i}}{\hbar c} \langle \mathrm{f} | e^{i\mathbf{q}\cdot\mathbf{r}_\mathrm{T}} | \mathrm{i} \rangle. \tag{5.107}$$

When using this expression in Eq. (5.106), we know that $q_z = q_0$ is fixed according to Eq. (5.105), so that

$$\begin{aligned} \langle \mathrm{f} | \alpha_z\, e^{i\mathbf{q}\cdot\mathbf{r}_\mathrm{T}} | \mathrm{i} \rangle &= \frac{1}{q_0} \langle \mathrm{f} | (\boldsymbol{\alpha} \cdot \mathbf{q} - \alpha_x q_x - \alpha_y q_y) e^{i\mathbf{q}\cdot\mathbf{r}_\mathrm{T}} | \mathrm{i} \rangle \\ &= \frac{1}{q_0} \left[ \frac{E_\mathrm{f} - E_\mathrm{i}}{\hbar c} \langle \mathrm{f} | e^{i\mathbf{q}\cdot\mathbf{r}_\mathrm{T}} | \mathrm{i} \rangle - \langle \mathrm{f} | (\alpha_x q_x + \alpha_y q_y)\, e^{i\mathbf{q}\cdot\mathbf{r}_\mathrm{T}} | \mathrm{i} \rangle \right], \end{aligned} \tag{5.108}$$

where the coefficient in front of the first matrix element of the second line turns out to be $\beta$ [see Eq. (5.105)]. Combining Eq. (5.108) with the contribution from the scalar potential in Eqs. (5.96) and (5.106), we get

$$\langle \mathrm{f}|(1-\beta\alpha_z)\, e^{i\mathbf{q}\cdot\mathbf{r}_{\mathrm{T}}}\mathrm{i}\rangle = \frac{1}{\gamma^2}\langle \mathrm{f}|e^{i\mathbf{q}\cdot\mathbf{r}_{\mathrm{T}}}|\mathrm{i}\rangle + \frac{\beta}{q_0}\langle \mathrm{f}|(\alpha_x q_x + \alpha_y q_y)\, e^{i\mathbf{q}\cdot\mathbf{r}_{\mathrm{T}}}|\mathrm{i}\rangle. \quad (5.109)$$

The term proportional to $1/\gamma^2$ arises from the near cancellation between the scalar potential $\Phi = A^0$ and the longitudinal part $A_z$ of the vector potential in the interaction. For high collision energies, this term gives a constant contribution, because the integration over $q_b$ in Eq. (5.96) introduces a factor $\gamma^2$. After having taken care of the near cancellation, it is now less critical to proceed to evaluating the matrix elements on the right-hand side of Eq. (5.109) with approximate wave functions.

Another reformulation of the $\alpha$-matrix element is given in Sec. 6.2.2, in which the $z$ and the spin quantization axes are chosen to lie along $\mathbf{q}$. The result looks similar to Eq. (5.109), but differs from it in subtle ways discussed there.

## 5.6 Many-electron systems

In the preceding discussion, we always assume that the collision system consists of two nuclei and a single electron. Experimentally, such a system cannot be prepared easily for $Z > 1$. In many cases, one has a beam of incompletely stripped ions and a neutral gas or solid-state target, that is, one has to consider a collision system with many electrons. A theoretical description of colliding many-electron atoms is extremely difficult, and for relativistic collision velocities, only the simplest attempts have been made.

Already in classical mechanics, a relativistic dynamical description of an interacting many-particle system meets conceptual problems. The rigorously valid no-interaction theorem (for an introduction, see [Roh79]) states the impossibility of formulating a theory that is (a) Lorentz invariant, (b) has a Hamiltonian description, (c) contains information on the particle trajectories, and (d) has an interaction depending only on the particle trajectories. There are, however, approximate approaches avoiding a conflict with the no-interaction theorem, see e.g. [ScK89].

We are concerned here with relativistic quantum approaches. They all assume independent electron motion. In other words, they take into account the electron-electron interaction only insofar as it can be represented by a self-consistent field. We give a brief outline of the framework for describing a system of independent electrons and show how the general treatment of transition amplitudes in Sec. 5.4 can be used to formulate the interaction of a bare projectile with a multi-electron target. This represents the basic treatment on which more elaborate theories have to be built, see Chap. 7. In particular, we show how the time evolution of a multiparticle state is related to the directly calculable time evolution of the single-particle states. We then discuss the case of both target and projectile nuclei carrying electrons and point out the fundamental difficulties encountered if one attempts to treat the electron-electron interaction relativistically.

### 5.6.1 Many-electron states in occupation number representation

In Sec.4.2.2, the occupation number representation is introduced as a device to formally overcome the difficulties associated with the Dirac sea and with the definition of a physical vacuum state. The same formalism of "second quantization" can be applied to treat a nonrelativistic or relativistic many-electron system subject to the Pauli principle.

*Slater determinants*

For simplicity of presentation, we confine ourselves to electrons and ignore the positrons which can be incorporated along the same lines. Having in mind electrons bound to one of the atoms, say the target, we assume that we have constructed a discrete basis of single-particle states labeled by sets of quantum numbers $j, k, l, \cdots$. We may think of a hydrogenic basis or a basis constructed from Hartree-Fock or Hartree-Fock-Dirac calculations. The stationary basis wave functions are denoted by $\psi_j(\mathbf{r},t) = \varphi_j(\mathbf{r})\, e^{-i\epsilon_j t/\hbar}$, where $\epsilon_j$ is the eigenenergy, so that the field operator (4.64) is expanded as

$$\widehat{\psi} = \sum_j \psi_j(\mathbf{r},t)\, \widehat{b}_j, \tag{5.110}$$

where $\widehat{b}_j$ is the time-independent destruction operator for an electron in the state $j$. The operators $\widehat{b}_j$ and the associated creation operators $\widehat{b}_k^\dagger$ obey the anticommutation relations (4.65), and the vacuum is defined by Eq. (4.68).

Before applying this expansion to the calculation of transition amplitudes (5.69), we first have to explicitly construct multi-electron states. In order to avoid the unessential difficulty of angular momentum coupling at this point, we assume that the system under consideration is given by a closed-shell state $\Phi_0$, in which all states $j$ with $j \leq n_{\mathrm{F}} \equiv n$, that is, up to the Fermi level, are occupied while all higher-lying states are vacant, see Fig. 5.2. The wave function of such a state is represented by the totally

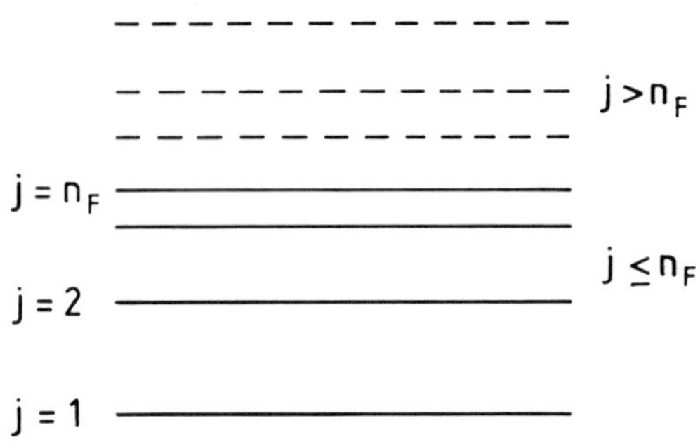


**Figure 5.2.** Definition of the Fermi level. Occupied states: solid lines; vacant states: dashed lines.

antisymmetric Slater determinant

$$\begin{aligned}\Phi_0 &= \frac{1}{\sqrt{n!}}\det|\,\varphi_k(\mathbf{r}_j)\,|, \qquad j,k = 1,2,\cdots,n \\ &= \frac{1}{\sqrt{n!}}\sum_P (-1)^P\, P[\varphi_{j_1}(\mathbf{r}_1),\cdots,\varphi_{j_n}(\mathbf{r}_n)], \end{aligned} \tag{5.111}$$

where $\mathbf{r}_j$ is the coordinate (including spin) of the electron, and $P$ denotes any one of all $n!$ possible permutations of the quantum numbers $j_1,\cdots,j_n$. The Slater determinant is the general representation of a state of independent (or uncorrelated) electrons. That is, the electrons interact only via a common self-consistent field.

In occupation number representation, the Slater determinant (5.111) corresponds to the normalized $n$-particle state

$$|\Phi_0\rangle = \prod_{j<n} \widehat{b}_j^\dagger |0\rangle \tag{5.112}$$

which, owing to Eqs. (4.65) and (4.68), has the property

$$\widehat{b}_k|\Phi_0\rangle = \begin{cases} (-1)^{s_k}\prod_{j\le n, j\neq k}\widehat{b}_J^\dagger|0\rangle & \text{if } k\le n \\ 0 & \text{if } k>n \end{cases} \tag{5.113}$$

and

$$\widehat{b}_k^\dagger|\Phi_0\rangle = \begin{cases} 0 & \text{if } k\le n \\ (-1)^{s_k}\prod_{j\le n, j=k}\widehat{b}_j^\dagger|0\rangle & \text{if } k>n \end{cases} \tag{5.114}$$

where $s_k$ is the number of transpositions needed to bring $\widehat{b}_k$ or $\widehat{b}_k^\dagger$ to the desired position. These rules, together with the anticommutation relations (4.65), enable us to calculate the matrix element between closed-shell states $|\Phi_0\rangle$ of any combination of operators $\widehat{b}_k$ and $\widehat{b}_l^\dagger$.

*Time evolution of Slater determinants*

Next we consider the change of an initial state $|\Phi_0\rangle$ caused by the collision with a bare projectile nucleus. At present, we just make use of the fact that a time evolution occurs, the detailed calculation being deferred to Chap. 7.

If the target is specified at time $t \to -\infty$ by $|\Phi_0\rangle$, the collision has the effect that the single-particle functions develop in time. This development is analogous to the procedure of Sec. 5.1.4 and has the same purpose. The evolving functions are represented by time-dependent linear combinations of the original basis functions,[4] expressed by

$$\psi_\alpha(\mathbf{r},t) = \sum_j a_{\alpha j}(t)\varphi_j(\mathbf{r}). \tag{5.115}$$

Here, the index $\alpha = 1, \cdots, n$ counts the orthogonal perturbed wave functions for each of the $n$ electrons. Equation (5.115) may be restated by allowing the creation (and annihilation) operators to develop in time. They are then denoted as Heisenberg operators[5] $\widehat{b}_\alpha^\dagger(t)$ with

$$\widehat{b}_\alpha^\dagger(t) = \sum_j a_{\alpha j}(t)\widehat{b}_j^\dagger; \tag{5.116}$$

conversely, because of the orthonormality of the sets of functions $\psi_\alpha$ and $\phi_i$, we have

$$\widehat{b}_{\mathrm{i}}^\dagger = \sum_\alpha a_{\alpha i}^*(t)\widehat{b}^\dagger\alpha(t). \tag{5.117}$$

Since the operators $\widehat{b}^\dagger(t), \widehat{b}(t)$ and $\widehat{b}^\dagger, \widehat{b}$ are connected by a unitary transformation, the anticommutation relations (4.65) also apply to the Heisenberg operators $\widehat{b}^\dagger(t), \widehat{b}(t)$ at any given time. With the single-particle states evolving according to Eq. (5.116), the Slater determinant will develop accordingly, so that at time $t$

$$|\Psi_0(t)\rangle = \prod_{\alpha\le n} \widehat{b}_\alpha^\dagger(t)|0\rangle \tag{5.118}$$

[4]For practical applications, it may be convenient to remove from $a_{\alpha_j}(t)$ the phase factor $\exp(-i\epsilon_j t/\hbar)$ describing the time oscillation of the basis state $j$ with energy $\epsilon_j$.

[5]We use the same symbol as for the time-independent operators, but distinguish them by explicitly displaying their time dependence.

represents the antisymmetric many-body state. Since the time evolution amounts to a redistribution of the $n$ particles upon the stationary basis states $j$, we may ask how the basis states are populated at a given time $t$. The answer is given by the expectation value of the number operator $\widehat{N}_j = \widehat{b}_j^\dagger \widehat{b}_j$ with respect to $|\Psi_0\rangle$, namely

$$\begin{aligned} \langle \Psi_0 | \widehat{N}_j | \Psi_0 \rangle &= \langle \Psi_0 | \widehat{b}_j^\dagger \widehat{b}_j | \Psi_0 \rangle \\ &= \sum_{\alpha\beta} a_{\alpha j}^* a_{\beta j} \langle \Psi_0 | \widehat{b}_\alpha^\dagger(t) \widehat{b}_\beta(t) | \Psi_0 \rangle \,. \end{aligned} \tag{5.119}$$

By noting that the rules (5.113) and (5.114) for $\widehat{b}^\dagger, \widehat{b}$ and $|\Phi_0\rangle$ translate into analogous rules for $\widehat{b}^\dagger(t), \widehat{b}(t)$ and $|\Psi_0\rangle$, we obtain

$$\langle \Psi_0 | \widehat{b}_\alpha^\dagger(t) \widehat{b}_\beta(t) | \Psi_0 \rangle = \begin{cases} \delta_{\alpha\beta} & \text{if } \alpha \leq n \\ 0 & \text{if } \alpha > n, \end{cases} \tag{5.120}$$

and by inserting into Eq. (5.119), we get the desired answer

$$\langle \Psi_0 | \widehat{N}_j | \Psi_0 \rangle = \sum_{\alpha \leq n} |a_{\alpha j}|^2, \tag{5.121}$$

expressing the fact that the particle occupation probability in an uncorrelated many-fermion system is an incoherent sum over single-particle occupation probabilities. This is one of the reasons why in this book we give major emphasis to the single-electron system.

It should be remarked that, according to Eqs. (5.113) and (5.114), one may also refer to the initially occupied state $|\Phi_0\rangle$ as a new vacuum state and redefine creation and annihilation operators $\widehat{b}^\dagger, \widehat{b}$ and $\widehat{d}^\dagger, \widehat{d}$ so that $\widehat{b}_j|\Phi_0\rangle = 0$ and $\widehat{d}_j|\Phi_0\rangle = 0$. Then $\widehat{b}_j^\dagger$ creates a particle above the Fermi level ($j > n$), and $\widehat{d}_j^\dagger$ creates a hole below the Fermi level ($j \leq n$). This is in complete analogy to the electron-positron problem discussed in Secs. 4.2.2 and 5.1.3. The occupied states below the Fermi level take the place of the Dirac sea.

### 5.6.2 Transition amplitudes between uncorrelated multi-electron states

As an illustration of the field-theoretic methods, we want to calculate the amplitude for excitation for an electron from a definite shell, say the K-shell of a closed-shell target atom. The result can be obtained from the general equation (5.69), if we specify the initial and final states accordingly. We assume that only the electron with quantum number $j_0$ is active while all other electrons $j \neq j_0$ remain frozen in their orbitals. The final state of the active electron, which can be in a bound or in a continuum state, is denoted by the quantum number $j_{\rm f}$. Then the initial and final multi-electron states are given by

$$\begin{aligned} |\Phi_{\rm i}\rangle &= \prod_{j<n} \widehat{b}_j^\dagger |0\rangle \\ |\Phi_{\rm f}\rangle &= \widehat{b}_{j_{\rm f}}^\dagger \widehat{b}_{j_0} |\Phi_{\rm i}\rangle. \end{aligned} \tag{5.122}$$

Inserting the expression (5.110) and the definition (5.122) into Eq. (5.69), we get the first-order transition amplitude

$$A_{\rm fi} = \frac{ie}{\hbar} \sum_{j,k} \langle \Phi_{\rm i} | \widehat{b}_{j_0}^\dagger \widehat{b}_{j_{\rm f}} \int d^4x \, \widehat{b}_j^\dagger \, \overline{\psi}_j(\mathbf{r},t) \gamma^\mu \psi_k(\mathbf{r},t) \widehat{b}_k \, A_\mu | \Phi_{\rm i} \rangle. \tag{5.123}$$

By applying the rules (5.113) and (5.114), we get the nondiagonal (i.e., $f \neq i$) transition amplitude in the form

$$A_{\rm fi} = \frac{ie}{\hbar} \int d^4x \, \overline{\psi}_{j_{\rm f}}(\mathbf{r},t) \gamma^\mu A_\mu \psi_{j_0}(\mathbf{r},t) \tag{5.124}$$

with the condition that the final state is unoccupied, or $j_{\rm f} > n$. This latter condition, imposed by the Pauli principle, is in fact the main difference as compared to the single-particle result (5.71). The preceding development serves only to demonstrate the power of the field-theoretic formulation in a particularly simple case.

We have assumed here that the external field produced by the moving projectile nucleus excites an uncorrelated target state. In a similar fashion, we may write down the amplitude for the excitation of a multi-electron projectile by the external field of the target nucleus as seen in the projectile system.

More complicated, i.e., correlated target or projectile, wave functions are usually described by a linear superposition of Slater determinants. The procedure outlined in this subsection can then be extended to this more complicated case in a straightforward manner.

We now discuss the case where target as well as projectile electrons are involved in the transition.

### 5.6.3 Transitions involving target and projectile electrons

In an actual experimental situation, it will often occur that the projectile as well as the target carries electrons. For an exact relativistic treatment, this leads to a very involved collision dynamics because we are dealing with perturbations that are no longer generated only by external fields. As discussed in Sec. 5.1.1, the electromagnetic fields produced by the nuclei can be regarded as external, because, owing to the large nuclear mass, the nuclear motion is almost unaffected by the electronic motion. Clearly, when projectile electrons collide with target electrons, this is no longer the case.

Only if we disregard the interaction between electrons belonging to different partners is it possible to write down a simple cross section formula. For a fast collision, when, except for a transient perturbation, the electrons may be regarded as moving solely under the influence of the field produced by their own parent nucleus, this assumption means that the target electrons are only excited by the projectile nucleus and the projectile electrons only by the target nucleus.

Let the target be excited from the state $i_{\mathrm{T}}$ to the state $f_{\mathrm{T}}$ and the projectile from $i_{\mathrm{P}}$ to $f_{\mathrm{P}}$. In the absence of electron-electron interactions between different partners, the excitation processes in the target and in the projectile are independent of each other, so that the transition probability factorizes into a product of individual transition probabilities $P_{f_{\mathrm{T}}i_{\mathrm{T}}}(b)$ and $P_{f_{\mathrm{P}}i_{\mathrm{P}}}$, which are assumed to be small compared to unity. Indeed, at high velocities, the Pauli principle acting between electrons belonging to different atoms can also be disregarded. Instead of Eq. (5.65), we then have the total cross section for double excitation

$$\sigma_{f_{\mathrm{T}}f_{\mathrm{P}},i_{\mathrm{T}}i_{\mathrm{P}}} = \int P_{f_{\mathrm{T}}i_{\mathrm{T}}}(b)P_{f_{\mathrm{P}}i_{\mathrm{P}}}(b)\,d^2b \tag{5.125}$$

between any set of initial states $i_{\mathrm{T}}i_{\mathrm{P}}$ and final states $f_{\mathrm{T}}f_{\mathrm{P}}$.

In many cases of interest, it is not justified to ignore the interaction between electrons belonging to different parent nuclei, for a discussion see Secs. 7.1 and 7.3. A nonrelativistic description of these interactions is still acceptable at collision energies of about 100 MeV/u or less. At higher energies, a relativistic description is needed.

The scattering of two free electrons at relativistic relative energy is commonly known as Møller scattering [BeL82]. A proper description of the mutual interaction requires the framework of quantum electrodynamics (QED), that is, it requires the solution of the coupled equations (5.1) and (5.2). In an approximate way, it is possible to eliminate the mediating photon field and to substitute *effective* electron-electron potentials, the so-called Breit interaction (see, e.g., [BeS57]). In addition to the instantaneous Coulomb interaction, it includes a current-current interaction between the convection currents of the electrons, a current-dipole interaction between the convection current of one electron and the magnetic moment of the other electron (a spin-orbit interaction), and finally a dipole-dipole interaction between the magnetic moments of both electrons. The Breit interaction is not Lorentz invariant and has other limitations. A covariant two-electron (in general, two-fermion) equation which does not explicitly refer to photon fields, has been given by Bethe and Salpeter (see [BeS57]). In this case, the interaction operator consists of an infinite series of terms, so that it is very difficult to apply.

While Møller scattering of free electrons has been fully treated within QED, and the Breit interaction has been successfully used in atomic structure calculations, there is, to our knowledge, no treatment available for the interaction of electrons bound to two centers moving with relativistic velocity against each other. This means that in the problem of mutual electron screening or electron excitation, one has to rely on nonrelativistic estimates, or one has to use the impulse approximation. In the latter approach, the electron would be considered as quasi-free with a momentum distribution dictated by the atomic binding. In order to get the bound-electron scattering amplitude, one would have to fold the known Møller scattering amplitude with the relative momentum distribution relativistically composed of the global atomic relative motion and the momenta of the two electrons in their respective parent atoms.

# Part II

# Elementary Atomic Processes

# Chapter 6

# Excitation and Ionization

Excitation and ionization of one-electron atoms by relativistic projectiles are treated simultaneously in this chapter and are not distinguished. Most applications refer to K-shell ionization which is the important process at high velocities. The theoretical methods used for relativistic collisions are very similar to those applied at nonrelativistic bombarding energies.

In Sec. 6.1 we start with an outline of first-order perturbation theory, and in Sec. 6.2, for the sake of illustration, we discuss the simple cases of the dipole approximation and of a quasirelativistic treatment of the matrix elements. Some of the methods used to evaluate the necessary interaction matrix elements more exactly are presented in Sec. 6.3, so that one is in a position to derive theoretical results and to compare them with the corresponding experimental data. In Sec. 6.4 we discuss the finite-difference method for solving the time-dependent Dirac equation of the collision system in a nonperturbative way. As an alternative nonperturbative method, we consider two-center coupled-channel methods in Sec. 6.5, and subsequently specialize to single-center coupled-channel calculations.

For definiteness, we assume in this chapter that a single *target* electron is excited or ionized by the projectile. If the active electron is carried by the *projectile*, excitation can occur by the target nucleus, and one has simply to interchange the role of target and projectile. Excitation is then influenced by the target electrons. These and other many-electron effects are discussed in Chap. 7.

# 6.1 First-order perturbation theory

For a fast collision and for impact parameters $b$ comparable with, or larger than the atomic K-shell radius, the transient perturbation of the target atom by the projectile is small enough that the first-order time-dependent perturbation theory is expected to be a good approximation, even for a high-$Z$ projectile.

## 6.1.1 General expressions

In first-order perturbation theory, by combining Eqs. (5.76) and (4.32), the transition amplitude for excitation and ionization of the target electron is written as

$$A_{\rm fi}^{\rm e}(b) = i\frac{\gamma Z_{\rm P} e^2}{\hbar} \int dt\, e^{\frac{i}{\hbar}(E_{\rm f}-E_{\rm i})t} \int d^3 r_{\rm T}\, \varphi_{\rm f}^{\dagger}(\mathbf{r}_{\rm T})(1-\beta\alpha_z)\frac{1}{r'_{\rm P}}\,\varphi_{\rm i}(\mathbf{r}_{\rm T}). \quad (6.1)$$

Here, $\gamma$ is the Lorentz factor, $Z_{\rm P}$ the projectile charge number, and $\varphi_{\rm i}$ and $\varphi_{\rm f}$ are the initial and final eigenfunctions of the target electron with energies $E_{\rm i}$ and $E_{\rm f}$. The coordinates are explained in Fig. 3.2, and $r'_{\rm P}$, given by Eq. (5.97), is the electron-projectile distance measured in the projectile rest system. The cross section for excitation is directly given by Eq. (5.65), while for the differential ionization cross section we have to replace $\sigma_{\rm fi}$ by $d\sigma_{\rm fi}/dE_{\rm f}$, so that for each specified initial and final state (see below) we have

$$\frac{d\sigma}{dE_{\rm f}} = 2\pi \int_0^{\infty} |A_{\rm fi}^{\rm e}|^2\, b\, db. \quad (6.2)$$

In this formula, it is assumed that the radial continuum wave function for each final partial wave $\kappa_{\rm f}$ or $j_{\rm f}, l_{\rm f}$ entering in the transition amplitude is normalized per unit energy interval, see Eq. (4.117). Since the emitted electron is not represented by a plane wave but a Coulomb wave, the approaches discussed here belong to the class of distorted-wave Born approximations (DWBA). Introducing the minimum momentum transfer $\hbar q_0$ through Eq. (5.105), the integration in Eq. (6.1) over the time dependence residing in the exponential factor and in the term $1/r'_{\rm P}$ can be directly carried out to yield [BaH59, AmA81]

$$|A_{\rm fi}^{\rm e}|^2 = 4\nu_{\rm P}^2 \left| \left\langle {\rm f} \left| (1-\beta\alpha_z) e^{iq_0 z_{\rm T}}\, K_0\left(\frac{q_0}{\gamma}\sqrt{(x_{\rm T}-b)^2+y_{\rm T}^2}\right) \right| {\rm i} \right\rangle \right|^2 . \quad (6.3)$$

Here and in the remaining part of this chapter, we use the bracket notation for the space integral. Furthermore, $\nu_{\rm P}$ is given by Eq. (5.52), and $K_0$ denotes the modified Bessel function. It is seen that the range parameter of the effective interaction is $q_0^{-1}$ along the path, i.e., parallel to the $z$-axis, but is $\gamma q_0^{-1}$ perpendicular to it, i.e., in the impact parameter plane. Hence, as $\gamma$ increases, the integrand falls off more slowly with $b$ than in the beam direction. This reflects the properties of the Liénard-Wiechert potential given by Eq. (3.34) and depicted in Fig. 3.3.

*Formulation in terms of the Fourier transform*

For an actual evaluation of the cross section, it is more convenient to return to the expression (6.1) and to rewrite the retarded interaction in terms of its Fourier transform. Using the relativistically generalized Bethe integral (5.103), the electronic transition matrix element in Eq. (6.1) can be cast into the form (5.104). By decomposing $\mathbf{q}$ as $\mathbf{q} = (\mathbf{q}_b, q_z)$ and performing the time integration, one obtains from Eq. (5.106) the transition amplitude

$$A^{\mathrm{e}}_{\mathrm{fi}}(b) = i\gamma v \nu_{\mathrm{P}}\, M_{\mathrm{fi}}(b) = \frac{i}{\pi}\nu_{\mathrm{P}} \int \frac{d^2 q_b}{q_b^2 + (1-\beta^2)q_0^2}\, e^{-i\,\mathbf{q}_b\cdot\mathbf{b}}\, M_{\mathrm{fi}}(\mathbf{q}) \tag{6.4}$$

and from Eq. (6.2) the cross section

$$\frac{d\sigma^{\mathrm{e}}_{\mathrm{fi}}}{dE_{\mathrm{f}}} = 8\pi\nu_{\mathrm{P}}^2 \int_0^\infty \frac{q_b\, dq_b}{[q_b^2 + (1-\beta^2)q_0^2]^2}\, |M_{\mathrm{fi}}(\mathbf{q})|^2, \tag{6.5}$$

where

$$M_{\mathrm{fi}}(\mathbf{q}) = \langle \mathrm{f}|(1-\beta\alpha_z)\, e^{i\,\mathbf{q}\cdot\mathbf{r}_{\mathrm{T}}}|\mathrm{i}\rangle. \tag{6.6}$$

The advantage of this formulation is that the complicated time dependence contained in the expression (5.97) for $r'_{\mathrm{P}}$ is removed and replaced with the more explicit wave number dependence (6.5) and (6.6). The expression (6.5), valid for each target electron and for specified initial and final states, agrees with the corresponding expression derived in a wave picture, in which certain simplifications, described in Sec. 5.2, are adopted [DaM78, Anh79].

*Specification of the initial and final states*

For *excitation*, the initial and final bound states $i$ and $f$ are specified by their quantum numbers $i \equiv \{n_{\mathrm{i}}, \kappa_{\mathrm{i}}, \mu_{j_{\mathrm{i}}}\}$ and $f \equiv \{n_{\mathrm{f}}, \kappa_{\mathrm{f}}, \mu_{j_{\mathrm{f}}}\}$, respectively, see Eqs. (4.84) and (4.85). In order to obtain the complete cross section, one averages over the initial angular momentum projections $\mu_{j_{\mathrm{i}}}$ and sums over the final angular momentum projections $\mu_{j_{\mathrm{f}}}$.

For *ionization*, Eqs. (6.2) and (6.5), the initial bound state is characterized in the same way. However, for the final state of a given continuum energy $E_{\rm f}$, one has to distinguish two possible situations determined by the experimental setup.

(a) The energy $E_{\rm f}$ and the direction $\hat{\mathbf{k}}$ of emission of the electron is detected. Then the final state $|{\rm f}\rangle = |E_{\rm f}, \hat{\mathbf{k}}, \sigma\rangle$ with spin projection $\sigma = \pm\frac{1}{2}$ is a coherent superposition of partial waves $\kappa_{\rm f}, \mu_{j_{\rm f}}$ given by Eq. (4.124). By inserting into (6.2) or (6.5) and summing over the (undetected) spin projections, one obtains the angle-differential cross section.

(b) The direction of emission of the electron is *not* detected. Then one integrates Eqs. (6.2) and (6.5) over the direction $\hat{\mathbf{k}}$ of the emitted electron in Eqs. (4.124) or (4.123) and sums over the spin projection $\sigma$ (or $m_s$). Owing to the orthogonality of the rotation matrices in (4.124) [or of the spherical harmonics in (4.123)] and of the Clebsch-Gordan coefficients, one obtains an *incoherent* summation over the partial waves $\kappa_{\rm f}$, $\mu_{j_{\rm f}}$.

The resulting total energy-differential cross section for ionization is then written explicitly as

$$\left.\frac{d\sigma_{\rm fi}^{\rm e}}{dE_{\rm f}}\right|_{\rm total} = 8\pi\nu_{\rm P}^2 \frac{1}{2j_{\rm i}+1} \sum_{\mu_{j_{\rm i}}} \sum_{\kappa_{\rm f},\mu_{j_{\rm f}}} \int_0^\infty \frac{q_b\, dq_b}{[q_b^2 + (1-\beta^2)q_0^2]^2}$$
$$\times \left| \langle E_{\rm f}, \kappa_{\rm f}, \mu_{j_{\rm f}} | (1 - \boldsymbol{\beta}\cdot\boldsymbol{\alpha})\, e^{i\,\mathbf{q}\cdot\mathbf{r}_{\rm T}} | n_{\rm i}, \kappa_{\rm i}, \mu_{j_{\rm i}} \rangle \right|^2, \qquad (6.7)$$

where $\boldsymbol{\beta} = \mathbf{v}/c$.

In the following, we keep the symbols $i$ and $f$ for fully specified states, i.e., $|{\rm i}\rangle = |n_{\rm i}, \kappa_{\rm i}, \mu_{j_{\rm i}}\rangle$ and $|{\rm f}\rangle = |E_{\rm f}, \kappa_{\rm f}, \mu_{j_{\rm f}}\rangle$. Only when calculating angular distributions, do we identify $f$ with a coherent superposition $|{\rm f}\rangle = |E_{\rm f}, \hat{\mathbf{k}}, \sigma\rangle$ of partial waves given by Eq. (4.124).

### 6.1.2 Long-range couplings and Coulomb boundary conditions

In the preceding subsection, we have assumed that the initial and final states are undistorted target eigenstates. However, according to the discussion in Sec. 5.3.3, we should use the exact solutions[1] (5.61)

$$\Phi_{\rm T}^\infty(\mathbf{r}_{\rm T}, t) = e^{-i\nu_{\rm P}\ln(R'-vt')}\psi_{\rm T}(\mathbf{r}_{\rm T}, t) \qquad (6.8)$$

of the asymptotic Dirac equation (5.49), that is, we should take into account the asymptotic distortion caused by the residual interaction $-S^2 Z_{\rm P} e^2/R'$. Inserting the solution (6.8) into Eq. (5.82), one finds from Eqs. (5.82) and (5.83) that, by construction, the residual interaction term proportional

[1] An irrelevant constant phase factor depending on the unit of length is discarded.

to $1/R'$ has to be subtracted from the interaction occurring in Eq. (6.1), see Eq. (6.9) below. The reason why one usually disregards the residual interaction term derives from nonrelativistic collisions, where, according to Eq. (5.47), $R = R'$, so that a term proportional to $1/R$ has vanishing matrix elements between orthogonal states.

For relativistic collisions, the situation is somewhat different. Let us examine the modified transition amplitude $\tilde{A}^{\mathrm{e}}$ calculated with phase-distorted states $\Phi_{\mathrm{T}}^{\infty}$ satisfying Coulomb boundary conditions [ToE90b]. Since the state-independent phase factors in Eq. (6.8) for initial and final states cancel each other, one has

$$\begin{aligned} \tilde{A}^{\mathrm{e}}_{\mathrm{fi}}(b) &= i\frac{\gamma Z_{\mathrm{P}} e^2}{\hbar} \int dt\, e^{\frac{i}{\hbar}(E_{\mathrm{f}}-E_{\mathrm{i}})t} \\ &\quad \times \left\langle \mathrm{f} \left| (1-\beta\alpha_z)\left(\frac{1}{r'_{\mathrm{P}}} - \frac{1}{R'}\right) \right| \mathrm{i} \right\rangle \end{aligned} \tag{6.9}$$

instead of (6.1). The additional term $1/R'$ in the perturbation operator with $R' = \sqrt{b^2 + v^2 t'^2}$ depends on the laboratory space coordinates through $t' = \gamma(t - v z_{\mathrm{T}}/c^2)$. In contrast to the nonrelativistic case, the space integral of the $1/R'$ term does *not* vanish. However, we still have to perform the time integration. Since the integration $\int d^4x$ over the complete four-dimensional space is Lorentz invariant, we may as well evaluate this integral in the projectile frame. In the moving coordinate system, the term $1/R'$ is a constant, therefore does not contribute to the matrix element between orthogonal (exact) eigenstates, and hence can be discarded. We may now transform back to the laboratory frame to obtain a result *without* the $1/R'$ term, that is, we recover Eq. (6.1). In other words, Eqs. (6.1) and (6.9) are equivalent, and Coulomb boundary conditions are without any influence. It should be emphasized here that the equality of the two treatments is valid only in first-order perturbation theory and only if we consider the occupation amplitude for the final state at time $t \to \infty$.

Obviously, the same reasoning is valid for any function $f(R')$ and the corresponding phase distortion. This means that the first-order transition amplitudes for excitation and ionization (and for pair production) are invariant with respect to gauge transformations. This has also been confirmed in [RuS93] by a different method.

As an illustration, we show in Fig. 6.1 the time evolution of occupation probabilities calculated from Eqs. (6.1) and (6.9), respectively, with the modification that the upper limit of the time integration is taken as $t$ rather than as $\infty$. In the upper part of the figure, calculated from the finite-time version of Eq. (6.1), we notice the effect of the long-range couplings which lead to a sizable population of $2p_{3/2}$ and $2p_{1/2}$ states in the

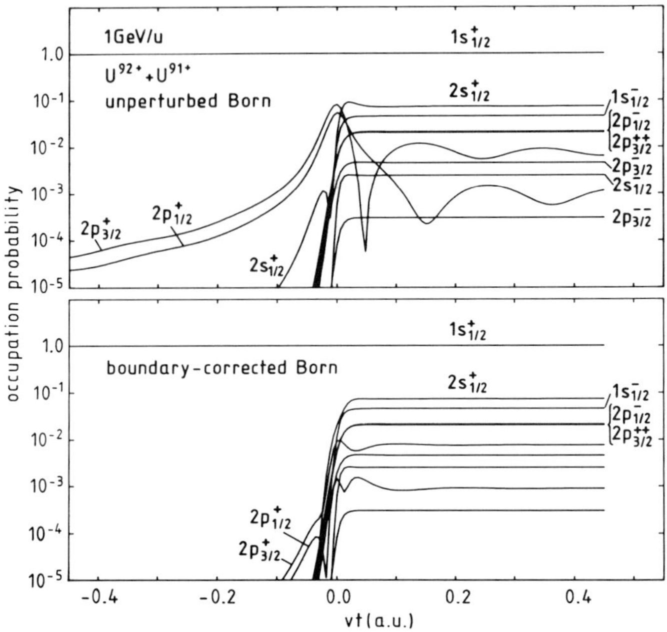


**Figure 6.1.** Time evolution of the occupation probabilities of target states in $U^{92+} + U^{91+}(1s_{1/2})$ collisions at 1 GeV/u. Calculations are performed in the Born approximation at $b = 0.01$ a.u. (about one K-shell radius). Upper part: unperturbed initial and final states, Eq. (6.1); lower part: boundary-corrected states, Eq. (6.9). The abscissa is the projection of the projectile-target separation on the beam direction. The projections of the angular momenta are indicated by $+, -, ++, --$ for $m_j = +\frac{1}{2}, -\frac{1}{2}, +\frac{3}{2}$, and $-\frac{3}{2}$, respectively. From [ToE90b].

entrance channel for projectile-target separations as large as 0.4 a.u. (about 37 K-shell radii). On the other hand, in the lower part of the figure, the finite-time version of Eq. (6.9) has been used. Here, as would be expected from the short-range potential in Eq. (6.9), the long-range couplings have disappeared. For large positive times, the occupation probabilities in the unperturbed and in the boundary-corrected Born approximation are exactly the same, thus verifying the reasoning given above. In a numerical evaluation of the time integrals, one expects that the boundary-corrected version converges much more rapidly.

## 6.2 Two simplified treatments

The treatment presented so far is exact within the first-order semiclassical approximation of Eq. (6.1). For the sake of an introductory illustration, we now present two different approximations, one valid at low values of $q_0$, which means small excitation energies or high projectile velocities, the other valid for $\alpha Z_T \ll 1$. These methods are called the dipole approximation and the quasirelativistic treatment, respectively.

### 6.2.1 Dipole approximation

In Bethe's dipole approximation [Bet30], one replaces in the matrix element (6.6) the factor $\exp(i\,\mathbf{q}\cdot\mathbf{r}_\mathrm{T})$ by $1+i\,\mathbf{q}\cdot\mathbf{r}_\mathrm{T}$. Unless the wave number $q$ becomes larger than the inverse K-shell radius $a_\mathrm{K}$ (for K-shell excitation or ionization) it is justified to neglect higher-order terms in the expansion because the exponential decay of the bound-state wave function cuts off contributions from larger $r$ values.

If we rewrite the Dirac matrix $\alpha_z$ with the aid of the identity (5.91), valid for exact eigenstates, and with the definition (5.105) for $q_0 = q_z$, the dipole approximation yields

$$M_\mathrm{fi} \simeq \langle \mathrm{f}|(1-\beta\alpha_z)e^{i\,\mathbf{q}\cdot\mathbf{r}_\mathrm{T}}|\mathrm{i}\rangle \approx i\,(\mathbf{q}\cdot\langle \mathrm{f}|\mathbf{r}_\mathrm{T}|\mathrm{i}\rangle - \beta^2 q_0\langle \mathrm{f}|z_\mathrm{T}|\mathrm{i}\rangle), \tag{6.10}$$

where each term obeys a different selection rule. Introducing spherical coordinates $r_s$ with the magnetic quantum number $s=0,\pm1$, so that $r_0 = z_\mathrm{T}$ and $r_{\pm1} = \mp\frac{1}{\sqrt{2}}(x_\mathrm{T}\pm i y_\mathrm{T})$, and subsequently defining reduced matrix elements [Ros57] through

$$\langle j_\mathrm{f}\mu_{j_\mathrm{f}}|r_s|j_\mathrm{i}\mu_{j_\mathrm{i}}\rangle = \left(\begin{array}{cc|c} j_\mathrm{i} & 1 & j_\mathrm{f} \\ \mu_{j_\mathrm{i}} & s & \mu_{j_\mathrm{f}} \end{array}\right)\langle \mathrm{f}\parallel \mathbf{r}_\mathrm{T}\parallel \mathrm{i}\rangle, \tag{6.11}$$

where $j_\mathrm{i},\mu_{j_\mathrm{i}}$ and $j_\mathrm{f},\mu_{j_\mathrm{f}}$ are the angular momenta and their projections of initial and final states, and $(\cdot\cdot|\cdot)$ is a Clebsch-Gordan coefficient, see Eq. (4.88), we get

$$|M_\mathrm{fi}|^2 = q_0^2(1-\beta^2)^2\left(\begin{array}{cc|c} j_\mathrm{i} & 1 & j_\mathrm{f} \\ \mu_{j_\mathrm{i}} & 0 & \mu_{j_\mathrm{i}} \end{array}\right)^2 |\langle \mathrm{f}\parallel \mathbf{r}_\mathrm{T}\parallel i\rangle|^2, \quad \text{if} \quad \mu_{j_\mathrm{f}} = \mu_{j_\mathrm{i}}, \tag{6.12}$$

and

$$|M_\mathrm{fi}|^2 = \tfrac{1}{2}(q_x^2+q_y^2)\left(\begin{array}{cc|c} j_\mathrm{i} & 1 & j_\mathrm{f} \\ \mu_{j_\mathrm{i}} & \pm1 & \mu_{j_\mathrm{i}}\pm1 \end{array}\right)^2 |\langle \mathrm{f}\parallel \mathbf{r}_\mathrm{T}\parallel i\rangle|^2, \quad \text{if} \quad \mu_{j_\mathrm{f}} = \mu_{j_\mathrm{i}}\pm1. \tag{6.13}$$

Both types of terms contribute to the cross section after summation over final angular momentum projections, provided initial and final states have different parity and the dipole selection rules embodied in the Clebsch-Gordan coefficients are satisfied. In writing down Eqs. (6.12) and (6.13) we still maintain a relativistic description of the electronic states. Expressions (6.12) and (6.13) have different $q$ dependencies and cannot be combined unless the squares of the Clebsch-Gordan coefficients are equal. For the particular transition $\mathrm{s}_{1/2} \to \mathrm{p}_{1/2}$ all $(\cdot\cdot|\cdot)^2 = 1/3$, but in general, e.g., for $\mathrm{s}_{1/2} \to \mathrm{p}_{3/2}$, this is not true.

In order to evaluate the cross sections for each of the two cases, we write $q^2 = q_0^2 + q_b^2$, where $\hbar q_b$ is the transverse momentum transfer. Since $q\,dq = \frac{1}{2}\,dq^2 = \frac{1}{2}\,dq_b^2$, we get from Eq. (6.1)

$$\frac{d\sigma_{\mathrm{fi}}^{\mathrm{e}}(\mu_{j_\mathrm{i}} \to \mu_{j_\mathrm{i}})}{dE_\mathrm{f}} = 4\pi\nu_\mathrm{P}^2 \left( \begin{array}{cc|c} j_\mathrm{i} & 1 & j_\mathrm{f} \\ \mu_{j_\mathrm{i}} & 0 & \mu_{j_\mathrm{i}} \end{array} \right)^2 |\langle \mathrm{f} \,\|\, \mathbf{r}_\mathrm{T} \,\|\, \mathrm{i} \rangle|^2 \times \int_0^\infty \frac{q_0^2(1-\beta^2)^2}{[q_b^2 + (1-\beta^2)q_0^2]^2}\, dq_b^2 \tag{6.14}$$

and

$$\frac{d\sigma_{\mathrm{fi}}^{\mathrm{e}}(\mu_{j_\mathrm{i}} \to \mu_{j_\mathrm{i}} \pm 1)}{dE_\mathrm{f}} = 2\pi\nu_\mathrm{P}^2 \left( \begin{array}{cc|c} j_\mathrm{i} & 1 & j_\mathrm{f} \\ \mu_{j_\mathrm{i}} & \pm 1 & \mu_{j_\mathrm{i}} \pm 1 \end{array} \right)^2 |\langle \mathrm{f} \,\|\, \mathbf{r}_\mathrm{T} \,\|\, \mathrm{i} \rangle|^2 \times \int_0^\infty \frac{q_b^2}{[q_b^2 + (1-\beta^2)q_0^2]^2}\, dq_b^2 . \tag{6.15}$$

According to Eq. (6.5), the integration over $q_b^2$ extends from zero to infinity. The dipole approximation breaks down for large values of $q$, giving rise to a term in Eq. (6.15) which diverges logarithmically. If, in the divergent term, we introduce a cut-off $q_{\mathrm{max}}$ as the upper integration limit, we have

$$\frac{d\sigma_{\mathrm{fi}}^{\mathrm{e}}(\mu_{j_\mathrm{i}} \to \mu_{j_\mathrm{i}})}{dE_\mathrm{f}} = 4\pi\nu_\mathrm{P}^2 \left( \begin{array}{cc|c} j_\mathrm{i} & 1 & j_\mathrm{f} \\ \mu_{j_\mathrm{i}} & 0 & \mu_{j_\mathrm{i}} \end{array} \right)^2 |\langle \mathrm{f} \,\|\, \mathbf{r}_\mathrm{T} \,\|\, \mathrm{i} \rangle|^2 \,(1-\beta^2) \tag{6.16}$$

and

$$\frac{d\sigma_{\mathrm{fi}}^{\mathrm{e}}(\mu_{j_\mathrm{i}} \to \mu_{j_\mathrm{i}} \pm 1)}{dE_\mathrm{f}} = 2\pi\nu_\mathrm{P}^2 \left( \begin{array}{cc|c} j_\mathrm{i} & 1 & j_\mathrm{f} \\ \mu_{j_\mathrm{i}} & \pm 1 & \mu_{j_\mathrm{i}} \pm 1 \end{array} \right)^2 |\langle \mathrm{f} \,\|\, \mathbf{r}_\mathrm{T} \,\|\, \mathrm{i} \rangle|^2 \times \left[ 2\ln\gamma + \ln\left( \frac{q_{\mathrm{max}}^2}{q_0^2} + \frac{1}{\gamma^2} \right) - 1 \right] . \tag{6.17}$$

In Eq. (6.17) we may neglect the term $\gamma^{-2}$ in the second logarithm. If, somewhat arbitrarily, we choose $q^2_{\rm max} = 2mE_{\rm i}/\hbar^2$ as the nonrelativistic inverse K-shell radius and, moreover, neglect $E_{\rm f}$ in the definition (5.105) for $q_0$ , we may replace $q^2_{\rm max}/q_0^2$ with $2mv^2/E_{\rm i}$.

Equations (6.16) and (6.17) illustrate a very important feature, namely that for $\mu_{j_{\rm i}} \to \mu_{j_{\rm i}}$ transitions, the cross section approaches a constant value[2] as the projectile energy increases, while for $\mu_{j_{\rm i}} \to \mu_{j_{\rm i}} \pm 1$ it grows proportional to $\ln\gamma$ for large values of $\gamma$. The increase of the cross section as $\ln\gamma$ in the extreme relativistic limit can be understood by considering the extension in the transverse direction of the Liénard-Wiechert potential (3.34) shown in Fig. 3.3. As a consequence, larger and larger impact parameters contribute, and the cross section keeps increasing with energy [AmA81]. This unbounded increase with energy is characteristic for all perturbation theory approaches, which do not conserve unitarity.

### 6.2.2 Quasirelativistic treatment

The cross section (6.5) is exact in first-order perturbation theory as far as relativistic effects are concerned. Here, we present a treatment which employs semirelativistic Darwin bound wave functions (Sec. 4.3.3) and semirelativistic plane waves to evaluate the matrix element (6.6) using a convenient coordinate system. The advantage of this treatment is that the final cross section is a sum of a nonrelativistic expression plus a relativistic modification. For the nonrelativistic part, many tabulations exist, given below, and for the relativistic part, approximate expressions are available.

For calculating the total cross section, we start from Eq. (6.7), using an integration over $q$ from $q_0$ to infinity rather than over $q_b$ from zero to infinity. It is advantageous to evaluate the matrix element in a $\tilde{x}, \tilde{y}, \tilde{z}$ coordinate system, in which the $\tilde{z}$-axis lies along $\mathbf{q}$ and the $\tilde{x}$-axis lies in the plane formed by $\mathbf{q}$ and $\boldsymbol{\beta}$. Then

$$\boldsymbol{\beta} \cdot \boldsymbol{\alpha} = \beta \cos\lambda \, \alpha_{\tilde{z}} + \beta \sin\lambda \, \alpha_{\tilde{x}}. \tag{6.18}$$

Here, $\lambda$ is the angle between $\mathbf{q}$ and $\boldsymbol{\beta}$, so that $\cos\lambda = q_z/q = q_0/q$, where the $z$-axis points along $\boldsymbol{\beta}$. Substituting this expression into Eq. (6.7), the matrix element of

$$\alpha_{\tilde{z}} \, e^{i\,\mathbf{q}\cdot\mathbf{r}_{\rm T}} = \alpha_{\tilde{z}} \, e^{i\,q\tilde{z}_{\rm T}}$$

[2]The cancellation between 1 and $\beta^2$ as $\gamma \to \infty$ is a result of the dipole approximation (6.10). Any deviation from it leads to a finite constant value in the extreme relativistic limit.

can be evaluated with the help of Eqs. (5.105) and (5.107) as

$$\langle \mathrm{f}|\alpha_{\tilde{z}}\, e^{i\,q\tilde{z}_{\mathrm{T}}}|\mathrm{i}\rangle = \frac{\beta q_0}{q}\,\langle \mathrm{f}|e^{i\,q\tilde{z}_{\mathrm{T}}}|\mathrm{i}\rangle. \tag{6.19}$$

This and Eq. (6.18) yield

$$\langle \mathrm{f}|(1-\boldsymbol{\beta}\cdot\boldsymbol{\alpha})\, e^{i\,\mathbf{q}\cdot\mathbf{r}_{\mathrm{T}}}|\mathrm{i}\rangle = (1-\beta^2\, q_0^2/q^2)F(q) - \beta\sqrt{1-q_0^2/q^2}\, G(q)\ , \tag{6.20}$$

with

$$\begin{aligned} F(q) &= \langle \mathrm{f}|e^{i\,q\tilde{z}_{\mathrm{T}}}|\mathrm{i}\rangle, \\ G(q) &= \langle \mathrm{f}|\alpha_{\tilde{x}} e^{i\,q\tilde{z}_{\mathrm{T}}}|\mathrm{i}\rangle\ . \end{aligned} \tag{6.21}$$

These definitions imply that the wave functions for the initial and final states are specified in the $\tilde{x}, \tilde{y}, \tilde{z}$ coordinate system.

Since the wave functions $|\mathrm{i}\rangle$ and $|\mathrm{f}\rangle$ are characterized by definite values of the magnetic quantum number $\mu_j$, the matrix elements $F(q)$ and $G(q)$ obey different selection rules: in $G$, the operator $\alpha_{\tilde{x}}$ changes the total angular momentum component along the $\tilde{z}$-axis, but not in $F$. For any specific combination of the magnetic quantum numbers $\mu_{j_\mathrm{i}}$ and $\mu_{j_\mathrm{f}}$, at most one of them contributes, so that the terms containing $F$ and $G$ add up incoherently in the total cross section (6.7). This equation now becomes

$$\left.\frac{d\sigma_{\mathrm{fi}}^{\mathrm{e}}}{dE_{\mathrm{f}}}\right|_{\mathrm{total}} = 8\pi\nu_{\mathrm{P}}^2 \sum \int_{q_0}^{\infty} \frac{dq}{q^3}\left(|F(q)|^2 + \frac{\beta^2(1-q_0^2/q^2)}{(1-\beta^2\, q_0^2/q^2)^2}|G(q)|^2\right)\ , \tag{6.22}$$

with the understanding that the sum is taken over all partial continuum waves, and that in Eq. (6.21) the wave function $|\mathrm{f}\rangle$ refers only to a single partial wave.

The term proportional to $\sum |F|^2$ is identical to that obtained for non-relativistic projectiles. The term proportional to $\sum |G|^2$ is the relativistic modification, as suggested by the factor $\beta^2$. As $\beta \to 1$, the first term becomes independent of $\beta$ (or $\gamma$), since $q_0$ becomes constant, see Eq. (5.105). To obtain the $\beta$ dependence of the second term, one notes that the main $\beta$ dependence arises near $q \simeq q_0$. Hence, one can perform the integration over $q$ assuming $G(q) \simeq G(q_0)$. One then finds that the second term is proportional to

$$-\left[\frac{1}{\beta^2}\ln(1-\beta^2)\right] - 1 = \frac{2}{\beta^2}\ln\gamma - 1\ , \tag{6.23}$$

which, as $\beta \to 1$, has the asymptotic form of Eq. (1.11).

The matrix elements $F$ and $G$ have been evaluated with 1s semirelativistic wave functions for the electron [DaM78,Anh79]. Averaging over the initial and summing over the final angular momentum projections of the electron, and integrating over the final electron energy, one obtains the cross section for ionizing a single K-shell electron. The expression is simplified if one makes the substitutions

$$W = 2\,\frac{E_\mathrm{f} - E_\mathrm{i}}{Z_\mathrm{T}^2 e^2/a_0} \tag{6.24}$$

and

$$Q = \frac{q^2 a_0^2}{Z_\mathrm{T}^2}\,. \tag{6.25}$$

One then finds [MeL58, KhC69, Anh79, AnG86][3]

$$\begin{aligned}\sigma_\mathrm{K} &= 4\pi a_0^2 \frac{Z_\mathrm{P}^2}{\eta_\mathrm{K}^2 Z_\mathrm{T}^4} \int_{W_\mathrm{min}}^{\infty} dW \int_{Q_\mathrm{min}}^{\infty} \frac{dQ}{Q^2} \\ &\times \left(|F_\mathrm{K}|^2 + \frac{(\beta Z_\mathrm{T}\alpha)^2(1 - Q_\mathrm{min}/Q)}{(1-\beta^2 Q_\mathrm{min}/Q)^2}\left(|G_\mathrm{K}|^2 + \tfrac{1}{4}Q|F_\mathrm{K}|^2\right)\right),\end{aligned} \tag{6.26}$$

where

$$\eta_\mathrm{K} = (\beta/\alpha Z_\mathrm{T})^2 \tag{6.27}$$

and

$$\begin{aligned} Q_\mathrm{min} &= W^2/4\eta_\mathrm{K}, \\ W_\mathrm{min} &= \theta_\mathrm{K} = \frac{2I_\mathrm{K}^\mathrm{exp}}{Z_\mathrm{T}^2 e^2/a_0}. \end{aligned} \tag{6.28}$$

Here, $I_\mathrm{K}^\mathrm{exp}$ is the experimental K-shell ionization energy, and for a neutral target atom, the target charge is taken as an effective charge $Z_\mathrm{T} \to Z_\mathrm{T}^\mathrm{bare} - 0.3$ according to the Slater screening rules [Sla60].

The factors $|F_\mathrm{K}|^2$, $|G_\mathrm{K}|^2$ in Eq. (6.26) are squares of quasirelativistic matrix elements calculated with hydrogenic Darwin and free-electron wave functions. They are given by

$$\begin{aligned} |F_\mathrm{K}|^2 &= (3Q + W)QA_3, \\ |G_\mathrm{K}|^2 &= \tfrac{1}{4}WA_2, \end{aligned} \tag{6.29}$$

[3] Since quasirelativistic wave functions are accurate only to order $\alpha Z_\mathrm{T}$, it is inconsistent to introduce normalization factors for these wave functions of the form $[1 + (\alpha Z_\mathrm{T}/2)^2]^{-1/2}$ and $[1 + (k\alpha Z_\mathrm{T}/2)^2]^{-1/2}$, respectively [DaM78, Anh79]. For this reason, these normalization factors are omitted in Eq. (6.26), as is in fact done also in [Anh85a, AnG86].

with

$$\begin{aligned} A_n &= \frac{2^7 \exp\{-2/k \arctan[2k/(Q+1-k^2)]\}}{3[1-\exp(-2\pi/k)][(Q+1-k^2)^2+4k^2]^n}, \\ k &= \sqrt{W-1}. \end{aligned} \tag{6.30}$$

The integration over $q$ in Eq. (6.26) has to be performed with some care (see the discussion preceding Eq. (6.22) and in [MeL58, Anh79,AnG86]). The first term with $|F_{\rm K}|^2$ arises from the longitudinal contribution present also in nonrelativistic collisions. The terms proportional to $|G_{\rm K}|^2$ and $\frac{1}{4}Q|F_{\rm K}|^2$ are interpreted as transverse and spin-flip contributions, respectively [Anh79]. The transverse and spin-flip terms are most important for high-$Z$ targets, for which the quasirelativistic approximations actually lose their justification. Tables for the longitudinal part of the cross section are available [KhC69, ChM73, RiB77, BeK78, JoB79], but rather than using the projectile energy entering in the tables, for relativistic collisions one must use the ion velocity entering through the parameter $\eta_{\rm K}$ given in Eq. (6.27).[4] Estimates for the transverse and spin-flip contributions are given in [Anh79, AnB87].

In Fig. 6.2, theoretical uranium K-vacancy production cross sections are shown for bare projectiles of charge $Z_{\rm P}$. According to Eq. (6.26), the total cross section is proportional to $Z_{\rm P}^2$. Hence, the cross section shown has been divided by $Z_{\rm P}^2$. For $\gamma \geq 10$, the transverse and spin-flip contributions account for more than one-half of the total cross section. Figure 6.9 shows that these contributions bring the theory into close agreement with the experimental data. Although, as noted previously, the wave functions used to evaluate the cross sections are valid only for $\alpha Z_{\rm T} \ll 1$, the agreement appears to stretch to $\alpha Z_{\rm T} \simeq 0.7$. This may be because the collision dynamics explores a part of the bound and continuum wave functions which is not sensitively dependent on $\alpha Z_{\rm T}$.

For general orientation, Fig. 6.3 shows reduced 1s-vacancy production cross sections by bare projectiles on one-electron targets as a function of $\ln\gamma$ for various values of $Z_{\rm T}$, calculated from Eq. (6.26).[5]

---

[4] In the theoretical expression (6.22), the projectile velocity enters through the parameter $q_0$ in the nonrelativistic term and through $q_0$ and $\beta$ in the relativistic term.

[5] One-electron (1s) loss cross sections for projectiles by bare targets are given by [AnB87]. By interchanging $Z_{\rm T}$ and $Z_{\rm P}$, the 1s-ionization cross section of a one-electron target by a bare projectile can be obtained. Figure 6.2 is taken from [Anh79] and refers to two electrons in the K-shell, whereas Fig. 6.3 is taken from [AnB87] and refers to a single 1s electron.

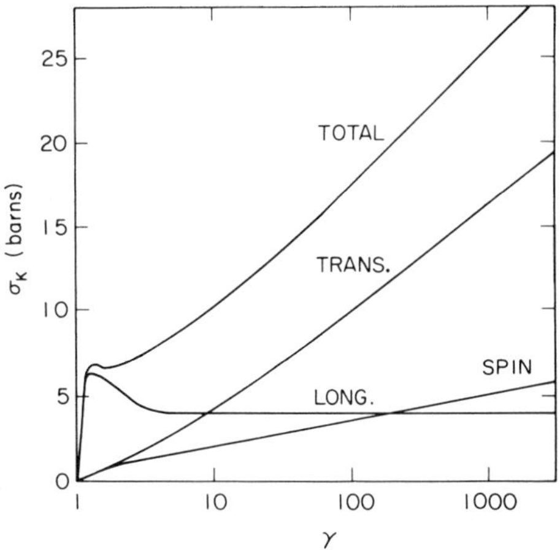


**Figure 6.2.** Calculated cross sections for U K-vacancy production by bare projectiles of charge $Z_P$, divided by $Z_P^2$, as a function of the Lorentz factor $\gamma$, using Eq. (6.26). The longitudinal, transverse, and spin-flip contributions are shown, as well as the total cross section. From [Anh79].

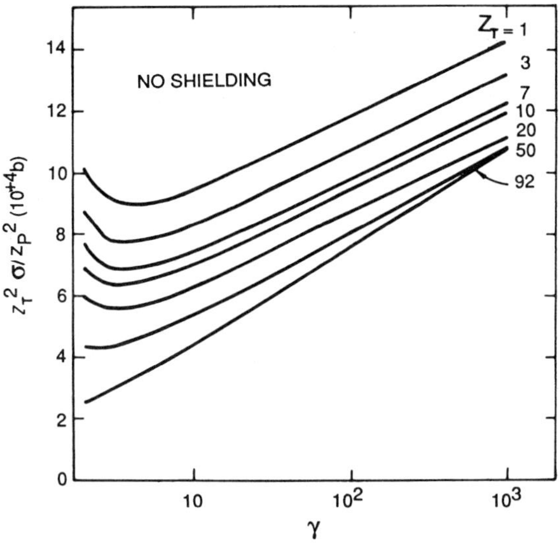


**Figure 6.3.** Calculated cross sections for one-electron target (1s) ionization by bare projectiles of charge $Z_P$, multiplied by $(Z_T/Z_P)^2$, as a function of the projectile Lorentz factor $\gamma$. The calculated curves are based on Eq. (6.26). The units are b/electron. Adapted from [AnB87].

## 6.3 Multipole expansion

For high-Z target atoms, it is mandatory to use exact relativistic wave functions for initial and final states. The factor $r^{s-1}$ in the Dirac $1s$ bound and continuum wave functions [see Eqs. (4.98), (4.103), and (4.114)] contracts the radial electronic density distribution. The resulting decrease of the radial matrix elements leads to smaller cross sections [AnG86] compared to the nonrelativistic treatment of electrons.

For calculating exact matrix elements, one may either use a multipole decomposition of the interaction in coordinate space [MeS89], or one may decouple the angular part of the electronic and projectile coordinates [VaB84, BeG85, Bec86] by introducing a multipole expansion for the exponential $\exp(i\,\mathbf{q}\cdot\mathbf{r}_\mathrm{T})$ occurring in Eq. (6.4). This amounts to an expansion in momentum space. In the following, we discuss both types of expansion in turn.

### 6.3.1 Multipole expansion in coordinate space

Returning to the starting equation (6.1), one may try to decouple the time integration associated with the nuclear motion from the integration over the electronic coordinates by performing a multipole decomposition of the interaction term $1/r'_\mathrm{P}$. Such an expansion is applicable not only in the perturbative formula (6.1) but also in a more rigorous coupled-channel treatment, see Sec. 6.5.

For a nonrelativistic interaction $1/|\mathbf{r}_1 - \mathbf{r}_2|$, the standard multipole expansion leads to a complete decoupling of the angular dependence of the vectors $\mathbf{r}_1$ and $\mathbf{r}_2$, i.e., a decoupling of the unit vectors $\hat{\mathbf{r}}_1$ and $\hat{\mathbf{r}}_2$. In a relativistic collision, the Lorentz contraction acts in the $z$-direction, so that the Cartesian coordinates are not treated on equal footing in Eq. (5.97). Some coupling in the polar angles will therefore remain. A partial decoupling [RuS93] can be achieved by rewriting the interaction term in the form

$$\frac{1}{r'_\mathrm{P}} = \frac{1}{\sqrt{(x_\mathrm{T}-b)^2 + y_\mathrm{T}^2 + \gamma^2(z_\mathrm{T}-vt)^2}} = \frac{1}{|\boldsymbol{\rho}_\mathrm{e} - \boldsymbol{\rho}_\mathrm{P}|}, \tag{6.31}$$

where we have formally introduced the vectors

$$\boldsymbol{\rho}_\mathrm{e} = (x_\mathrm{T}, y_\mathrm{T}, \gamma z_\mathrm{T}) \quad \text{with} \quad \rho_\mathrm{e} = \sqrt{x_\mathrm{T}^2 + y_\mathrm{T}^2 + \gamma^2 z_\mathrm{T}^2} \tag{6.32}$$

and

$$\boldsymbol{\rho}_\mathrm{P} = (b, 0, \gamma v t) \quad \text{with} \quad \rho_\mathrm{P} = \sqrt{b^2 + \gamma^2 v^2 t^2}. \tag{6.33}$$

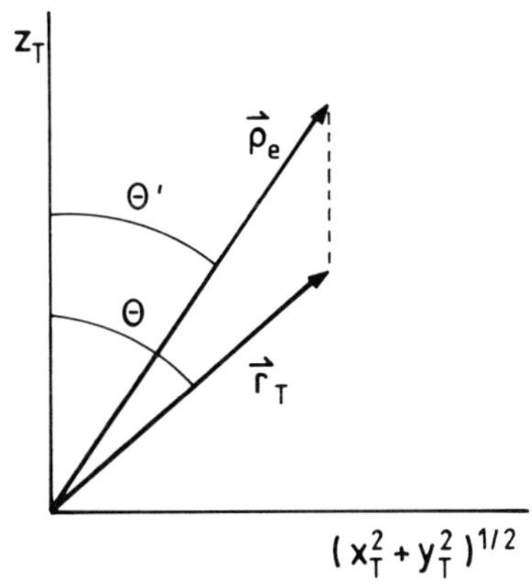


**Figure 6.4.** Definition of the vector $\rho_e$.

These vectors are not physical position vectors for the electron and projectile, respectively, nevertheless they are formally useful in partly separating the electron motion from the projectile motion. In Fig. 6.4, we show the relation of the modified vector $\boldsymbol{\rho}_e$ to the position vector $\mathbf{r}_T = (x_T, y_T, z_T)$. Introducing the polar angles $\theta$ of $\mathbf{r}_T$ and $\theta'$ of $\boldsymbol{\rho}_e$, we can rewrite

$$\rho_e = r_T u(\theta) \tag{6.34}$$

with

$$u(\theta) = \sqrt{1 + (\gamma^2 - 1)\cos^2\theta}, \tag{6.35}$$

so that

$$\cos\theta' = \frac{\gamma z_T}{\rho_e} = \frac{\gamma\cos\theta}{u(\theta)}. \tag{6.36}$$

We now decompose the inverse distance $1/|\boldsymbol{\rho}_e - \boldsymbol{\rho}_P|$ of Eq. (6.31) into spherical harmonics $Y_{LM}$, depending on the directions $\widehat{\boldsymbol{\rho}}_e$ and $\widehat{\boldsymbol{\rho}}_P$, in the form

$$\frac{1}{|\boldsymbol{\rho}_e - \boldsymbol{\rho}_P|} = \begin{cases} \sum_{L=0}^{\infty} \frac{4\pi}{2L+1} \frac{[r_T u(\theta)]^L}{\rho_P^{L+1}} \sum_{M=-L}^{L} Y_{LM}^*(\widehat{\boldsymbol{\rho}}_P) Y_{LM}(\widehat{\boldsymbol{\rho}}_e), & \text{if} \quad \rho_e \le \rho_P, \\ \sum_{L=0}^{\infty} \frac{4\pi}{2L+1} \frac{\rho_P^L}{[r_T u(\theta)]^{L+1}} \sum_{M=-L}^{L} Y_{LM}^*(\widehat{\boldsymbol{\rho}}_P) Y_{LM}(\widehat{\boldsymbol{\rho}}_e), & \text{if} \quad \rho_e \ge \rho_P, \end{cases} \tag{6.37}$$

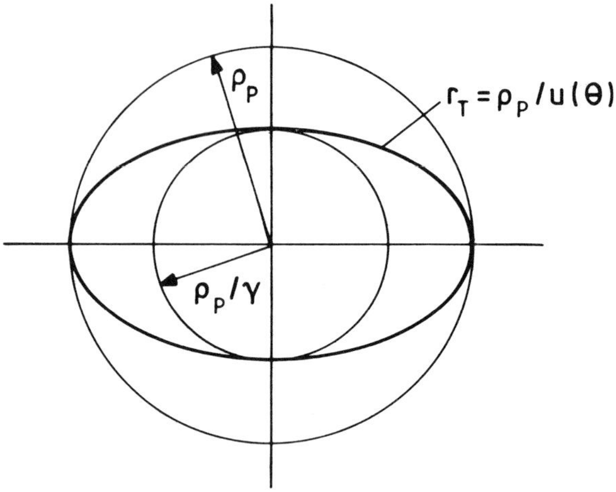


**Figure 6.5.** Illustration of the inner and outer regions in which the angular and the radial integration decouple and of the intermediate regions in which they do not decouple.

or, equivalently,

$$\frac{1}{|\boldsymbol{\rho}_{\rm e}-\boldsymbol{\rho}_{\rm P}|}=\sum_{LM}\frac{4\pi}{2L+1}v_L(r_{\rm T},\theta,\rho_{\rm P})Y^*_{LM}(\widehat{\boldsymbol{\rho}}_{\rm P})Y_{LM}(\widehat{\boldsymbol{\rho}}_{\rm e}) \tag{6.38}$$

with

$$v_L(r_{\rm T},\theta,\rho_{\rm P})=\begin{cases}\dfrac{[r_{\rm T}u(\theta)]^L}{\rho_{\rm P}^{L+1}}, & \text{if} \quad r_{\rm T}\le\rho_{\rm P}/u(\theta),\\ \dfrac{\rho_{\rm P}^L}{[r_{\rm T}u(\theta)]^{L+1}}, & \text{if} \quad r_{\rm T}\ge\rho_{\rm P}/u(\theta).\end{cases} \tag{6.39}$$

The distinction, which one of $\rho_{\rm e}$ or $\rho_{\rm P}$ is larger, is seen to depend on the angle $\theta$, unlike for the nonrelativistic case. Since, according to Eq. (6.35), $u(\theta)$ varies between 1 and $\gamma$, the separation between the radial and the angular integration is complete only in the range $r_{\rm T}\le\rho_{\rm P}/\gamma$ and $r_{\rm T}\ge\rho_{\rm P}$. The dividing surface between the inner and the outer region is given by the ellipsoid $\rho_{\rm e}=r_{\rm T}u(\theta)=\rho_{\rm P}$. The inner, outer, and intermediate regions are illustrated in Fig. 6.5.

The first-order transition amplitude (6.1) is now expressed as

$$A^{\rm e}_{\rm fi}(b)=i\frac{\gamma Z_{\rm P}e^2}{\hbar}\int dt\, e^{i(E_{\rm f}-E_{\rm i})t/\hbar}M_{\rm fi}(b,t). \tag{6.40}$$

Here, the matrix element is

$$M_{\mathrm{fi}}(b,t) = \sum_{L=0}^{\infty} \frac{4\pi}{2L+1} \langle \mathrm{f}|(1-\beta\alpha_z)$$
$$v_L(r_{\mathrm{T}},\theta,\rho_{\mathrm{P}}) Y_{L,\mu_{j_\mathrm{f}}-\mu_{j_\mathrm{i}}}(\widehat{\boldsymbol{\rho}}_\mathrm{e})|\mathrm{i}\rangle \, Y^*_{L,\mu_{j_\mathrm{f}}-\mu_{j_\mathrm{i}}}(\widehat{\boldsymbol{\rho}}_\mathrm{P}). \tag{6.41}$$

The quantities $\mu_{j_\mathrm{i}}$ and $\mu_{j_\mathrm{f}}$ are the angular momentum projections of the initial and final states, respectively. Formula (6.40) is a possible point of departure for the accurate evaluation of excitation, ionization, and pair production cross sections using Eqs. (5.64) and (5.65) within first-order perturbation theory. The matrix elements (6.41) can, of course, also be employed for detailed single-center coupled-channel calculations [RuS93].

In evaluating the matrix elements (6.41), one has to observe that, owing to (6.39), not only the radial integration over $r_\mathrm{T}$ but also the integration over the polar angle $\theta$ within the interval $\rho_\mathrm{P}/\gamma \le r_\mathrm{T} \le \rho_\mathrm{P}$ depends on the particular value of $b$ and $t$ through the relation $\rho_\mathrm{P} = \sqrt{b^2+\gamma^2 v^2 t^2}$ of Eq. (6.33). Only the integral over the azimuthal angle can be carried out easily.

### 6.3.2 Multipole expansion in momentum space

Let us now turn to the calculation of the transition amplitude in momentum space which allows for a more complete separation of the integrations associated with the electron and the projectile coordinates.

If we start from the right-hand side of Eq. (5.104) [which leads to Eq. (6.4)] rather than from the left-hand side, a decoupling is achieved by decomposing the exponential $\exp(i\mathbf{q}\cdot\mathbf{r}_\mathrm{T})$. In terms of a spherical representation $\hat{\mathbf{q}} = \mathbf{q}/q = (\theta_q,\phi_q)$, where $q^2 = q_b^2 + q_0^2$, and $\hat{\mathbf{r}}_\mathrm{T} = \mathbf{r}_\mathrm{T}/r_\mathrm{T} = (\theta,\phi)$, we can write the expansion of the plane wave as

$$e^{i\,\mathbf{q}\cdot\mathbf{r}_\mathrm{T}} = 4\pi \sum_{L=0}^{\infty} \sum_{M=-L}^{L} i^L\, j_L(qr_\mathrm{T}) Y^*_{LM}(\hat{\mathbf{q}})\, Y_{LM}(\hat{\mathbf{r}}_\mathrm{T}). \tag{6.42}$$

While in each term of the sum, the angular dependencies factorize, the radial parts remain coupled by the spherical Bessel function $j_L(qr_\mathrm{T})$. Using $\mathbf{q}_b\cdot\mathbf{b} = q_x b = qb\sin\theta_q\cos\phi_q$ and the integral representation of the Bessel function $J_M$, one may carry out the $\phi_q$ integration within the angular integral on the right-hand side of the matrix element (5.104) to get

$$\int_0^{2\pi} d\phi_q \int_{-1}^{1} d\cos\theta_q \, \frac{e^{-iqb\sin\theta_q\cos\phi_q}}{q^2(1-\beta^2\cos^2\theta_q)}\, Y^*_{LM}(\theta_q,\phi_q) =$$

$$2\pi\, i^{-M} \int_{-1}^{1} d\left(\frac{q_z}{q}\right) \frac{1}{q^2 - \beta^2 q_z^2}\, Y_{LM}\left(\arccos\frac{q_z}{q}, 0\right)\, J_M(b\sqrt{q^2 - q_z^2}\,). \tag{6.43}$$

By substituting Eq. (6.42) into (6.4) and employing Eq. (6.43), one obtains the transition amplitude

$$\begin{aligned} A_{\mathrm{fi}}^{\mathrm{e}}(b) &= 8i\pi\nu_{\mathrm{P}} \sum_{L,M} i^{L-M} \int_0^\infty \frac{q\, dq}{q^2 - \beta^2 q_0^2} \\ &\quad \times \langle \mathrm{f}|(1-\beta\alpha_z) j_L(q r_{\mathrm{T}}) Y_{LM}(\hat{\mathbf{r}}_{\mathrm{T}})|\mathrm{i}\rangle\; B_{LM}(b, q_0, q), \end{aligned} \tag{6.44}$$

where

$$B_{LM}(b, q_0, q) = \Theta(q - q_0)\, Y_{LM}\left(\arccos\frac{q_0}{q}, 0\right) J_M\left(b\sqrt{q^2 - q_0^2}\right) \tag{6.45}$$

is the straight-line path factor [AmA81, VaB84, BeG85] that already occurs in the SCA for nonrelativistic collisions [BaH59]. The Heaviside step function $\Theta(q - q_0)$ assures that $q$ is greater than its projection $q_0$.

The total cross section is derived either from Eq. (6.44) using the orthogonality relation for the Bessel functions or, more directly, from Eq. (6.5). The final result for the ionization cross section per electron can be written as

$$\frac{d\sigma_{\mathrm{fi}}^{\mathrm{e}}}{dE_{\mathrm{f}}} = 8\pi\nu_{\mathrm{P}}^2 \int_{q_0}^\infty \frac{q\, dq}{(q^2 - \beta^2 q_0^2)^2}\, |M_{\mathrm{fi}}(q)|^2, \tag{6.46}$$

with

$$\begin{aligned} M_{\mathrm{fi}}(q) &= 4\pi \sum_{L=0}^{\infty} i^L\, \langle \mathrm{f}|(1-\beta\alpha_z) j_L(q r_{\mathrm{T}}) Y_{L,\mu_{j_{\mathrm{f}}}-\mu_{j_{\mathrm{i}}}}(\hat{\mathbf{r}}_{\mathrm{T}})|\mathrm{i}\rangle \\ &\quad \times Y_{L,\mu_{j_{\mathrm{f}}}-\mu_{j_{\mathrm{i}}}}\left(\arccos\frac{q_0}{q}, 0\right), \end{aligned} \tag{6.47}$$

where $\mu_{j_{\mathrm{i}}}$ and $\mu_{j_{\mathrm{f}}}$ are the angular momentum projections of the initial and final state, respectively. This formula is the starting point for the accurate evaluation of excitation and ionization cross sections within first-order perturbation theory [VaB84, BeG85].

Before moving on to discuss some details and generalizations, a few comments may be appropriate.

(a) First, we observe that in contrast to Eq. (6.15) in the dipole approximation, there is no convergence problem at the upper limit of the $q$-integration in Eq. (6.46). The cross section is well-defined without ambiguous cut-off procedures.

(b) The behavior of the cross section (6.46) in the extreme relativistic limit $\gamma \gg 1$ can be investigated by expanding the integrand around $q = q_0$. In this way, it is shown by Becker [Bec86] that for $\mu_{j_\mathrm{f}} = \mu_{j_\mathrm{i}}$ (and also for $|\mu_{j_\mathrm{f}} - \mu_{j_\mathrm{i}}| \geq 2$), the integral tends to a constant value as $\gamma$ increases. However, for $\mu_{j_\mathrm{f}} = \mu_{j_\mathrm{i}} \pm 1$, the integral increases with $\gamma$ as $\ln \gamma$. This is in accordance with the results (6.17) or (6.23), derived more simply in the dipole approximation or the quasirelativistic treatment, and with the asymptotic behavior given in Eqs. (1.10) and (1.11).

(c) The choice of the gauge for the electromagnetic potentials is discussed in some detail in [AmA81]. Here, we have used the Lorentz condition (3.3) which is Lorentz invariant and hence allows us to derive the electromagnetic potentials $\mathbf{A}$ and $\Phi$ by a Lorentz transformation from the static Coulomb potential. This renders it possible to use Eq. (6.4) to derive rather simple path factors (6.45) which actually are the same as for nonrelativistic projectiles. On the other hand, the Coulomb gauge defined by $\nabla \cdot \mathbf{A}^{\mathrm{Coul}} = 0$ leads to more complicated path factors. A comparison of numerical results for cross sections calculated in the Lorentz and Coulomb gauges in the plane-wave Born approximation as well as in the SCA shows agreement with one another on the 1% level.

(d) The electronic matrix elements in Eq. (6.47) have to be calculated with relativistic wave functions. The use of nonrelativistic approximations either leads to a violation of gauge invariance (since the electromagnetic coupling $j_\mu A^\mu / c$ , Eq. (5.31) is not compatible with the Schrödinger equation) or leads to second-order terms in the potentials and hence to a spurious projectile charge dependence as $Z_\mathrm{P}^4$. Only by a careful analysis [AmA81] is it possible to remove these spurious terms.

In evaluating the matrix elements (6.47) rigorously, it is no longer useful to reduce the terms containing the Dirac matrix $\alpha_z$ by employing Eq. (5.90). Rather, with the definitions (4.86) and (4.87), one may decompose the matrix elements as [VaB84, BeG85]

$$\begin{aligned}\langle \mathrm{f}|(1-\beta\alpha_z)\, j_l(qr_\mathrm{T})Y_{lm}(\hat{\mathbf{r}}_\mathrm{T})|\mathrm{i}\rangle &= \\ A_1F_L^{(1)}(q) + i\beta A_2F_L^{(2)}(q) &- i\beta A_3F_L^{(3)}(q) \qquad (6.48)\end{aligned}$$

where the angular integrals are defined by

$$\begin{aligned} A_1 &= \langle \chi_{\kappa_\mathrm{f}}^{\mu_{j_\mathrm{f}}}|Y_{LM}|\chi_{\kappa_\mathrm{i}}^{\mu_{j_\mathrm{i}}}\rangle \\ A_2 &= \langle \chi_{-\kappa_\mathrm{f}}^{\mu_{j_\mathrm{f}}}|Y_{LM}\sigma_z|\chi_{\kappa_\mathrm{i}}^{\mu_{j_\mathrm{i}}}\rangle \\ A_3 &= \langle \chi_{\kappa_\mathrm{f}}^{\mu_{j_\mathrm{f}}}|Y_{LM}\sigma_z|\chi_{-\kappa_\mathrm{i}}^{\mu_{j_\mathrm{i}}}\rangle \qquad (6.49)\end{aligned}$$

and are evaluated with standard techniques [Ros57, Edm57]. The radial integrals are defined as

$$\begin{aligned} F_L^{(1)}(q) &= \int_0^\infty j_L(qr)(g_\mathrm{i}g_\mathrm{f} + f_\mathrm{i}f_\mathrm{f})\, r^2\, dr \\ F_L^{(2)}(q) &= \int_0^\infty j_L(qr) g_\mathrm{i} f_\mathrm{f}\, r^2\, dr \\ F_L^{(3)}(q) &= \int_0^\infty j_L(qr) f_\mathrm{i} g_\mathrm{f}\, r^2\, dr. \end{aligned} \tag{6.50}$$

The integrals can be rewritten in terms of hypergeometric functions [JaZ57, Amu78, JaA80, TrB83, VaB84] if exact Coulomb bound-state wave functions (4.98) and continuum partial waves (4.114) are inserted. The actual evaluation is performed numerically [BeG85].

The calculation of the total and energy-differential ionization cross section according to Eq. (6.7) requires an averaging over all $\mu_{j_\mathrm{i}}$ and a summation over all values of $\kappa_\mathrm{f}$ and $\mu_{j_\mathrm{f}}$. In practice, the summation has to be truncated at a maximum value $\kappa_\mathrm{max}$ because of computer limitations. For the angle-differential cross section, one has to use the wave functions (4.123) or (4.124) with well-defined asymptotic momenta instead of the spherical partial waves.

### 6.3.3 Perturbative results and comparison with experiment

In an experiment, we usually have a solid or gas target, that is, we have to take into account many-electron effects. Deferring a more detailed discussion to Chap. 7, we assume for the present purpose that the K-shell wave function of a complex target atom can be approximated by an effective hydrogenic wave function. The generally accepted method uses an effective Slater-screened target charge $Z_\mathrm{T}^* = Z_\mathrm{T}^\mathrm{bare} - 0.3$ in conjunction with experimental K-shell binding energies. These binding energies are not consistent with the wave functions used. One may expect, however, that mainly the inner part of the wave function is responsible for the matrix elements. This part is well described by the effective nuclear charge $Z_\mathrm{T}^*$. The outer part is related to the binding energy but does not enter sensitively into the results. Nevertheless, it is not clear whether the inconsistency may lead to sizable effects in some cases. In most results to be discussed in this chapter, it is justified to neglect the effective-charge correction.

The first step in calculating differential (in electron energy and angle) and total cross sections consists in studying the impact-parameter dependence of the ionization probability $P(b)$. The weighted K-shell ionization probability $2\pi b\, P(b)$ usually has a maximum in the vicinity of the K-shell radius $a_K$. This is similar as for the excitation probabilities in Fig. 6.13 for coupled-channel calculations.

Figures 6.6 and 6.7 show calculated double-differential ionization cross sections $d^2\sigma/dE_e\,d\Omega_e$ for fixed electron energies and fixed electron emission angles, respectively, for the system $Z_P$ + U [MoG89]. In Fig. 6.6, we can see that electron emission occurs mainly at smaller electron energies. As Fig. 6.7 demonstrates, the angular anisotropy of the emitted electrons becomes peaked in the forward direction as their kinetic energy increases. Since, owing to limitations in computing power, $|\kappa| \leq 10$ has been used, the convergence of the partial-wave contributions in Eq. (4.123) becomes unsatisfactory at backward angles, so that this angular range is not included in Fig. 6.7.

In Figure 6.8, the total K-shell ionization cross section for p + U collisions is plotted as a function of the Lorentz factor $\gamma$. As already seen explicitly in the dipole approximation of Eq. (6.17), the cross section increases monotonically with $\gamma$. The quasirelativistic calculation of Anholt [Anh79] and the multipole expansion calculation of Becker et al. [BeG85] show the same qualitative behavior as the collision energy increases. For $\gamma > 2$, the absolute difference between the calculations increases slowly with $\gamma$. For large values of $\gamma$, Becker et al. have the problem that a growing number of terms contribute in the multipole expansion (calculations have been carried up to $l = 6$); however, the authors estimate the resulting inaccuracy to be less than 5 %. On the other hand, differences in the wave function will affect the matrix elements changing the magnetic quantum number by one unit and hence will increase as $\ln\gamma$ for large energies. It is therefore quite natural that discrepancies slowly build up as the energy increases.

Numerical calculations have established an approximate scaling rule [AmA81, Bec86] for the ionization probability at high relativistic energies ($\gamma > 5$) and fixed impact parameters. For $b = 0$, the ionization probability

$$P(b = 0) = 1.8 \times 10^{-4} Z_P^2 \tag{6.51}$$

is independent (within $\pm 10\%$) of the target charge up to the heaviest elements. This empirical finding shows that for $Z_P \geq 75$ the ionization probability exceeds unity, and hence perturbation theory breaks down.

There are few experimental data to compare with theoretical results. At present, only measurements for total cross sections of K-shell ionization in various target atoms are available. In Fig. 6.9, the total K-shell ionization cross section is shown for 4.88 GeV protons impinging on target atoms with charge numbers $Z_T$. For the calculations of [DaM78, Anh79], quasirelativistic wave functions are used. For high-$Z$ targets, the deviation

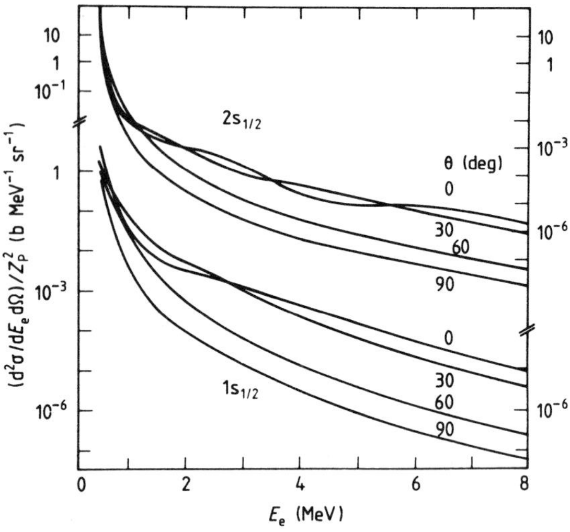


**Figure 6.6.** Double-differential cross section for the ionization of the $1s_{1/2}$ and $2s_{1/2}$ target states in $Z_P$ + U collisions at $E_{lab}$ = 1 GeV/u as a function of the energy of the ionized electron. The parameter at each curve gives the laboratory emission angle $\theta$. From [MoG89].

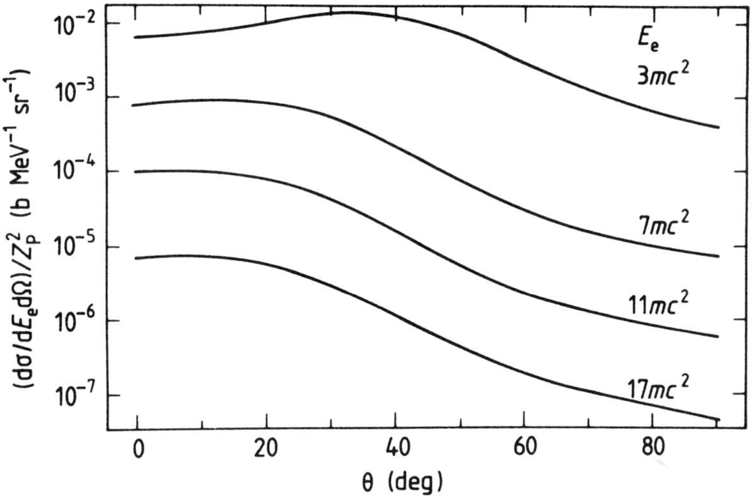


**Figure 6.7.** Double-differential cross section for target K-shell ionization in $Z_P$ + U collisions at $E_{lab}$ = 1 GeV/u as a function of the electron laboratory emission angle $\theta$ for various electron energies $E_e$. From [MoG89].

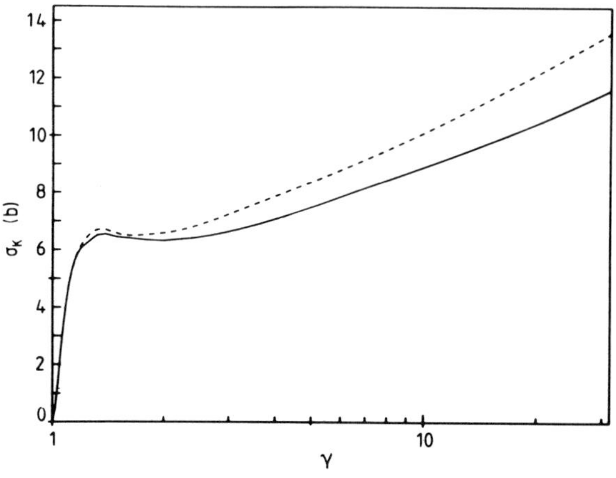


**Figure 6.8.** K-shell ionization cross section for p + U collisions as a function of the Lorentz factor $\gamma$. Dashed curve: [Anh79]; solid curve: [BeG85]. In both curves, experimental binding energies have been used.

from exact relativistic wave functions (used in [BeG85]) should be most significant. In fact, the results for the two types of calculations differ by only 10%. Unfortunately, the calculations using more accurate electronic wave functions lie farther away from the experimental data. Considering the experimental errors, it certainly would not be justified to prefer any of the calculations just on the basis of a comparison with the data. For further discussion see [Eic90].

In view of the scarcity of ionization data for relativistic atomic projectiles, it is tempting to turn to other projectiles. Ionization measurements have been made with electrons up to 23 GeV, corresponding to $\gamma = 4.4 \times 10^4$ [SpM94]. At these extremely high values of $\gamma$, the relevant impact parameters are so large that they lie in the medium surrounding the target atom.[6] This leads to a shielding of the projectile's electromagnetic field [Fer40], which causes the cross section to saturate [Sor87, ChE89]. Such "many-body" effects lie outside the scope of this book. Below the onset of the saturation, though, typically for $\gamma \leq 10^3$, one can expect that high-energy electrons have ionization cross sections similar to those of protons. This is true, as long as the trajectory of the electron projectile can be approximated by a straight line for the relevant collisions. Calculations by Davidović et al. [DaM78], based on a relativistic first-order wave theory, show that even

[6] At these high values of $\gamma$, the de Broglie wavelength $h/(\gamma m_e c)$ is small enough ($\sim 10^{-2}$ fm) that the concept of impact parameter is applicable even to electrons.

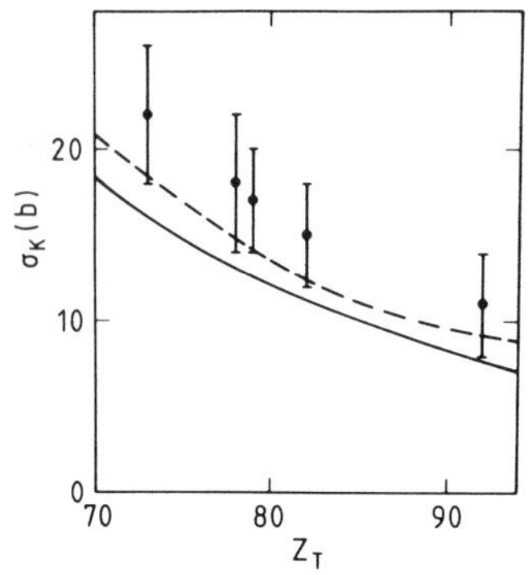


**Figure 6.9.** K-shell ionization cross section for 4.88 GeV protons versus the target charge. Experimental data and dashed curve: Anholt et al. [Anh79], see also Davidović et al. [DaM78]; solid curve: Becker et al. [BeG85].

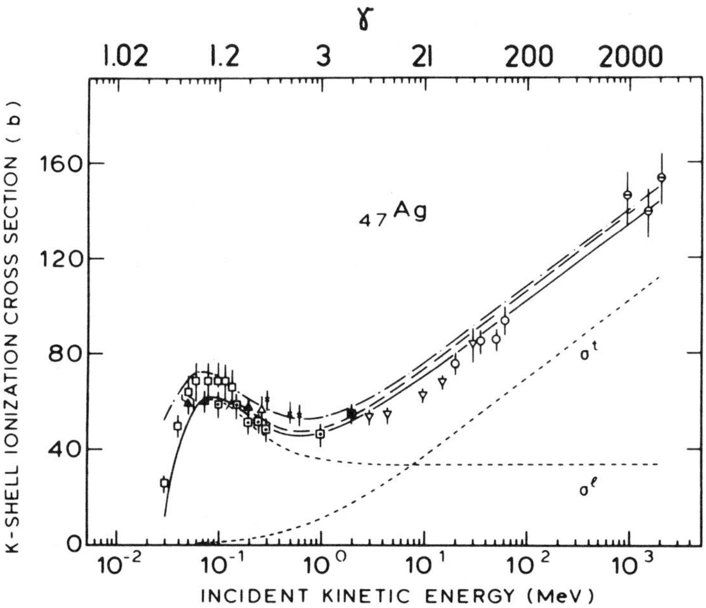


**Figure 6.10.** K-shell ionization cross section for Ag bombarded by electrons. The dash-dotted curve gives the result of a semiclassical calculation [Kol75], the dashed curve a fully quantum-mechanical calculation [Sco78], and the solid line a calculation based on structure functions with one-photon exchange [EsM82], together with a decomposition into longitudinal ($\sigma^l$) and transverse plus spin-flip ($\sigma^t$) contributions. Note that the abscissa is not linear in $\ln \gamma$. From [Gen84].

for $\gamma$ as low as 5 (2-MeV electrons), electron and proton cross sections differ by less than 5% up to $Z_P = 50$. Therefore, we show in Fig. 6.10 the K-shell ionization cross section for electrons on Ag. A comparison is made with various theories. One of these, by Eschwey and Manakos [EsM82], allows a decomposition into longitudinal ($\sigma^l$) and transverse plus spin-flip ($\sigma^t$) components, shown in dotted lines on Fig. 6.10. The former goes to a constant, and the latter increases as $\ln\gamma$, as $\gamma$ increases to high values, illustrating the expected behavior of these cross section components. Comparisons of other relativistic electron data with Eq. (6.26) can be found in [Anh79].

# 6.4 Nonperturbative approaches

The empirical scaling rule Eq. (6.51) of the ionization probability at impact parameter $b = 0$ shows that perturbation theory breaks down at large projectile charges. Already for $Z_\mathrm{P} \geq 75$ the probability $P(b = 0) \geq 1.0$, so that for medium-$Z$ projectiles, the method will start to become questionable. Although this conclusion is weakened for the more important larger impact parameters, it is nevertheless mandatory to examine nonperturbative approaches. We first consider the finite difference method and in Secs. 6.5 and 6.6 proceed to coupled-channel methods.

We would like to mention another relativistic nonperturbative approach, namely the classical-trajectory Monte-Carlo (CTMC) method [TeT80, HoH92]. The nonrelativistic CTMC version, while difficult to justify theoretically, is known to be successful in low-velocity collisions. The relativistic version treats the nuclear as well as the electronic motion by solving the classical relativistic equations of motion for the three-body system assuming a relativistic microcanonical distribution of initial positions and momenta and ignoring the electron spin.

## 6.4.1 Finite-difference methods

Let us now turn to finite-difference calculations. Here, the time-dependent Dirac equation in its Hamiltonian form (5.40) is solved numerically. Writing, for a small time step $\Delta t$,

$$\Psi(\mathbf{r}_\mathrm{T}, t + \Delta t) = U(t,\, t + \Delta t)\Psi(\mathbf{r}_\mathrm{T}, t), \tag{6.52}$$

the time evolution operator is evaluated as

$$U(t,\, t + \Delta t) = \left(1 + \frac{i}{2\hbar}\,\Delta t\, H(t + \frac{1}{2}\Delta t)\right)^{-1} \left(1 - \frac{i}{2\hbar}\,\Delta t\, H(t + \frac{1}{2}\Delta t)\right) \tag{6.53}$$

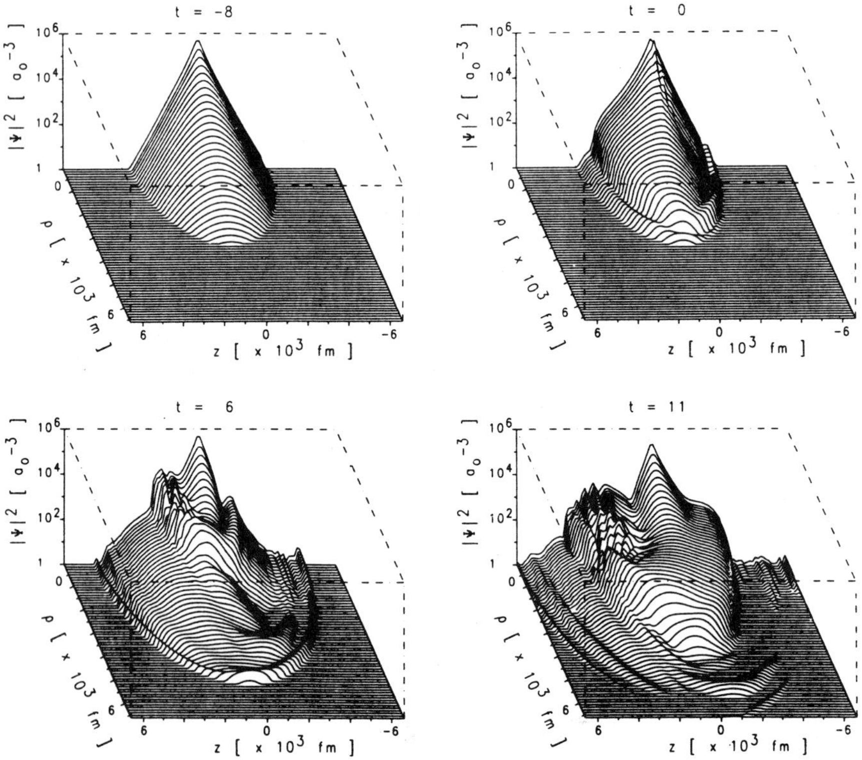


**Figure 6.11** Time evolution of the electron density distribution in the $z_T - \rho$-plane $[\rho = (x_T^2 + y_T^2)^{1/2}]$ for a $U^{92+} + U^{91+}$ collision at 10.0 GeV/u. The target remains fixed at $z_T = \rho = 0$ while the projectile moves at $b = 0$ from right to left along the $z_T$-axis in the positive $z$-direction. It is located at $z_P = -3.1 \times 10^3$ fm for $t = -8.0$ [in units of $\hbar/(m_e c^2) = 1.28 \times 10^{-21}$ s], at $z_P = 0$ for $t = 0$, at $z_P = 2.3 \times 10^3$ fm for $t = 6$, and at $z_P = 4.2 \times 10^3$ fm for $t = 11$. The densities are given in units of $a_0^{-3}$ ($a_0 = 5.3 \times 10^4$ fm). From [ThB92].

up to second order in $\Delta t$. After discretizing the wave function in space as well, one may follow its time development starting from a suitable initial condition. The numerical technique is described in detail by Becker et al. [BeG83].

In order to render the numerical procedure manageable, one has to confine oneself to an impact parameter $b = 0$, thus enforcing cylindrical symmetry about the beam direction and keeping the problem at two dimensions in space [BeG83, ThB92]. While this choice of the impact parameter carries zero weight for the cross section, it is instructive to study this particular case.

Since the trajectory passes right through the singular points of the Coulomb potentials, it is necessary to replace the Coulomb potentials in Eq. (5.40) by nonsingular potentials produced by a finite-size nucleus. The simplest choice for the nuclear charge distribution is a homogeneously charged sphere with radius $R_0$, separately fixed for the target and the projectile nucleus. In other words, one has to substitute in Eq. (5.40) for both collision partners

$$\frac{1}{r} \longrightarrow f_0(r) = \begin{cases} 1/r & \text{for } r \geq R_0 \\ \frac{1}{2}(3 - r^2/R_0^2)/R_0 & \text{for } r \leq R_0 \end{cases} . \tag{6.54}$$

Figure 6.11 conveys a vivid picture of the time development of the probability distribution for $U^{92+} + U^{91+}$ collisions at 10 GeV/u [ThB92]. In the calculations leading to these results, a grid of $600 \times 300$ meshes with a spacing of $4.6 \times 10^{-2}\ a_K = 26.5$ fm (where $a_K$ is the uranium K-shell radius) is used. The nuclear radius $R_0$ is arbitrarily chosen to be twice this amount in order to guarantee the stability of the numerical procedure. The calculations are initialized at the time $t = -8.5$ [in natural relativistic units $t_0 = \hbar/(m_e c^2) = 1.28 \times 10^{-21}$s], when the projectile has a distance of about $5.7\ a_K = 3.3 \times 10^{-10}$ cm from the target. At $t_0$, the density of the electron distribution is assumed to be determined by the *unperturbed* $1s_{1/2}$ ground state of $U^{91+}$. Figure 6.11 shows the electron density at $t = -8.0$ and separation $5.4\ a_K$. At the next stage, $t = 0$, when the nuclear charge distributions completely overlap, the density distribution still is almost unaltered. The main change takes place while the projectile recedes and leaves the target atom behind, at a separation of about $4a_K$ at $t = 6$ and of about $7.3a_K$ at $t = 11$. The numerically calculated wave functions can be projected onto exact target eigenstates and thus be analyzed with respect to their content of various excited and continuum states.

While these calculations are very instructive in permitting one to visualize the collision process, they are not sufficient to calculate cross sections.

In Sec. 6.1 it is also shown that it is not realistic to assume that in the entrance channel the target wave function is still unperturbed at a projectile separation of 5.7 $a_K$. In reality, owing to the long range of the Coulomb interaction, changes in the target wave function start to accumulate when the projectile is still a distance of (500-2000)$a_K$ away [ToE88b], see Fig. 6.12. Unfortunately, numerical calculations on a discretized grid within such a large time span are presently hardly possible.

Following the arguments outlined in Sec. 6.1.2, one may expect a much better convergence if one uses a phase-distorted wave function (5.62) for the initialization of the time evolution and, correspondingly, the short-range interaction occurring in Eq. (6.9).

## 6.5 Two-center coupled-channel methods

The direct solution of the time-dependent Dirac equation (5.39, 5.40) in the laboratory frame on a discrete space-time grid is still not feasible with sufficient accuracy for a general collision geometry. However, one may solve the problem within a limited space of basis functions. For this purpose, one expands the wave function $\Psi(\mathbf{r}_T, t)$ in terms of a suitable set of basis states $\psi_k(\mathbf{r}_T, t)$. The degree of approximation then is essentially determined by the choice and the truncation of the basis set. A basis set embracing the most important reaction channels should not only allow for excitation and ionization but also for charge transfer.

In this section, we discuss two-center coupled-channel calculations, and proceed to single-center coupled-channel methods in Sec. 6.6. Both methods are based on the same idea, namely an expansion of the time-dependent wave function in terms of atomic orbitals. In the former, more symmetrical approach, one uses a subset of basis functions centered around the target and another subset centered around the projectile nucleus. In the latter method, which is computationally simpler, the basis functions are attached only to the target nucleus, with the projectile merely acting as a source of perturbation.

### 6.5.1 Coupled-channel equations

A treatment of the time-dependent Dirac equation (5.39, 5.40) for relativistic ion-atom collisions requires the expansion of the time-dependent wave function $\Psi(\mathbf{r}_T, t)$ in terms of target and projectile eigenstates $\psi_k(\mathbf{r}_T, t)$ and $\psi_{k'}(\mathbf{r}'_P, t')$, respectively, as

$$\Psi(\mathbf{r}_T, t) = \sum_k a_k(t)\psi_k(\mathbf{r}_T, t) \; + \; \sum_{k'} a_{k'}(t)\, S^{-1}\psi_{k'}(\mathbf{r}'_P, t'), \tag{6.55}$$

where we have referred the total wave function to the laboratory system. The first sum represents an expansion in terms of target states; the second sum is an expansion in terms of projectile states originally defined in the projectile system and subsequently transformed to the laboratory system with the aid of the operator $S^{-1}$ defined in Eqs. (4.29) and (4.31).

The use of two centers for the expansion of the wave function $\Psi(\mathbf{r}, t)$ has three important implications: (a) In the interaction region, linear combinations of atomic orbitals around the target and around the projectile nucleus provide flexibility to approximate transient molecular orbitals which may play a role during the collision. (b) Charge transfer is explicitly included as a reaction channel. This not only renders charge exchange at all possible, but also will affect other reactions by entering as an intermediate state. (c) Specifically, ionization of an electron may occur as a direct ionization of the target atom as well as capture into the continuum of the projectile. In this way, one will get a better representation of the true continuum by a finite set of basis states.

In fact, the only approximation within the method consists in the restriction to a finite set of atomic basis states. Since each of the atomic basis sets is truncated, there is no problem with overcompleteness that would arise from two complete (or nearly complete) sets. In practical calculations [ToE88a, ToE88b], basis sets have been confined to 20 or 36 substates, namely those of the $1s_{1/2}$, $2s_{1/2}$, $2p_{1/2}$, $2p_{3/2}$, $3s_{1/2}$, $3p_{1/2}$, and $3p_{3/2}$ shells. The use of exact hydrogenic eigenstates calculated from Eqs. (4.86) and (4.98) guarantees that the atomic basis functions attached to the *same* center are orthogonal in *both* the target frame and the projectile frame according to Eqs. (4.41) and (4.44). This would no longer be true for variational wave functions or basis states obtained by diagonalization in *one* of the two frames. The difficulties associated with pseudostate calculations are discussed in Sec. 6.5.3.

In Eq. (6.55) the Lorentz transformation of the time factor [Eq. (4.45)] included in the projectile states automatically implies the relativistic counterparts of the nonrelativistic translation factors, see Eq. (1.6).

*Coupled equations*

In order to determine the time-dependent expansion coefficients $a_k$ and $a_{k'}$ in Eq. (6.55), we have to insert $\Psi(\mathbf{r}_\mathrm{T}, t)$ into the Dirac equation (5.39). Observing that the atomic eigenfunctions satisfy

$$\left(H - i\hbar\frac{\partial}{\partial t}\right)\psi_k = -S^2\,\frac{Z_\mathrm{P}e^2}{r'_\mathrm{P}}\,\psi_k, \tag{6.56}$$

and using Eq. (5.42), we get

$$\left(H - i\hbar\frac{\partial}{\partial t}\right) S^{-1}\psi_{k'} = -\frac{Z_{\mathrm{T}}e^2}{r_{\mathrm{T}}} S^{-1}\psi_{k'}. \tag{6.57}$$

With these two relations, the time-dependent two-center Dirac equation can be rewritten as a set of coupled differential equations for the occupation amplitudes $a_k(t)$ and $a_{k'}(t)$. Combining the amplitudes into a column vector $\mathbf{a} = (a_k, \cdots; a_{k'}, \cdots)$, we may write the coupled equations in a compact matrix form as

$$i\hbar\mathbf{N}\,\dot{\mathbf{a}} = \mathbf{V}\,\mathbf{a} \tag{6.58}$$

or

$$\dot{\mathbf{a}} = -\frac{i}{\hbar}\mathbf{N}^{-1}\,\mathbf{V}\,\mathbf{a}. \tag{6.59}$$

The overlap matrix $\mathbf{N}$ and the interaction matrix $\mathbf{V}$ are each built from four submatrices corresponding to target-target, target-projectile, projectile-target, and projectile-projectile transitions.

The overlap matrices are

$$\begin{aligned}
N_{ik} &= \delta_{ik}\\
N_{ik'} &= \int \varphi_{\mathrm{i}}^{\dagger}(\mathbf{r}_{\mathrm{T}})S^{-1}\varphi_{k'}(\mathbf{r}'_{\mathrm{P}})\, e^{i\gamma E_{k'}(v/\hbar c^2)z_{\mathrm{T}}}\; d^3r_{\mathrm{T}}\; e^{-i(\gamma E_{k'}-E_{\mathrm{i}})t/\hbar}\\
N_{i'k} &= \int \varphi_{i'}^{\dagger}(\mathbf{r}'_{\mathrm{P}})S^{-1}\varphi_{k}(\mathbf{r}_{\mathrm{T}})\, e^{-i\gamma E_{i'}(v/\hbar c^2)z_{\mathrm{T}}}\; d^3r_{\mathrm{T}}\; e^{-i(E_k-\gamma E_{i'})t/\hbar}\\
N_{i'k'} &= \delta_{i'k'}.
\end{aligned} \tag{6.60}$$

The last equation, valid for exact atomic eigenstates only, is derived in Eq. (4.44). The single-center overlap matrix elements (target-target and projectile-projectile) are time independent while the two-center matrix elements (target-projectile and projectile-target) are time dependent owing to the relative motion of the collision partners.

In analogy to the overlap, we have four types of interaction matrix elements, namely

$$\begin{aligned}
V_{ik} &= \int \varphi_{\mathrm{i}}^{\dagger}(\mathbf{r}_{\mathrm{T}})\left(-S^2\frac{Z_{\mathrm{P}}e^2}{r'_{\mathrm{P}}}\right)\varphi_k(\mathbf{r}_{\mathrm{T}})\, d^3r_{\mathrm{T}}\;\; e^{-i(E_k-E_{\mathrm{i}})t/\hbar}\\
V_{ik'} &= \int \varphi_{\mathrm{i}}^{\dagger}(\mathbf{r}_{\mathrm{T}})\left(-S^{-1}\frac{Z_{\mathrm{T}}e^2}{r_{\mathrm{T}}}\right)\varphi_{k'}(\mathbf{r}'_{\mathrm{P}})\, e^{i\gamma E_{k'}(v/\hbar c^2)z_{\mathrm{T}}}\, d^3r_{\mathrm{T}}\\
&\qquad \times e^{-i(\gamma E_{k'}-E_{\mathrm{i}})t/\hbar}
\end{aligned}$$

$$\begin{aligned} V_{i'k} &= \int \varphi_{i'}^{\dagger}(\mathbf{r}_{\mathrm{P}}') \left( -S \frac{Z_{\mathrm{P}} e^2}{r_{\mathrm{P}}'} \right) \varphi_k(\mathbf{r}_{\mathrm{T}})\, e^{-i\gamma E_{i'} (v/\hbar c^2) z_{\mathrm{T}}}\, d^3 r_{\mathrm{T}} \\ &\qquad \times e^{-i(E_k - \gamma E_{i'})t/\hbar} \\ V_{i'k'} &= \int \varphi_{i'}^{\dagger}(\mathbf{r}_{\mathrm{P}}') \left( -S^{-2} \frac{Z_{\mathrm{T}} e^2}{r_{\mathrm{T}}} \right) \varphi_{k'}(\mathbf{r}_{\mathrm{P}}')\, e^{-i\gamma (E_{k'} - E_{i'})(v/\hbar c^2) z_{\mathrm{T}}}\, d^3 r_{\mathrm{T}} \\ &\qquad \times e^{-i\gamma(E_{k'} - E_{i'})t/\hbar}. \end{aligned} \tag{6.61}$$

One verifies that in the nonrelativistic limit, all matrix elements merge into the corresponding nonrelativistic expressions including translation factors.

As long as the expansion (6.55) is not truncated, the set of coupled equations (6.59) is equivalent to the original two-center Dirac equation (5.39). The choice and the truncation of the basis set constitutes the only approximation made in solving the problem. Of course the truncation implies that some of the symmetry properties of the original Dirac equation get lost in the coupled-channel equations. With an incomplete basis set, Eq. (6.59) is neither Lorentz invariant nor gauge invariant. An explicit numerical example for the loss of gauge invariance is given in Sec. 6.5.4.

*Initial conditions*

In order to actually solve the coupled equations (6.59), one has to specify the initial conditions. If we assume that the target electron is initially in the state $k_0$, we have

$$\begin{aligned} a_{k_0}(t \to -\infty) &= 1 \\ a_k(t \to -\infty) &= 0 \quad \text{for} \quad k \neq k_0 \\ a_{k'}(t \to -\infty) &= 0 \end{aligned} \tag{6.62}$$

Equation (6.60) shows that an expansion in terms of atomic eigenfunctions has the advantage that at asymptotic separations, i.e., for $t \to \pm\infty$, all basis states are mutually orthogonal. This means that the expansion coefficients $a_k, a_{k'}$ can be interpreted unambiguously as occupation amplitudes in this limit. In particular, there is no overlap with the negative-energy continuum which might lead to spurious results if there were a finite overlap. In the interaction region, where target and projectile states are nonorthogonal, and where each bound state of one center has some overlap with bound states and positive- as well as negative-energy continuum states of the other center, an interpretation in terms of probability amplitudes is not needed. It suffices that the overlap matrix (6.60) is rigorously taken into account.

*Cross sections*

The final probabilities are

$$P_k(b) = |a_k(t \to \infty)|^2 \quad \text{for excitation} \tag{6.63}$$

and

$$P_{k'}(b) = |a_{k'}(t \to \infty)|^2 \quad \text{for charge transfer.} \tag{6.64}$$

The integrated cross section for exciting state $k$ from the initial state $k_0$ according to Eq. (5.65) is given by

$$\sigma_k = 2\pi \int_0^\infty P_k(b) b \, db \tag{6.65}$$

while for transfer, $k$ is replaced with $k'$.

*Numerical procedure*

The main numerical task consists in evaluating the matrix elements defined in Eqs. (6.60) and (6.61). In contrast to the analogous nonrelativistic problem, there is no symmetry about the internuclear axis because of the Lorentz contraction in the beam direction. A fully numerical integration is required.

The details of the numerical procedure are discussed in [ToE88b]. The task is facilitated by a symmetry relation

$$N_{\alpha\beta}(t) = (-1)^{l_A^\alpha + l_A^\beta + \mu_j^\alpha + \mu_j^\beta + 1} \, N_{\alpha\beta}^*(-t) \tag{6.66}$$

for the overlap matrix and a corresponding relation for the potential matrix elements, where $\alpha, \beta$ denote any of the labels $i, k, i', k'$ and $l_A^{\alpha,\beta}$ are the orbital angular momenta for the large components of the states $\alpha, \beta$, while $\mu_j^{\alpha,\beta}$ are the projections of the total angular momenta $j = l_A \pm \frac{1}{2}$. Estimating the radius of the interaction region by $\bar{r} = 4a_0/(Z_\mathrm{T} + Z_\mathrm{P})$, the limits of time integration are extended to $|vt| \leq 250\ \bar{r}$, and, when needed for convergence, to $500\ \bar{r}$ or $1000\ \bar{r}$. Beyond $|vt| = 20\ \bar{r}$, the two-center matrix elements are negligibly small, so that only single-center matrix elements have to be calculated. Since the calculation of the matrix elements takes the overwhelming part of the computing time, it is convenient to solve the coupled time-dependent equations simultaneously with all the different initial conditions that might be of interest.

### 6.5.2 Application to $U^{92+} + U^{91+}$ collisions

For the collision of $U^{92+}$ ions with $U^{91+}$ targets at 0.5 and 1.0 GeV/u, experimental data are not available but, on the other hand, the collision is well suited to serve as a model case for which effects of relativistic electron and relativistic projectile motion can be studied. According to Fig. 5.1, the matching energy for $1s_{1/2}$ electrons is 240 MeV/u, so that at least for 500 MeV/u we still have rather "slow" relativistic collisions for which the two-center coupled-channel method is most appropriate.

Figure 6.12 shows some results for the time evolution of the occupation probabilities for target states (excitation) and projectile states (charge transfer) for the impact energy of 500 MeV/u using 36 atomic basis states up to the $3p_{3/2}$ shell of target and projectile and assuming an impact parameter of 0.01 a.u. This distance is approximately equal to the K-shell radius $a_K$ and roughly corresponds to the region of maximum contribution to the excitation cross section, see Fig. 6.13. The interval $|vt| \leq 40\ a_K$, displayed in Fig. 6.12, represents a narrow window out of the total time range $|vt| \leq (500 - 2000)\, a_K$ actually treated in the calculation.

In conjunction with the time developments shown, it should be noted that the interpretation of the expansion coefficients as probability amplitudes near $t = 0$ is only approximately valid because of the nonorthogonality in the reaction zone of basis states belonging to different centers. Therefore a transient "probability" exceeding unity occurs in Fig. 6.12(a).

All figures reveal certain qualitative features: (a) The importance of multistep processes is shown by the rapid fluctuations of some of the occupation probabilities. (b) Electron transfer, both during the collision as well as in the exit channel, is comparable to excitation and cannot be neglected. (c) With increasing shell size, the interaction region extends much farther out, see Fig. 6.12(b). (d) In this treatment, certain target states are affected by the projectile long before and long after it reaches its distance of closest approach at $t = 0$. In Sec. 6.5.4, we come back to these long-range couplings, also observed for first-order perturbation theory in Fig. 6.1.

Figure 6.13 shows weighted excitation and transfer probabilities $P(b)\,b$ for the leading transitions as a function of the impact parameter $b$. As already apparent from Fig. 6.12, the probabilities for excitation reach out to much larger impact parameters than for transfer. A particular role is played by the spin-flip transition $1s_{1/2}(\frac{1}{2}) \rightarrow 1s_{1/2}(-\frac{1}{2})$. Being degenerate within the coupled-channel treatment, the two states are strongly coupled even at large impact parameters so that the total cross section diverges logarithmically. This divergence is spurious and can be remedied by including higher-order QED corrections or hyperfine interactions removing the degeneracy.

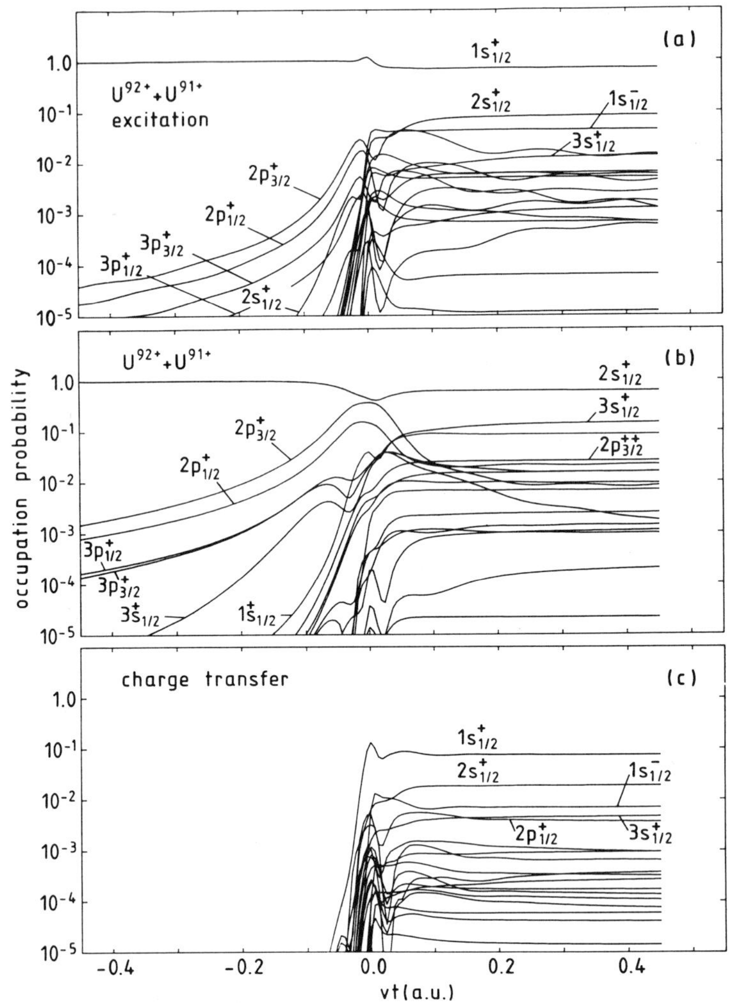


**Figure 6.12.** Time evolution of the occupation probabilities of target states in excitation and of projectile states in charge transfer in $U^{92+} + U^{91+}$ collisions at 500 MeV/u laboratory energy and impact parameter $b = 0.01$ a.u. $\approx a_K$. The abscissa plotted is the projection of the projectile-target separation on the beam direction. The projection of the angular momentum is indicated by $+, -$, and $++$ for $\mu_j = \frac{1}{2}, -\frac{1}{2}$, and $\frac{3}{2}$, respectively. (a) Excitation from an initial $1s_{1/2}$ state; (b) excitation from an initial $2s_{1/2}$ state; (c) charge transfer from an initial $1s_{1/2}$ state. Adapted from [ToE88b].

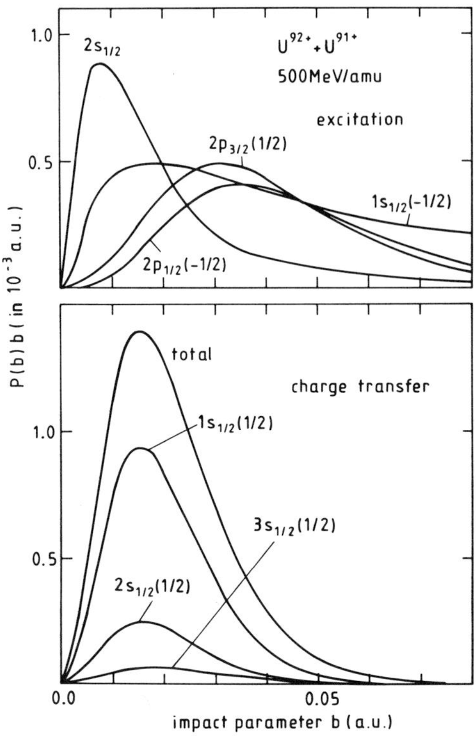


**Figure 6.13.** Weighted excitation and capture probabilities $P(b)\,b$ for $U^{92+}$ + $U^{91+}(1s_{1/2})$ collisions at 500/u MeV laboratory energy as a function of the impact parameter $b$. Only the leading probabilities into specific final states $nl_j(\mu_j)$ are plotted. From [ToE88b].

For detailed state-to-state cross sections within a 36-state fully relativistic coupled-channel treatment of 500 MeV/u $U^{92+}$ + $U^{91+}$ collisions see [ToE88b, Eic90].

### 6.5.3 Pseudostates for the continuum

The coupled-channel treatment discussed in the preceding subsections uses a basis of bound states in the target and the projectile atom and hence is confined to excitation and charge transfer into bound states. However, in order to describe ionization processes on the same footing as excitation and capture, it is mandatory to include target and projectile continuum states in the basis set. It is known from nonrelativistic collisions [FrL91] that by augmenting the atomic orbital expansion by appropriate "pseudostates" attached to target and projectile, one may simulate intermediate and final continuum states and at the same time improve the representation of molecular states at small internuclear separations. Pseudostates are arbitrary basis functions chosen so as to improve the flexibility of the basis set in approximating the true two-center wave function throughout the collision. Not being eigenstates of the unperturbed atomic Hamiltonian, pseudostates, if properly chosen, have an overlap (a) with bound highly excited atomic orbitals not explicitly included in the basis set, (b) with atomic or molecular continuum states, and (c) with united-atom wave functions (for $R = 0$ and $Z = Z_{\rm T} + Z_{\rm P}$) needed to improve the representation of molecular orbitals. This method is a powerful tool for simultaneously describing excitation, ionization, and charge transfer in nonrelativistic collisions [FrL91]. For relativistic collisions, see [ToE89].

The use of approximate wave functions such as pseudostates to describe relativistic collisions entails two problems that are absent in the nonrelativistic situation. One of them is the occurrence of negative-energy states arising from the Dirac theory for the relativistic electron motion. The other one is the loss of the orthogonality property of approximate eigenfunctions under Lorentz transformations.

*Negative-energy components*

We first address the problem of negative-energy components in pseudostates. The Dirac equation, while formally appearing as a single-particle equation, is fully interpretable only if considered as an equation for the field operator, see Secs. 4.2.2 and 5.1. As a consequence, many-electron–many-positron systems are equally represented by the same equation. For most applications in atomic and molecular physics, however, we wish to confine ourselves to a system with a *fixed number* of electrons. Therefore, one wants to avoid "negative-energy solutions" of the Dirac equation, that is solutions with eigenvalues $E < -m_{\rm e}c^2$ (defining $E = m_{\rm e}c^2$ for a free electron at rest). Indeed, for variational problems, negative-energy solutions may always occur since the Dirac Hamiltonian is not bounded from below. In spite of these problems, techniques have been developed to avoid negative-energy states and to perform reliable relativistic many-electron calculations for atoms and molecules. The most important feature common to many of the recipes appears to be that the ratio between upper and lower Dirac components has to be fixed to a reasonable value and is not left open to variation. It is thus a pragmatic procedure to augment the set of exact Dirac eigenstates as basis functions in a linear variational procedure by hydrogenic wave functions in Eq. (6.55) generated by effective charges

$Z^\star$, see e.g. [ToE89, Eic90].

*Loss of orthogonality*

When pseudostates are included in the basis set, a difficulty arises in the calculation of overlap and potential matrix elements that is peculiar to relativistic projectile motion. As discussed in Sec. 4.1.4, the operator $S$ represents a Lorentz transformation only for rigorous eigenstates and represents an *approximate* Lorentz transformation for *approximate* eigenstates such as those obtained by diagonalizing the single-center Hamiltonian appearing in Eqs. (5.53) and (5.54) in a basis containing pseudostates. For example, approximate projectile eigenstates which, by construction, are orthogonal in the projectile system are no longer orthogonal in the target frame, in which the collision is described. This means that the last of Eqs. (6.60) is no longer valid, and instead, we have in the target system

$$N_{i'k'} \begin{cases} = \delta_{i'k'} & \text{if } i', k' \text{ are both real states} \\ \neq \delta_{i'k'} & \text{if at least one of } i', k' \text{ is a pseudostate} \end{cases} \tag{6.67}$$

A similar relation holds for the Hamiltonian matrices. For example, if the projectile Hamiltonian has been diagonalized in the projectile frame, only the submatrix spanned by real states remains diagonal under Lorentz transformations.

The loss of the diagonal property of overlap and Hamiltonian matrices under Lorentz transformations poses serious problems to calculation and interpretation within a framework of coupled channels involving pseudostates. Then, no matter whether the projectile Hamiltonian is diagonalized in the projectile or in the target frame, an observer in the target frame will always find that for $t \to \infty$ the coupling matrix $(H'_{\text{P}} - i\hbar \frac{\partial}{\partial t'})$ in the projectile frame is nondiagonal and hence the occupation amplitudes keep oscillating among the various projectile basis states. Such a behavior is physically meaningless. Several approximate ways to treat these problems are discussed in [ToE89]

The loss of orthogonality is generally unavoidable if basis states are used that are not eigenstates of target or projectile Hamiltonian. Discretized wave packets for the continuum, see Sec. 6.6.1, are expected to give rise to similar problems.

### 6.5.4 Long-range couplings and Coulomb boundary conditions

In Figs. 6.12(a) and (b) we present calculated results for $U^{92+} + U^{91+}$ collisions which exhibit long-range couplings among the excited target states.

In this subsection, we analyze this effect and show how to remove it. The anomalous long-range couplings are of relativistic origin and are absent in nonrelativistic coupled-channel calculations.

According to Eqs. (6.56) and (5.97), the interaction leading to target excitation is

$$S^2 \frac{Z_{\rm P} e^2}{r'_{\rm P}} = \gamma Z_{\rm P} e^2 (1 - \beta \alpha_z) \left[ \frac{1}{R'_0} + \frac{b x_{\rm T} + \gamma^2 v t z_{\rm T}}{(R'_0)^3} + \cdots \right] \tag{6.68}$$

where $R'_0 = (b^2 + \gamma^2 v^2 t^2)^{1/2}$ and higher multipole terms have been omitted. The main term in (6.68) giving rise to long-range couplings between initial and final states is the term proportional to $\beta \alpha_z / R'_0$, which is of magnetic origin; for a more detailed analysis, see [ToE88a, ToE88b].

*Basis states satisfying Coulomb boundary conditions*

In coupled-channel calculations, the choice of the basis states is a matter of convenience, both for the desired interpretation of the results and for the ease of calculations. In Eq. (6.55), we have chosen the conventional view that the basis states are stationary atomic orbitals. In light of the discussion above, it appears more appropriate to use a set of basis states that satisfy Coulomb boundary conditions (5.61) and (5.62). We therefore replace Eq. (6.55) with the expansion [ToE90b]

$$\begin{aligned} \Psi(\mathbf{r}_{\rm T}, t) &= \sum_k a_k(t) e^{-i\nu_{\rm P} \ln(R' - vt')} \psi_k(\mathbf{r}_{\rm T}, t) \\ &+ \sum_{k'} a_{k'}(t) e^{i\nu_{\rm T} \ln(R + vt)} \, S^{-1} \psi_{k'}(\mathbf{r}'_{\rm P}, t'). \end{aligned} \tag{6.69}$$

The first term represents an expansion in terms of target states asymptotically phase-distorted by the projectile, while the second sum is an expansion in terms of projectile states asymptotically phase-distorted by the target.

When writing down the equations corresponding to (6.56) and (6.57), we find that on the right-hand side, the time differentiation of the logarithmic phase factors yields additional terms so that we get the replacement

$$\begin{aligned} \frac{Z_{\rm P} e^2}{r'_{\rm P}} &\rightarrow \frac{Z_{\rm P} e^2}{r'_{\rm P}} - \frac{Z_{\rm P} e^2}{R'} \\ \frac{Z_{\rm T} e^2}{r_{\rm T}} &\rightarrow \frac{Z_{\rm T} e^2}{r_{\rm T}} - \frac{Z_{\rm T} e^2}{R}. \end{aligned} \tag{6.70}$$

The phase factors themselves drop out in the one-center integrals since they are state independent, but they remain in the two-center integrals. With these substitutions it is obvious how to modify the overlap matrix elements (6.60) and the potential matrix elements (6.61). We emphasize that the multiplication of the basis states with a phase factor and the simultaneous replacement (6.70) of the interaction simply constitute a gauge transformation.

It becomes apparent that with the basis expansion of (6.70), the long range of the Coulomb interaction is explicitly taken into account and hence the long-range couplings discussed in Sec. 6.5.2 disappear. Indeed, if the replacement (6.70) is inserted into the left-hand side of Eq. (6.68) the term $1/R_0'$ in the square brackets responsible for these couplings is cancelled.

*Elimination of long-range couplings*

The elimination of the long-range couplings by a suitable gauge transformation is illustrated in Fig. 6.14. Here, the time evolution during a $U^{92+}$ + $U^{91+}$ collision at 1 GeV/u is shown for the occupation probabilities of the target states. The impact parameter is 0.01 a.u. or about one uranium K-shell radius. Unperturbed basis states (6.55) are used in the upper part of the figure, while the lower part is obtained for boundary-corrected basis states (6.69). Here, the long-range couplings visible for unperturbed basis states have disappeared. The $2p_{3/2}(\frac{1}{2})$, $2p_{1/2}(\frac{1}{2})$, $3p_{3/2}(\frac{1}{2})$, and $3p_{1/2}(\frac{1}{2})$ states that are excited for unperturbed states already long before the distance of closest approach is reached begin to be populated for boundary-corrected basis states only at about $vt = -10a_K$. Similarly, after the collision, the boundary-corrected basis states attain their asymptotic occupation probabilities at much smaller values of $vt$ than the unperturbed basis states. Most importantly, the asymptotic values themselves are changed, in sharp contrast to the Born approximation discussed in Sec. 6.1.2. A similar behavior has been confirmed for single-center calculations [ToE90b]. Generally speaking, Fig. 6.14 explicitly illustrates that for an incomplete basis set the coupled-channel equations (6.59) are not invariant with respect to gauge transformations.

One can conclude that for relativistic single-center and two-center coupled-channel calculations with unperturbed basis states, the coupling among states is of long range and acts long before and long after the encounter takes place, while for basis states satisfying Coulomb boundary conditions, the couplings are localized around the distance of closest approach. However, in contrast to the Born approximation, Sec. 6.1.2, where there is no backcoupling, the multiple couplings occurring in a coupled-channel calculation eventually lead to transition probabilities and cross sections that are distinctly different for unperturbed and boundary-corrected basis states. This shows that it is mandatory for a description of direct reactions to take

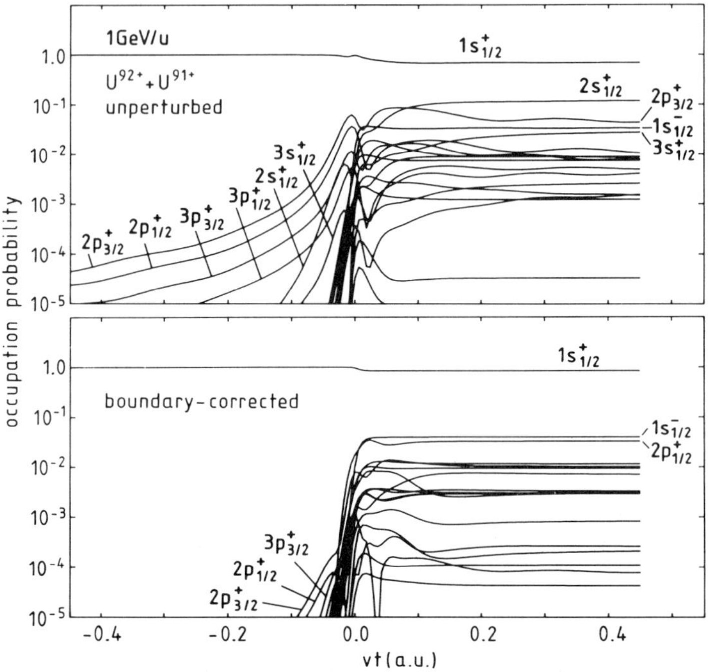


**Figure 6.14.** Time evolution of the occupation probabilities of target states in $U^{92+}$ + $U^{91+}(1s_{1/2})$ collisions at 1 GeV/u laboratory energy. Calculations are performed with 18 hydrogenic basis states at each of target and projectile and for an impact parameter $b = 0.01a.u.$ The upper set of curves represents results obtained from unperturbed basis states (6.55), while the lower set of curves corresponds to boundary-corrected basis states, Eq. (6.69). For the notation, see Fig. 6.1. From [ToE90b].

care of the long-range Coulomb interaction from the beginning. It is to be expected that this also improves the convergence behavior with respect to the number of basis states.

## 6.6 Single-center coupled-channel methods

While two-center coupled-channel calculations have the merit of allowing for charge exchange, one obtains great simplifications in a single-center coupled-channel calculation, where one regards the projectile merely as the source of a time-dependent electromagnetic field. In this way, one discards charge exchange not only as a final but also as an intermediate channel and moreover does not permit the formation of transient quasimolecular states. On the other hand, by renouncing a second set of basis states and the associated complications, one may handle larger sets centered around the target and carry the calculations to higher energies.

By dropping the projectile-centered expansion in Eq. (6.55), we may write

$$\Psi(\mathbf{r}_\mathrm{T}, t) = \sum_k a_k(t)\psi_k(\mathbf{r}_\mathrm{T}, t), \tag{6.71}$$

where the $\psi_k(\mathbf{r}_\mathrm{T}, t) = \varphi_k(\mathbf{r}_\mathrm{T}) \exp(-iE_k t/\hbar)$ are mutually orthogonal and normalized bound-state and continuum ($E > m_\mathrm{e}c^2$) wave functions. This means that in each of Eqs. (6.60) and (6.61) only the first lines apply, in particular one gets rid of the unpleasant time-dependent overlap terms. Dropping the primed terms, the initial conditions (6.62), the expressions for the probability (6.63), and the cross section (6.65) are valid in exactly the same way.

## 6.6.1 Wave packets for the continuum

The main problem of all coupled-channel calculations consists in treating the continuum. In Sec. 6.5.3, we discuss pseudostates. Another method which allows one to understand better which part of the continuum is included is provided by the use of discretized continuum wave functions constructed as superpositions of continuous eigenstates.

*Weyl wave packets*

One first segments a certain energy range of the continuum into finite intervals centered around $E_k$ and with the width $\Delta E_k$. By integrating the exact continuum wave function $\varphi_E \exp(-iEt/\hbar)$ over the energy interval $\Delta E_k$, one obtains Weyl wave packets in the form

$$\psi_{E_k}(\mathbf{r}_\mathrm{T}, t) = \frac{1}{\sqrt{\Delta E_k}} \int_{E_k - \Delta E_k/2}^{E_k + \Delta E_k/2} \varphi_E(\mathbf{r}_\mathrm{T}) e^{-i(E-E_k)t/\hbar}\, dE. \tag{6.72}$$

These wave packets travel radially into the reaction region for $t < 0$ and outwards for $t > 0$ with a peak momentum $p_k = (E_k^2 - m_\mathrm{e}^2 c^4)^{\frac{1}{2}}/c$ [HeG81, MeR85]. Since the continuum wave functions are normalized on the energy scale, see Eq. (4.117), and the energy intervals $\Delta E_k$ belonging to different discretized states do not overlap, the Weyl wave packets are orthonormal, that is

$$\int \psi_{E_k}^\dagger(\mathbf{r}_\mathrm{T}, t)\psi_{E_l}(\mathbf{r}_\mathrm{T}, t)\, d^3 r_\mathrm{T} = \delta_{k,l}. \tag{6.73}$$

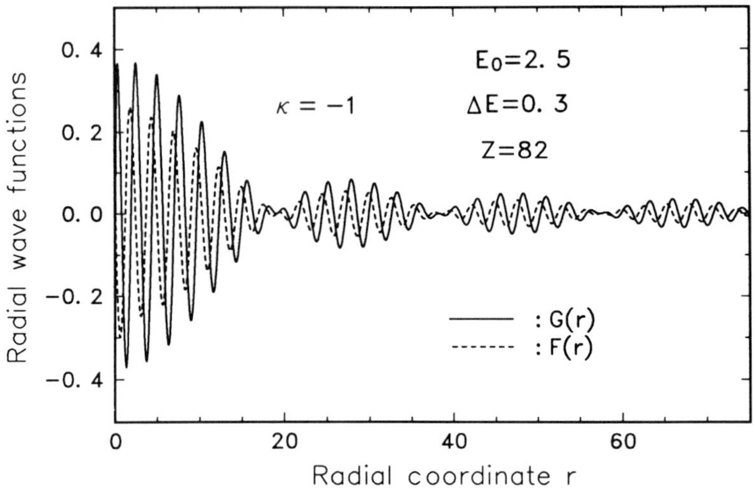


**Figure 6.15.** The radial components $G(r) = r\,g(r)$ and $F(r) = r\,f(r)$ of relativistic wave packets in relativistic units ($\hbar = c = m_e = 1$) versus the radial coordinate $r$. The s-state ($\kappa = -1$) with the total energy $E_0 = 2.5$ for $Z = 82$ is considered. The integration over continuum states is performed assuming a width of $\Delta E = 0.3$.

On the other hand, they are no exact solutions of the target Dirac equation.

*Stationary wave packets*

If one superimposes not time-dependent but stationary continuum solutions of the Dirac equation, one obtains stationary wave packets, see e.g. [BaS85], $\psi_{E_k}(\mathbf{r}_\mathrm{T}, t) = \varphi_{E_k} \exp(-iE_k t/\hbar)$, where

$$\varphi_{E_k}(\mathbf{r}_\mathrm{T}) = \frac{1}{\sqrt{\Delta E_k}} \int_{E_k-\Delta E_k/2}^{E_k+\Delta E_k/2} \varphi_E(\mathbf{r}_\mathrm{T})\, dE. \tag{6.74}$$

Owing to the superposition of adjacent continuum functions, the fall-off of the radial wave functions $g(r)$ and $f(r)$ is much faster than the decrease $1/r$ of the functions $g(r)$ and $f(r)$ themselves, see Fig. 6.15. In the evaluation of matrix elements containing these wave packets, it is sometimes necessary to take slowly varying quantities outside of the energy integrals.

*Gaussian wave packets*

As is known from quantum chemistry, it is often advantageous to expand atomic or molecular orbitals in terms of Gaussian-type orbitals (GTO)

$$\varphi_{nlm}(\mathbf{r}) = \sum_{\nu} c_{\nu}^{(nl)} e^{-\alpha_\nu r^2} r^l Y_{lm}(\hat{\mathbf{r}}), \tag{6.75}$$

where the solid harmonics $r^l Y_{lm}(\hat{\mathbf{r}})$ are written in Cartesian coordinates, and the Gaussian width parameters are chosen in a suitable (e.g., geometric progression). It is the merit of this expansion that the single-center and multi-center matrix elements entering in the Hamiltonian matrix can be evaluated analytically. Collision problems, however, demand the accurate numerical calculation of a complex auxiliary function defined as a one-dimensional integral, which arises from the complex oscillatory factors in Eqs. (6.60) and (6.61). It has been shown [ToE91, ToE92] that the method of GTO wave packets is very powerful in nonrelativistic coupled-channel calculations, even for energetic collisions when high-lying continuum states are needed. The Cartesian decomposition, which comes naturally with the GTO method, should be particularly appropriate for relativistic collisions, where, because of the Lorentz contraction, the $z$-coordinate has to be treated separately. Actual calculations have not yet been performed.

### 6.6.2 Numerical results

In order to solve the coupled equations (6.59), where $\mathbf{N}$ is the unit matrix for a single center with orthogonal basis states, one evaluates the coupling matrix elements (6.61) (first line only). Instead of a direct numerical quadrature, one may use the multipole decomposition discussed in Sec. 6.3.

Adopting the multipole expansion in momentum space, Sec. 6.3.2, and Weyl wave packets (6.72), Becker [Bec86] considered collisions of $Z_\mathrm{P}$ + U at 1 GeV/u and $b = 0$. The calculated differential cross sections for K-shell ionization are very close to the corresponding cross sections calculated in first-order perturbation theory (see Sec. 6.3.3). However, for higher projectile charges (see Fig. 6.16), the calculations show that the cross sections do not scale anymore with $Z_\mathrm{P}^2$ as they do in perturbation theory and hence do not coincide in the diagram. All these calculations require considerable computer time and cannot be easily extended to arbitrary impact parameters in order to provide complete cross sections.

More extended calculations of excitation and ionization cross sections in collisions of $Z_\mathrm{P}$ + $\mathrm{U}^{90+}$ at 430 MeV/u and of $Z_\mathrm{P}$ + $\mathrm{Xe}^{52+}$ at 82 MeV/u are performed by Momberger et al. [MoG90] within a one-center coupled-channel formalism. While there is good qualitative agreement with the experimental data, discrepancies at small values of $Z_\mathrm{P}/Z_\mathrm{T}$ still remain, see Fig. 7.6 for projectile ionization. Differential ionization cross sections are calculated in [MoG89].

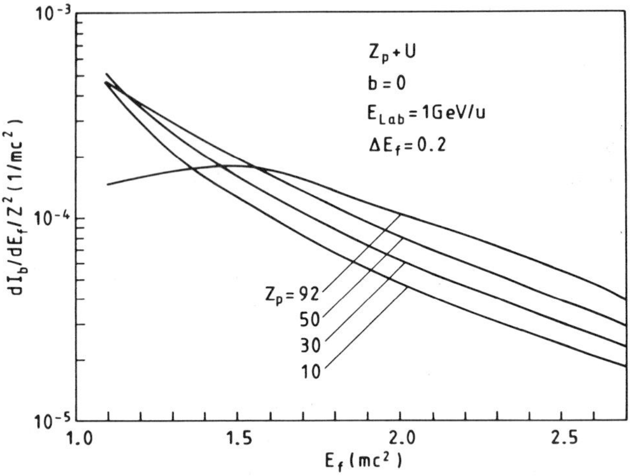


**Figure 6.16.** Differential K-shell ionization probabilities in $Z_P$ + U collisions, divided by $Z_P^2$, at 1.0 GeV/u and $b = 0$ versus the energy of the ionized electron. The curves are obtained from coupled-channel calculations by Becker et al. [BeG87b].

*Nonperturbative behavior at very small impact parameters*

Figure 6.16 and the earlier discussions reveal that single-center and two-center coupled-channel calculations yield significant deviations from the results of perturbation theory, Sec. 6.1. These deviations are expected to be largest if the interaction is extremely strong, that is, for high-$Z$ collision partners at very small impact parameters close to the limit of grazing collisions between the nuclei.

In order to investigate this interesting behavior, Rumrich et al. [RuM91, RuS93] have performed single-center coupled-channel calculations using stationary wave packets (6.74) for describing electronic continuum states (and also positron states, see Sec. 10.2.3) and a multipole decomposition of the interaction in coordinate space, see Sec. 6.3.1. Since they start from the expansion (6.71) without boundary corrections (Sec. 6.5.4), they introduce a damping procedure to avoid the time oscillations of the occupation amplitudes at the internuclear separation where the initial conditions are specified.

In part of their calculations, they include negative-energy basis states. The basis includes continuum states between 1.1 and 3.5 $m_e c^2$ and angular momentum quantum numbers $\kappa$ between −2 and +2. It is found that for the system $Z_P + Pb^{81+}$ at the bombarding energy of 1.2 GeV/u and impact

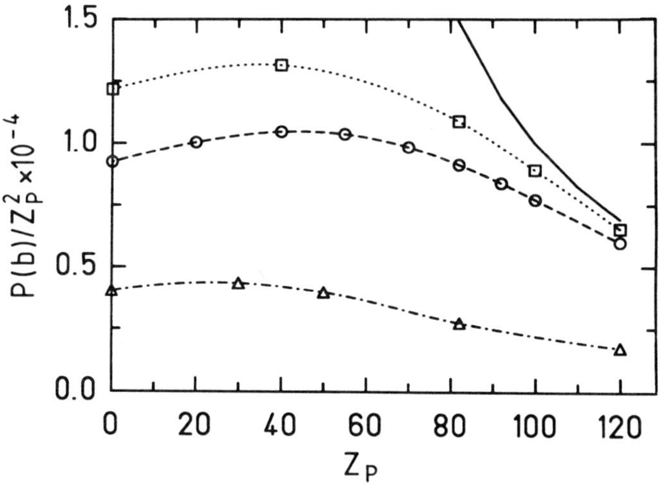


**Figure 6.17.** $1s_{1/2}$ and $2s_{1/2}$ ionization rates versus $Z_P$ for the system $Z_P$ + $Pb^{81+}$ at an impact parameter of $b = 20$ fm. The notation is: □, $1s_{1/2}$ ionization for 0.5 GeV/u ; ○, $1s_{1/2}$ ionization for 1.2 GeV/u; and △, $2s_{1/2}$ ionization for 1.2 GeV/u. Full line: unitarity limit. From [RuS93].

parameter $b = 20$ fm, the ionization rate is reduced by up to 30% if negative-energy states are included.

Figure 6.17 shows $1s_{1/2}$ and $2s_{1/2}$ ionization rates divided by $Z_P^2$ as a function of the projectile charge $Z_P$ for the impact parameter $b = 20$ fm. If perturbation theory were valid one would have horizontal straight lines. The calculations show that the $2s_{1/2}$ ionization rate is lower than the $1s_{1/2}$ ionization rate by a factor of 3.

In accordance with the previous remarks regarding the influence of boundary conditions, it has been pointed out by Baltz et al. [BaR93b] that the nonperturbative behavior seen in coupled-channel calculations is strongly gauge dependent, in particular for small basis sizes. Insofar, such results have to be viewed with some caution.

In Chap. 10, in the context of electron-positron pair production, we return to the problem of discrepancies between calculated perturbative and nonperturbative cross sections. In this case, the estimated effects are by far larger.

# Chapter 7

# Ionization – Many Electrons

## 7.1 Introduction

The previous chapter discussed ionization or excitation of a one-electron ion by a fully ionized projectile. Here, we concentrate on "many-electron" situations, where either the target carries more than one electron (structured target) or there are electrons on the projectile (structured projectile). The need for treating these more complicated situations becomes clear if we consider briefly the experimental situation.

Inner-shell ionization cross sections are typically determined from target x-ray yields (Chap. 13). Unfortunately, target vacancies (and x-rays) are also produced if a target inner-shell electron is captured by a stripped projectile, a process that increases strongly with $Z_P$ (Chap. 8). The target ionization and the capture processes cannot be distinguished experimentally, except by a coincidence measurement with the projectile charge state: target ionization generally leaves the projectile charge state unchanged, whereas capture decreases the charge state by unity (unless simultaneous projectile electron loss occurs). Such experiments require targets thin enough to assure single-collision conditions.

A clean way to test the theory is to study the ionization of a relativistic one-electron *projectile* by a thin solid or gaseous target. Capture of the projectile electron by a neutral target then is forbidden by the Pauli exclusion principle (negative-ion formation is completely negligible at relativistic velocities). A complication arises, though, because the target electrons can also affect the electron loss of the projectile. (Henceforth, ionization of a projectile electron is called *electron loss* and the word *ionization* is reserved for target ionization.) But, already the simplest case of electron loss by a relativistically moving one-electron projectile interacting with a one-electron target cannot be treated rigorously for the reasons mentioned in Sec. 5.6.3. Only a nonrelativistic treatment of this simplest "many-electron" problem is possible, within the framework of the plane-wave Born approximation (PWBA) or the SCA.

Despite this experimental preference for electron loss studies, below we first discuss target ionization by a structured projectile in order to retain as much as possible the notation and the point of view of the preceding chapters. The influence of capture is ignored here, but is treated in Chaps. 8 and 9. Only toward the end of the chapter, when comparison with experiment is made, the notation is changed to that appropriate for electron loss.

For general orientation, it is useful to start with the simplest many-electron model, the free-electron model [Boh48]: one assumes that each projectile electron acts as if it were free, but moves with the projectile velocity. If there are $N_\mathrm{P}$ electrons on the projectile, the total target ionization cross section is given by

$$\sigma_\mathrm{T}^\mathrm{i} = Z_\mathrm{P}^2 \sigma_\mathrm{T}^\mathrm{i}(\mathrm{p}) + N_\mathrm{P} \sigma_\mathrm{T}^\mathrm{i}(\mathrm{e}), \tag{7.1}$$

where $Z_\mathrm{P}$ is the projectile atomic number and $\sigma_\mathrm{T}^\mathrm{i}(\mathrm{p})$ and $\sigma_\mathrm{T}^\mathrm{i}(\mathrm{e})$ are the target ionization cross sections induced by a proton and an electron, respectively, both moving with the projectile velocity. Both cross section could be obtained from the theoretical expressions given in Chap. 6. In Eq. (7.1), the first term on the right represents the contribution of the projectile nucleus. Its form follows from any of the perturbative theories presented in Chap. 6. The second term is the contribution of the projectile electrons. The largest relative electron contribution occurs for $\mathrm{H}^0$ projectiles ($N_\mathrm{P} = Z_\mathrm{P} = 1$); it then decreases rapidly with increasing $Z_\mathrm{P}$, so that for $Z_\mathrm{P} \geq 20$ the effect of the electrons can be neglected even if $N_\mathrm{P} \simeq Z_\mathrm{P}$.

In Eq. (7.1), the free-electron ionization cross section $\sigma_\mathrm{T}^\mathrm{i}(\mathrm{e})$ is zero below a certain projectile (threshold) velocity, because a free electron needs to have a kinetic energy high enough to ionize the target, say its K-shell. The corresponding threshold projectile-electron kinetic energy is given by Eq. (5.63) and the corresponding projectile energy can be looked up in Fig. 5.1. The Lorentz $\gamma$ factor at threshold is given by

$$\gamma_\mathrm{thr} = 1 + I_\mathrm{K}^\mathrm{T}/m_e c^2 \,, \tag{7.2}$$

where $I_\mathrm{K}^\mathrm{T}$ is the target K-electron binding energy. If the projectile velocity lies in the nonrelativistic regime, one finds for the threshold velocity the simple result

$$v_{\rm thr} = v_{\rm K}^{\rm T} = Z_{\rm T}\,\alpha c, \tag{7.3}$$

where $v_{\rm K}^{\rm T}$ is the target K-electron Bohr velocity. For a U target, the threshold kinetic energy given by Eq. (7.2) is 240 MeV/u, i.e., below such a projectile energy, a "free" projectile electron could not contribute to the uranium K-shell ionization cross section.

In reality, the projectile electrons are bound. As a result, the threshold in the energy dependence of the projectile-electron-induced ionization cross section is spread out by the momentum distribution of the projectile electrons. One finds, though, that the cross section decreases rapidly below $v_{\rm thr}$. The purpose of the present chapter is to work out this problem in some detail.

The chapter is organized as follows. The nonrelativistic theory for a projectile with one electron is given in Sec. 7.2. The influence of additional *target* electrons can be taken into account through internal and external screening parameters. Additional *projectile* electrons typically are treated as if they act independently of each other (Sec. 7.3). By comparison of the quasirelativistic theory of Sec. 6.2.2 and the nonrelativistic theory for bare projectiles, the nonrelativistic theory for a one-electron projectile suggests how the quasirelativistic theory might be modified if the projectile carries one or more electrons. There is no *ab initio* theoretical treatment for this case. In Sec. 7.4, the notation is switched to that appropriate for electron loss and comparisons with experiment are made. Multiple target ionization and multiple projectile stripping are discussed in Sec. 7.5.

## 7.2 Nonrelativistic one-electron projectile

To progress toward the many-electron case, we sketch the nonrelativistic theory of ionization of a one-electron target by a one-electron projectile. For continuity, we follow the impact parameter method used in Chap. 6, although the original PWBA treatment by Bates and Griffing is easy to follow [BaG53, BaG54, BaG55]. The result is similar to Eq. (6.26) without the transverse and spin-flip parts, but with the projectile squared charge $Z_{\rm P}^2$ replaced by an effective squared charge $S_{\rm P}(Q)$ in the integral over $dQ$.

Figure 7.1 gives the coordinates applying to the present collision system. It is similar to Fig. 3.2, except that, besides the omission of relativistic effects, an electron has been added which originally is associated with the projectile. Its coordinate with respect to the projectile nucleus is denoted by $\boldsymbol{\xi}_{\rm P}$. For the present collision system, the (nonrelativistic) perturbing

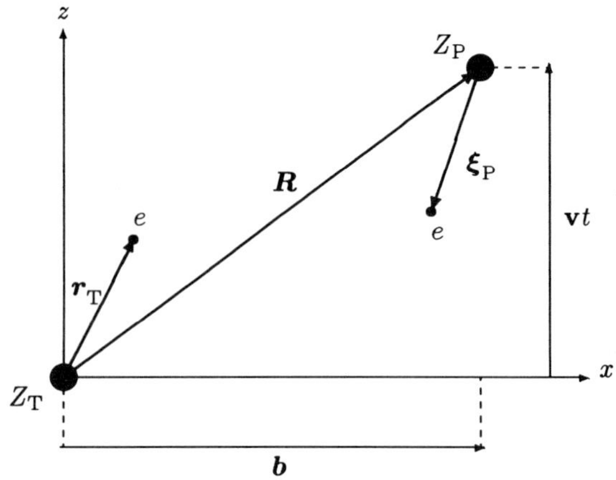


**Figure 7.1.** Nonrelativistic coordinate system for the two-electron case. The origin of the coordinate system is chosen to be at the target nucleus ($Z_T$). The projectile nucleus ($Z_P$) moves with constant velocity $v$ parallel to the $z$-axis, at an impact parameter $b$. One electron has the coordinate $\mathbf{r}_T$ with respect to $Z_T$, the other electron has the coordinate $\xi_P$ with respect to $Z_P$.

potential in Eq. (6.1)

$$-\frac{Z_P e^2}{r_P} = -\frac{Z_P e^2}{|\mathbf{R} - \mathbf{r}_T|}$$

is replaced by

$$V(\mathbf{R}, \mathbf{r}_T, \boldsymbol{\xi}_P) = -\frac{Z_P e^2}{|\mathbf{R} - \mathbf{r}_T|} - \frac{Z_T e^2}{|\mathbf{R} + \boldsymbol{\xi}_P|} + \frac{e^2}{|\mathbf{R} - \mathbf{r}_T + \boldsymbol{\xi}_P|} \tag{7.4}$$

and the wave functions $\varphi_i$ and $\varphi_f$ are now functions of $\mathbf{r}_T$ and $\boldsymbol{\xi}_P$. Correspondingly, the integral over $d^3 r_T$ has to be replaced by an integral over $d^3 r_T\, d^3 \xi_P$. If the collision is fast with respect to the Bohr velocities of the projectile and target electrons, electron exchange can be neglected and the initial and final wave functions can be written as products of target and projectile wave functions $\varphi_i(\mathbf{r}_T)\,\varphi_0(\boldsymbol{\xi}_P)$ and $\varphi_f(\mathbf{r}_T)\,\varphi_n(\boldsymbol{\xi}_P)$, respectively. Here, the quantum numbers of the initial and final target states are denoted by i and f, respectively, and those of the projectile states by 0 and $n$.

In a process describing target excitation or ionization, one always has f$\neq$i, but there is no *a priori* requirement that the projectile state has to change. In fact, one finds that the cross sections for $n = 0$ and for $n \neq 0$ can be equally important. In most experimental situations, the final state of the projectile electron is not observed, so that a sum over cross sections with

$n \neq 0$ has to be executed. It is convenient to treat the $n = 0$ and $\sum_{n\neq 0}$ cross sections separately. For this reason, they have been given different names, such as "(target) inelastic – (projectile) elastic" and "(target) inelastic – (projectile) inelastic", respectively, or "screening" and "antiscreening." Here, they are called "single inelastic (sinel)" and "double inelastic (dinel)."[1]

One further alteration needs to be made in Eq. (6.5). The energy change $E_\mathrm{f} - E_\mathrm{i}$ which occurs in the minimum momentum transfer (5.105) has to be replaced by the total energy change of the system:

$$\Delta E = E_\mathrm{f}^\mathrm{T} - E_\mathrm{i}^\mathrm{T} + E_n^\mathrm{P} - E_0^\mathrm{P}, \tag{7.5}$$

where the superscripts refer to the target and the projectile, respectively. Correspondingly, the minimum momentum transfer is replaced by

$$\hbar q_0(n) = \Delta E/v, \tag{7.6}$$

where, for later use, the dependence of $q_0$ on $n$ is noted.

The derivation of the cross section now follows the development in Sec. 6.1.1 with the following differences. The matrix element of the perturbing potential (7.4)

$$\begin{aligned} \langle n,\mathrm{f}|V|\mathrm{i},0\rangle &= -e^2 \int\!\!\int d^3r_\mathrm{T}\, d^3\xi_\mathrm{P}\, \varphi_n^*(\boldsymbol{\xi}_\mathrm{P})\, \varphi_\mathrm{f}^*(\mathbf{r}_\mathrm{T}) \\ &\quad \times \left( \frac{Z_\mathrm{P}}{|\mathbf{R}-\mathbf{r}_\mathrm{T}|} + \frac{Z_\mathrm{T}}{|\mathbf{R}+\boldsymbol{\xi}_\mathrm{P}|} - \frac{1}{|\mathbf{R}-\mathbf{r}_\mathrm{T}+\boldsymbol{\xi}_\mathrm{P}|} \right) \\ &\quad \times \varphi_\mathrm{i}(\mathbf{r}_\mathrm{T})\, \varphi_0(\boldsymbol{\xi}_\mathrm{P}) \end{aligned} \tag{7.7}$$

has contributions only from the first and third terms because of the orthogonality of the initial and final target wave functions. If $n = 0$, the first and third terms both contribute, but if $n \neq 0$, only the third term contributes because of the orthogonality of the projectile wave functions $\varphi_n(\boldsymbol{\xi}_\mathrm{P})$ and $\varphi_0(\boldsymbol{\xi}_\mathrm{P})$. Hence, in the present approach, the double-inelastic process is solely due to the projectile-electron – target-electron interaction.

In detail, one has

$$\langle 0,\mathrm{f}|V|\mathrm{i},0\rangle = -e^2 \int d^3r_\mathrm{T}\, \varphi_\mathrm{f}^*(\mathbf{r}_\mathrm{T}) \left[ \frac{Z_\mathrm{P}}{|\mathbf{R}-\mathbf{r}_\mathrm{T}|} - \int d^3\xi_\mathrm{P} \right.$$

[1] Physically, one must distinguish in the double-inelastic process the ionization (or excitation) of both electrons by the electron-(other) electron interaction and that by the electron-(other) nucleus interaction. Only the former is nonzero in the PWBA or SCA, but the latter has been observed and has been called "two-center double ionization" [MoM92b].

$$\times \varphi_0^*(\boldsymbol{\xi}_\mathrm{P}) \frac{1}{|\mathbf{R} - \mathbf{r}_\mathrm{T} + \boldsymbol{\xi}_\mathrm{P}|} \varphi_0(\boldsymbol{\xi}_\mathrm{P}) \Big] \varphi_\mathrm{i}(\mathbf{r}_\mathrm{T}) ,$$

$$\begin{aligned} \langle n \neq 0, \mathrm{f}|V|\mathrm{i}, 0\rangle &= e^2 \int\int d^3r_\mathrm{T}\, d^3\xi_\mathrm{P}\, \varphi_n^*(\boldsymbol{\xi}_\mathrm{P})\, \varphi_\mathrm{f}^*(\mathbf{r}_\mathrm{T}) \\ &\quad \times \frac{1}{|\mathbf{R} - \mathbf{r}_\mathrm{T} + \boldsymbol{\xi}_\mathrm{P}|} \varphi_\mathrm{i}(\mathbf{r}_\mathrm{T})\, \varphi_0(\boldsymbol{\xi}_\mathrm{P}) . \end{aligned}$$

Application of the nonrelativistic approximation to Eq. (5.103), obtained by the replacements $r'_\mathrm{P} \to r_\mathrm{P}, \gamma \to 1$, and $\beta \to 0$, converts these expressions to

$$\begin{aligned} \langle 0, \mathrm{f}|V|\mathrm{i}, 0\rangle &= -\frac{e^2}{2\pi^2} \int \frac{d^3q}{q^2} e^{-i\mathbf{q}\cdot\mathbf{R}} \langle \mathrm{f}|e^{i\mathbf{q}\cdot\mathbf{r}_\mathrm{T}}|\mathrm{i}\rangle \\ &\quad \times \left( Z_\mathrm{P} - \langle 0|e^{-i\mathbf{q}\cdot\boldsymbol{\xi}_\mathrm{P}}|0\rangle \right) , \end{aligned} \tag{7.8}$$

$$\begin{aligned} \langle n \neq 0, \mathrm{f}|V|\mathrm{i}, 0\rangle &= \frac{e^2}{2\pi^2} \int \frac{d^3q}{q^2} e^{-i\mathbf{q}\cdot\mathbf{R}} \langle \mathrm{f}|e^{i\mathbf{q}\cdot\mathbf{r}_\mathrm{T}}|\mathrm{i}\rangle \\ &\quad \times \langle n|e^{-i\mathbf{q}\cdot\boldsymbol{\xi}_\mathrm{P}}|0\rangle . \end{aligned} \tag{7.9}$$

The remaining development follows Eqs. (6.4) and (6.5), leading to the differential cross sections for ionization, with $d^2q_b$ replaced by $d^2q = 2\pi q\, dq$,

$$\frac{d\sigma_\mathrm{fi}(\mathrm{sinel})}{dE_\mathrm{f}} = \frac{8\pi a_0^2}{\eta_\mathrm{K}^2} \int_{q_0(0)}^{\infty} \frac{dq}{q^3} |F_\mathrm{T}|^2\, |Z_\mathrm{P} - F_\mathrm{P}^0|^2 , \tag{7.10}$$

$$\frac{d\sigma_\mathrm{fi}(\mathrm{dinel})}{dE_\mathrm{f}} = \frac{8\pi a_0^2}{\eta_\mathrm{K}^2} \sum_{n\neq 0} \int_{q_0(n)}^{\infty} \frac{dq}{q^3} |F_\mathrm{T}|^2\, |F_\mathrm{P}^n|^2 , \tag{7.11}$$

where $\eta_\mathrm{K}$ is defined in Eq. (6.27) and the form factors $F$ are given by

$$\begin{aligned} F_\mathrm{T} &= \langle \mathrm{f}|e^{-i\mathbf{q}\cdot\mathbf{r}_\mathrm{T}}|\mathrm{i}\rangle, \\ F_\mathrm{P}^0 &= \langle 0|e^{-i\mathbf{q}\cdot\boldsymbol{\xi}_\mathrm{P}}|0\rangle \\ &= \left( 1 + \frac{q^2 a_0^2}{4Z_\mathrm{P}^2} \right)^{-2} , \\ F_\mathrm{P}^n &= \langle n|e^{-i\mathbf{q}\cdot\boldsymbol{\xi}_\mathrm{P}}|0\rangle. \end{aligned} \tag{7.12}$$

The expression for $F_{\mathrm{T}}$ is identical to that for $F$ in Eq. (6.21). Equation (7.12) is valid for the ground state of a hydrogenic ion. Closed expressions for $F_{\mathrm{P}}^{n}$ for excited and continuum states of such ions are given by [BaG53,ChK66].

Without going further, one can see from Eq. (7.10) that in the single-inelastic cross section the projectile electron frozen in its ground state effectively shields the projectile nuclear charge and makes the target ionization less effective. On the other hand, Eq. (7.11) shows that the double-inelastic cross section, which in the present treatment is solely due to the interaction between the target electron and an excited (or ionized) projectile electron, enhances the target ionization.

Although there is no difficulty in evaluating the single-inelastic cross section (7.10), the double-inelastic cross section in general cannot be evaluated. The difficulty lies in the summation over excited projectile states $n$ in Eq. (7.11), in which the lower limit $q_0(n)$ of the integral depends on the energy $E_n^{\mathrm{P}}$ of each excited state. Only for hydrogenic ions, can the sum be computed exactly, using closed expressions for $F_{\mathrm{P}}^{n}$. For all other ions, some approximation has to be made.

We discuss briefly five approximations which have been proposed. (1) Ignore the $n$ dependence of $q_0$ by replacing the projectile excitation energies by a suitably chosen mean energy $\bar{E}^{\mathrm{P}}$. The summation sign in Eq. (7.11) can then be moved under the integral sign and $\sum |F_{\mathrm{P}}^{n}|^2$ can be evaluated by closure. The resultant total cross section fits very poorly with experiment near the free-electron threshold. (2) A correction factor for the double-inelastic cross section is suggested by the free-electron model to remove this defect. (3) The actual projectile states are represented by approximate hydrogenic states, so that the sum over projectile states can be evaluated. (4) The sum over projectile excited states is partly executed by using a higher-order sum rule due to Bethe and Jakiw [BeJ68]. (5) The impulse approximation is used to evaluate the double-inelastic cross section, by assuming that the projectile electrons act as free electrons, but have an initial momentum distribution corresponding to their initial state. [In the free-electron model expression (7.1), the projectile electrons are assumed to be initially at rest with respect to the projectile.]

*Closure approximation*

If $E_n^{\mathrm{P}}$ and $q_0(n)$ [Eq. (7.6)] are replaced by $\bar{E}^{\mathrm{P}}$ and $\bar{q}_0$, respectively, in Eq. (7.11) the summation sign can be moved inside the integral over $dq$ and closure can be invoked [Sch68]:

$$\sum_n |F_{\mathrm{P}}^n|^2 = 1 \quad \text{or} \quad \sum_{n\neq 0} |F_{\mathrm{P}}^n|^2 = 1 - |F_{\mathrm{P}}^0|^2, \tag{7.13}$$

leading to

$$\frac{d\sigma_{\mathrm{fi}}(\text{dinel})}{dE_{\mathrm{f}}} = \frac{8\pi a_0^2}{\eta_{\mathrm{K}}^2} \int_{\bar{q}_0}^{\infty} \frac{dq}{q^3} |F_{\mathrm{T}}|^2 (1 - |F_{\mathrm{P}}^0|^2) \,. \tag{7.14}$$

In the integration over final electron energies $E_{\mathrm{f}}$ in Eqs. (7.10) and (7.14) it is convenient to make substitutions equivalent to Eqs. (6.24) and (6.25), such as

$$W = 2\frac{E_{\mathrm{f}}^{\mathrm{T}} - E_{\mathrm{i}}^{\mathrm{T}} + \bar{E}^{\mathrm{P}} - E_0^{\mathrm{P}}}{Z_{\mathrm{T}}^2 e^2/a_0} \,. \tag{7.15}$$

One then arrives at a result analogous to the nonrelativistic part of Eq. (6.26), with $F_{\mathrm{K}}$ replaced by $F_{\mathrm{T}}$:

$$\begin{aligned} \sigma_{\mathrm{fi}} &= \sigma_{\mathrm{fi}}(\text{sinel}) + \sigma_{\mathrm{fi}}(\text{dinel}) \\ &= \frac{4\pi a_0^2}{\eta_{\mathrm{K}}^2 \, Z_{\mathrm{T}}^4} \int_{W_{\min}}^{\infty} dW \int_{Q_{\min}}^{\infty} \frac{dQ}{Q^2} |F_{\mathrm{T}}|^2 S_{\mathrm{P}}(Q), \end{aligned} \tag{7.16}$$

where

$$S_{\mathrm{P}}(Q) = |Z_{\mathrm{P}} - F_{\mathrm{P}}^0|^2 + (1 - |F_P^0|^2). \tag{7.17}$$

One sees that the essential difference between the cross section induced by a bare projectile and a one-electron projectile is the replacement in the former of the projectile squared charge $Z_{\mathrm{P}}^2$ by an effective squared charge $S_{\mathrm{P}}(Q)$.

In Eq. (7.15), one usually sets $\bar{E}^{\mathrm{P}} - E_0^{\mathrm{P}} = 0$, especially if $Z_{\mathrm{P}} \ll Z_{\mathrm{T}}$, although other proposals have been made which are discussed in [HaW87]. As noted above, though, as soon as a constant value for $q_0$ is used in Eq. (7.11), the projectile velocity dependence of the total ionization cross section fits very poorly with experiment in the neighborhood of the free-electron threshold (Sec. 7.1). One can see the reason for this difficulty by realizing that if one applies Eq. (7.17) to the case of negligible projectile-electron binding, so that $Z_{\mathrm{P}}(\text{effective}) \simeq 0$ in Eq. (7.12) and $F_{\mathrm{P}}^0 \simeq 0$, one obtains

$$S_{\mathrm{P}}(\text{nobind}) = Z_{\mathrm{P}}^2 + 1, \tag{7.18}$$

so that

$$\sigma_{\mathrm{fi}}(\text{nobind}) = (Z_{\mathrm{P}}^2 + 1)\sigma_{\mathrm{T}}^{\mathrm{i}}(\mathrm{p}), \tag{7.19}$$

where $\sigma^{\rm i}_{\rm fi}({\rm p})$ is the target ionization cross section by protons moving with the projectile velocity. But, the free-electron model (7.1), which is valid if binding is negligible, predicts

$$\sigma^{\rm i}_{\rm T} = Z_{\rm P}^2 \sigma^{\rm i}_{\rm T}({\rm p}) + \sigma^{\rm i}_{\rm T}({\rm e}). \tag{7.20}$$

Here, $\sigma^{\rm i}_{\rm T}({\rm e})$ is the electron-induced ionization cross section which has a threshold at the projectile velocity (7.3), whereas there is no threshold in expression (7.19). Hence the latter expression must be incorrect near threshold.

*Correction factor approximation*

A remedy for the preceding defect is suggested by modifying expression (7.17) [Anh86]

$$S_{\rm P}^{\rm mod}(Q) = \frac{\left|Z_{\rm P} - F_{\rm P}^0\right|^2 \sigma^{\rm i}_{\rm T}({\rm p}) + \left(1 - |F_{\rm P}^0|^2\right) \sigma^{\rm i}_{\rm T}({\rm e})}{\sigma^{\rm i}_{\rm T}({\rm p})}. \tag{7.21}$$

The cross sections $\sigma^{\rm i}_{\rm T}({\rm e})$ and $\sigma^{\rm i}_{\rm T}({\rm p})$ are to be evaluated at the projectile velocity and can be calculated with Eq. (6.26) or the equivalent nonrelativistic expressions. This procedure improves the fit of the cross section with experiment above the free-electron threshold, but introduces an artificially sharp kink in the total ionization cross section since $\sigma^{\rm i}_{\rm T}({\rm e})$ has an abrupt threshold at $v_{\rm thr}$.

*Approximate hydrogenic spectrum*

It has been suggested that in evaluating the sum in Eq. (7.11), it is not necessary to use the exact excitation spectrum, but that it is sufficient to mimic the spectrum by a hydrogenic one. Also, in calculating the form factors, one may be able to use 1s hydrogenic wave functions for the dominant terms. The effective value of the nuclear charge is obtained from the actual ionization energy of each state under consideration. We refer to [HaW87] for details. The results are in reasonably good agreement with an accurate PWBA calculation for the case of electron loss of $H^0$ incident on a He target.

*Use of an extended sum rule*

One can evaluate the sum over $n$ in Eq. (7.11) more accurately than provided by the closure relation if one breaks up

$$q_0(n) \quad = \quad \frac{E^{\rm T}_{\rm f} - E^{\rm T}_{\rm i}}{\hbar v} + \frac{E^{\rm P}_n - E^{\rm P}_0}{\hbar v}$$

$$= \quad q_{\mathrm{T}} + q_{\mathrm{P}}(n). \tag{7.22}$$

The integral in Eq. (7.11) is now written as $\int_{q_{\mathrm{T}}}^{\infty} + \int_{q_{\mathrm{P}}(n)}^{q_{\mathrm{T}}}$. Summing over projectile states, the sum over the first integral can be evaluated by closure and the sum over the second integral can be evaluated approximately by using a higher-order sum rule. The results are in reasonably good agreement with an accurate PWBA calculation for H + H ionization [MoM91a].

*Impulse approximation*

The impulse approximation is essentially an improvement of the free-electron model, Eq. (7.1).[2] It takes into account the initial motion of the electron(s) in the projectile by averaging the target ionization cross section over the momentum distribution of these electrons. The double-inelastic cross section is computed by the impulse approximation as follows.

Referring to Fig. 7.1 for the coordinate system, we denote the momentum of the projectile electron with respect to the projectile nucleus by $\mathbf{q}_{\mathrm{P}}$. Then, nonrelativistically, the momentum of the projectile electron with respect to the target is given by $m_{\mathrm{e}}\mathbf{v} + \mathbf{q}_{\mathrm{P}}$, and its kinetic energy by

$$\begin{aligned} T_{\mathrm{eP}} &= \frac{|m_{\mathrm{e}}\mathbf{v} + \mathbf{q}_{\mathrm{P}}|^2}{2m_{\mathrm{e}}} \\ &= \tfrac{1}{2}m_{\mathrm{e}}v^2 + \mathbf{q}_{\mathrm{P}}\cdot\mathbf{v} + \frac{q_{\mathrm{P}x}^2}{2m_{\mathrm{e}}} + \frac{q_{\mathrm{P}y}^2}{2m_{\mathrm{e}}} + \frac{q_{\mathrm{P}z}^2}{2m_{\mathrm{e}}}. \end{aligned} \tag{7.23}$$

The terms $(q_{\mathrm{P}x}^2 + q_{\mathrm{P}y}^2)/(2m_{\mathrm{e}})$ can be neglected in most collisions, but retention of the term $q_{\mathrm{P}z}^2/(2m_{\mathrm{e}})$ and subtraction of the projectile electron binding energy $I_{\mathrm{P}}$ from $T_{\mathrm{eP}}$ improves agreement with experiment. This gives

$$T_{\mathrm{eP}} = \tfrac{1}{2}m_{\mathrm{e}}v^2 + q_{\mathrm{P}z}\,v + \frac{q_{\mathrm{P}z}^2}{2m_{\mathrm{e}}} - I_{\mathrm{P}}. \tag{7.24}$$

Each projectile electron has a momentum distribution with respect to the projectile nucleus given by the square of its momentum wave function $\tilde{\varphi}(\mathbf{q}_{\mathrm{P}})$. In the impulse approximation, the contribution of each projectile electron to the target double-inelastic ionization cross section is then given by an expression of the form

$$\sigma_{\mathrm{T}}^{\mathrm{i}}(\mathrm{dinel}) \quad = \quad \int d^3q_{\mathrm{P}}\,\sigma_{\mathrm{T}}^{\mathrm{i}}(\mathrm{e}, T_{\mathrm{eP}})|\tilde{\varphi}(\mathbf{q}_{\mathrm{P}})|^2$$

[2]The approximation used here is an impulse-like approximation but cannot be justified by the derivations of the usual impulse approximation [GoW64,McC70]. These derivations typically assume a rigid projectile impinging on a target with loosely bound electrons. In the present case, though, the projectile is structured and *it* is assumed to have loosely bound electrons [MoM94].

$$= \int_0^\infty \sigma_{\mathrm{T}}^{\mathrm{i}}(\mathrm{e}, T_{\mathrm{eP}}) \mathcal{J}(_{\mathrm{P}z}) dq_{\mathrm{P}z}, \tag{7.25}$$

where the second expression makes use of the definition of the Compton profile function of the projectile electron:

$$\mathcal{J}(q_{Pz}) = \int_0^\infty dq_{\mathrm{P}x} \int_0^\infty dq_{Py} |\tilde{\varphi}(\boldsymbol{q}_{\mathrm{P}})|^2. \tag{7.26}$$

The cross section $\sigma_{\mathrm{T}}^{\mathrm{i}}(e, T_{\mathrm{eP}})$ is the same free-electron target ionization cross section as occurs in Eq. (7.1), but here is evaluated at the electron kinetic energy $T_{\mathrm{eP}}$, Eq. (7.24), rather than at the energy $\frac{1}{2} m_{\mathrm{e}} v^2$ as in the free-electron model. For the evaluation of Eq. (7.25), semiempirical formulas for electron-induced ionization cross sections of neutral atoms [JaG87, LeB88] or of hydrogenic ions [MoG80], or tabulations [Sco78] are useful. Compton profile functions can be looked up in [BiM75, Lee77]. The results, when added to the single-inelastic cross section, are in good agreement with experiment [LeZ92].

In surveying the various approximations, (4) and (5) appear to be best founded theoretically, although (2) is convenient and is often used for the evaluation of relativistic cross sections, as described in the next sections.

## 7.3 More electrons. Relativistic projectiles

If the partner to be ionized, in the present discussion the target, has many electrons, each shell or subshell which is ionized can be treated separately. One introduces an inner screening constant which reduces the nuclear charge [Sla60], and an outer screening constant $\theta$ defined similar to the K-shell expression in Eq. (6.28), as the ratio of the experimental electron-binding energy divided by the hydrogenic-binding energy of the shell or subshell [MeL58].

The (nonrelativistic) method sketched in the preceding section can be extended to a projectile with more than one electron. To make use of the orthogonality relations similar to those used in obtaining Eqs. (7.10) and (7.11), one has to assume very simple independent-electron product wave functions for the target and the projectile. Making again use of closure to evaluate the double-inelastic cross section, the modified effective squared projectile charge equivalent to Eq. (7.21), for a projectile with $N_{\mathrm{P}}$ electrons, is [BrT78,McS81,Anh85a]

$$S_{\mathrm{P}}^{\mathrm{mod}}(Q) = \left| Z_{\mathrm{P}} - \sum_{i=1}^{N_{\mathrm{P}}} F_{\mathrm{P}}^0(i) \right|^2 + \left( N_{\mathrm{P}} - \sum_{i=1}^{N_{\mathrm{P}}} |F_{\mathrm{P}}^0(i)|^2 \right) \frac{\sigma_{\mathrm{T}}^{\mathrm{i}}(\mathrm{e})}{\sigma_{\mathrm{T}}^{\mathrm{i}}(\mathrm{p})}. \tag{7.27}$$

Here, the correction factor expression (7.21) has been used. Tabulations by [HuV75] and analytic expressions by [SaM87] are useful in evaluating the form factor sums.

In principle, Eq. (7.27) can be substituted only into the nonrelativistic expression (7.16). If one wants to consider target ionization by relativistic projectiles, one is on secure theoretical grounds for bare projectiles, if one uses one of the "relativistically exact" expressions derived in Chap. 6. Since these equations can also be used to compute the ionization cross section for relativistic electrons, under the neglect of exchange, one is also on good theoretical grounds in applying the free-electron expression (7.1) to a relativistic projectile with $N_{\mathrm{P}}$ electrons. Of course, the kinetic energy of the projectile electron is only $\sim m_{\mathrm{e}}/M_{\mathrm{P}}$ times the kinetic energy of the projectile, e.g., for a 1-GeV/u projectile, the kinetic energy of a free projectile electron is only 0.5 MeV. In that case, an expression like Eq. (6.26) might not be sufficiently accurate, and it might be better to use one of the references cited after Eq. (7.26) to evaluate $\sigma_{\mathrm{T}}^{i}(e)$.

A more complete many-electron expression for relativistic projectiles is suggested by the substitutions, in Eq. (7.16),

$$S_{\mathrm{P}}(Q) \rightarrow S_{\mathrm{P}}^{\mathrm{mod}}(Q), \tag{7.28}$$

$$|F_{\mathrm{T}}|^2 \rightarrow |F_{\mathrm{T}}|^2 + \frac{(\beta Z_{\mathrm{T}}\alpha)^2(1 - Q_{\min}/Q)}{(1 - \beta^2 Q_{\min}/Q)^2}\,(|G_{\mathrm{T}}|^2 + \tfrac{1}{4}Q|F_{\mathrm{T}}|^2), \tag{7.29}$$

from Eqs. (7.27) and (6.26), respectively [Anh85a]. In the second expression, which is valid only for target K-shell ionization, $|F_{\mathrm{T}}|^2$ and $|G_{\mathrm{T}}|^2$ are given by Eqs. (6.29) and the other quantities are also defined in Sec. 6.2.2. Below, we confront the resulting cross section with experiment.

## 7.4 Electron loss. Comparison with experiment

To facilitate the comparison with experiment, cross section expressions for K-electron loss of a projectile induced by a neutral target atom are derived from the above expressions for target ionization, by interchanging the roles of the target and the projectile. It is not difficult to extend the results to L-shell loss [AnM84b], but the expressions given below apply only to K-shell loss, except for the free-electron model expression which is general. We start by rewriting the latter, Eq. (7.1):

$$\sigma_{\mathrm{P}}^{1} = Z_{\mathrm{T}}^{2}\sigma_{\mathrm{P}}^{1}(\mathrm{p}) + Z_{\mathrm{T}}\sigma_{\mathrm{P}}^{1}(\mathrm{e}), \tag{7.30}$$

where $\sigma^1_{\rm P}({\rm p})$ and $\sigma^1_{\rm P}({\rm e})$ denote the projectile loss cross sections induced by protons and free electrons, respectively, both moving with the projectile velocity $v$. The target is assumed to be neutral, with $Z_{\rm T}$ "free" electrons.

For the more complete, many-electron expression, we use the one proposed in the preceding section, which gives, per target K electron,

$$\sigma^1_{\rm K} = \frac{4\pi a_0^2}{(\eta^{\rm P}_{\rm K})^2 Z_{\rm P}^4} \int_{W_{\rm min}}^{\infty} dW \int_{Q_{\rm min}}^{\infty} \frac{dQ}{Q^2} |F^{\rm rel}_{\rm P}|^2 \, S^{\rm mod}_{\rm T}(Q), \tag{7.31}$$

where

$$S^{\rm mod}_{\rm T}(Q) = \left| Z_{\rm T} - \sum_{i=1}^{Z_{\rm T}} F^0_{\rm T}(i) \right|^2 + \left( Z_{\rm T} - \sum_{i=1}^{Z_{\rm T}} |F^0_{\rm T}(i)|^2 \right) \frac{\sigma^1_{\rm P}({\rm e})}{\sigma^1_{\rm T}({\rm p})}, \tag{7.32}$$

$$\begin{aligned}
|F^{\rm rel}_{\rm P}|^2 &= |F_{\rm P}|^2 + \frac{(\beta Z_{\rm P}\alpha)^2 (1 - Q_{\rm min}/Q)}{(1-\beta^2 Q_{\rm min}/Q)^2} \left( |G_{\rm P}|^2 + \tfrac{1}{4} Q |F_{\rm P}|^2 \right), \\
W &= 2\,\frac{E^{\rm P}_{\rm f} - E^{\rm P}_{\rm i} + \langle E^{\rm T}_n - E^{\rm T}_0 \rangle}{e^2 Z_{\rm P}^2/a_0}, \\
Q_{\rm min} &= \frac{W^2}{4\eta^{\rm P}_{\rm K}}, \quad W_{\rm min} = \theta^{\rm P}_{\rm K} = \frac{2E^{\rm P}_{\rm K}}{e^2 Z_{\rm P}^2/a_0}, \\
\eta^{\rm P}_{\rm K} &= \left( \frac{\beta}{\alpha Z_{\rm P}} \right)^2, \quad Z_{\rm P} = Z^{\rm bare}_{\rm P} - 0.3,
\end{aligned}$$

where $\langle E^{\rm T}_n - E^{\rm T}_0 \rangle$ is a suitably chosen target excitation energy. The form factors $|F_{\rm P}|^2$ and $|G_{\rm P}|^2$ are given by Eq. (6.29) applied to the projectile electron. In the expression for $W$, the designations i and f for the initial and final states of the ionized partner, and 0 and $n$ for the ionizing partner, have been kept. In the calculations shown below, $\langle E^{\rm T}_n - E^{\rm T}_0 \rangle$ has been set equal to zero.

In examining experimentally the validity of a theoretical loss cross section expression like Eq. (7.31), one can ask the following questions of physical interest. (1) Is there experimental evidence for the role of the double-inelastic cross section, i.e., for the increase of the projectile loss cross section due to the target electrons? (2) Is there evidence for the screening effect of the target electrons in the single-elastic cross section, i.e., for the reduction of the effective squared target charge due to the form factors $F^0_{\rm T}(i)$? Considering the relative sizes of the electron distributions in the projectile and the target, one would expect target screening effects to be large in projectile loss if $Z_{\rm P} \ll Z_{\rm T}$ and small if $Z_{\rm P} \gg Z_{\rm T}$. (3) If screening effects are small, is there any evidence that perturbation theory is valid in relativistic collisions, which implies that $\sigma^1(\text{sinel}) \propto Z_T^2$? So far, only points (1) and (3) have been examined experimentally, but we give calculated results for (2).

*Role of double-inelastic cross section*

Figure 7.2 shows measurements of the one-electron loss cross section of 100- and 380-MeV/u $Au^{52+}$ ions in (a) $H_2$ and (b) He targets. The dashed curves are the single-inelastic and the solid curves are the sums of single- and double-inelastic cross sections. Below the "threshold" at 10 MeV/u, only the former is effective. The cross section calculation for the $H_2$ target includes a small correction for the molecular nature of the target [MeH91]. The theoretical results agree with experiment only if the double-inelastic cross section is included. This is seen more strikingly in Fig. 7.3 which gives the ratio of the $H_2$ target to the He target cross section (a) for the above ions and (b) for 405-MeV/u $U^{86+}$ ions. Without the double-inelastic process, the ratio would follow the dashed line.

*Effect of target-electron screening*

The role of screening in the total loss cross section of various one-electron ions colliding with a Al, Ag, and U targets is shown in Fig. 7.4. For projectiles with $Z_P < Z_T$, the target screening effect in the single-inelastic cross section is quite dramatic (for the targets shown, the double-inelastic cross section plays only a minor role). It effectively removes the transverse and spin-flip contributions to the cross section, so that instead of increasing as $\ln\gamma$, the loss cross section becomes independent of $\gamma$.

*Validity of perturbation theory*

The validity of perturbation theory has been examined in measurements of 1s-electron loss for 82- to 200-MeV/u Xe and 105- to 955-MeV/u U ions incident on various solid targets between Be and Au. Figure 7.5 shows the quantity

$$\sigma^{\text{corr}}/Z_T^2 = [\sigma_P^1 - Z_T\sigma_P^1(\text{e})]/Z_T^2, \qquad (7.33)$$

which according to the free-electron model expression (7.30) should be equal to $\sigma_P^1(\text{p})$. The data points have been obtained from single K-electron loss measurements using $Xe^{53+}$ and $Xe^{52+}$ ions (the latter data have been corrected for the influence of double K-electron loss). One sees that in the energy region examined, perturbation theory breaks down for the higher-$Z_T$ targets, but increasingly less so, as the projectile velocity increases.

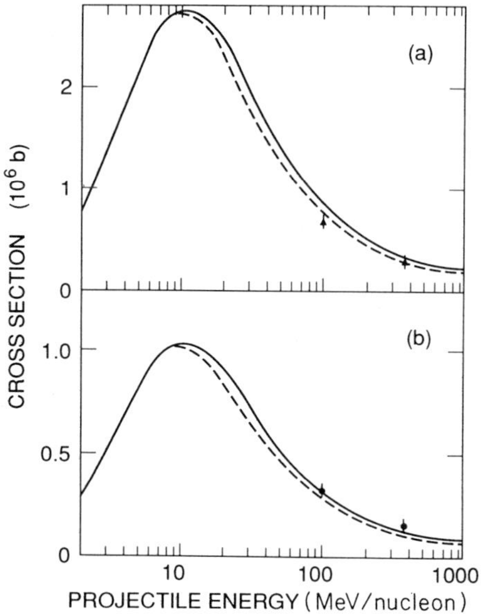


**Figure 7.2.** Projectile one-electron loss cross section of $Au^{52+}$ in (a) $H_2$ and (b) He as a function of the projectile energy. The solid curves are the sums of the single- and double-inelastic cross sections, the dashed curves are only the former. From [HuF91].

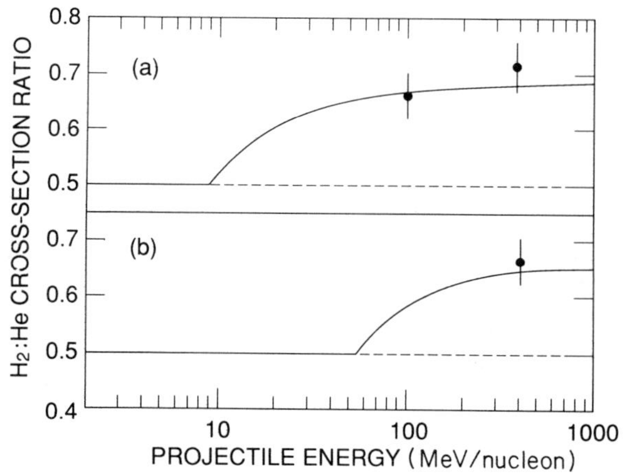


**Figure 7.3.** Ratio of $H_2$ to He one-electron loss cross section for (a) $Au^{52+}$ and (b) $U^{86+}$ ions, as a function of the projectile energy. Without a double-inelastic process, the ratio would follow the dashed line. From [HuF91].

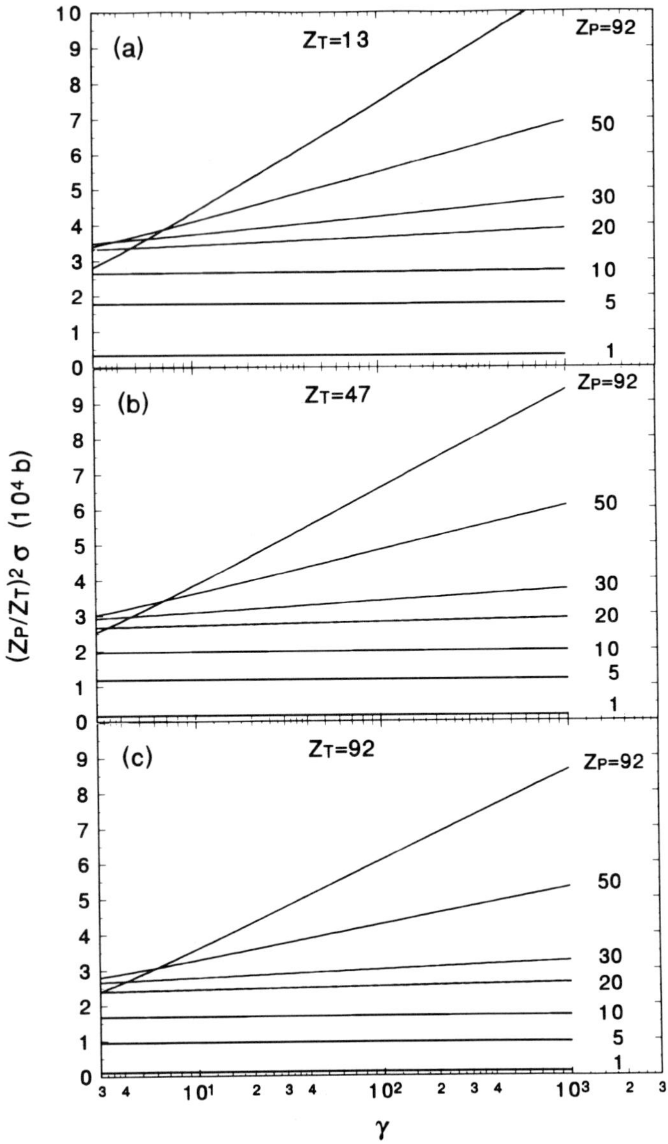


**Figure 7.4.** Loss cross sections multiplied by $(Z_P/Z_T)^2$ for various one-electron projectiles colliding with (a) Al, (b) Ag, and (c) U targets as a function of the projectile energy. Adapted from [AnB87].

Qualitatively, one expects a smaller perturbation of the projectile electron wave function by the target (nucleus) with increasing projectile velocity. The data are compared with the Glauber approximation [Mcg82], in which the distortion of the projectile wave function by the target is partially taken into account by means of an eikonal phase (Sec. 8.4.1).

It is somewhat disturbing that for $Z_{\mathrm{T}} \leq 13$, where perturbation theory is expected to be valid, the experimental results do not agree better with the theoretical ones. This is not only the case for Xe projectiles, as is shown by Fig. 7.6, which gives the ratio of experimental to theoretical cross sections for relativistic Xe, Bi, and U projectiles. At present, there is no explanation for these effects, even though sometimes they have been blamed on experimental uncertainties [MoG90].

# 7.5 Multiple ionization

With relativistic projectiles of high atomic number, there is always the possibility of multiple-electron target ionization and projectile loss in a single collision, in addition to the single-electron processes. (Multiple projectile electron loss is often called *stripping*.) In the present section, these processes are discussed briefly.

## 7.5.1 Multiple target ionization

Multiple ionization of gaseous targets (Ne, Ar, Kr, $I_2$) by 420-MeV/u $U^{91+}$ ions under single-collision conditions has been investigated to find out whether highly charged, slow (eV) target recoil ions can be produced copiously enough so that these recoil ions could be used as secondary beams for other studies. Figure 7.7 gives an overview of the production cross sections of the heavier ions as a function of their charge $q$. Charge states up to $q \sim 0.5 Z_{\mathrm{T}}$ are indeed produced with reasonable cross sections.

In these collisions, $Z_{\mathrm{P}} > Z_{\mathrm{T}}$, so that perturbation approaches of the kind presented in Chap. 6 and in this chapter are not applicable. Horbatsch [Hor86, HoD86] has made a calculation in which the time evolution of the target many-electron density is obtained from a classical equation for the phase-space distribution function. The calculation is statistical in nature and uses an average over shell structure, similar to the Thomas-Fermi distribution [Sch68]. The probability of ionization per electron, $P_1(b)$, for a given nuclear trajectory with impact parameter $b$ is obtained by a Monte-Carlo method which uses a time-dependent classical Hamiltonian

$$H = \frac{p^2}{2m_{\mathrm{e}}} - \frac{Z_{\mathrm{T}}^{\mathrm{eff}}(t)\, e^2}{r} - \frac{Z_{\mathrm{P}}\, e^2}{|\mathbf{R}(t) - \mathbf{r}|} \tag{7.34}$$

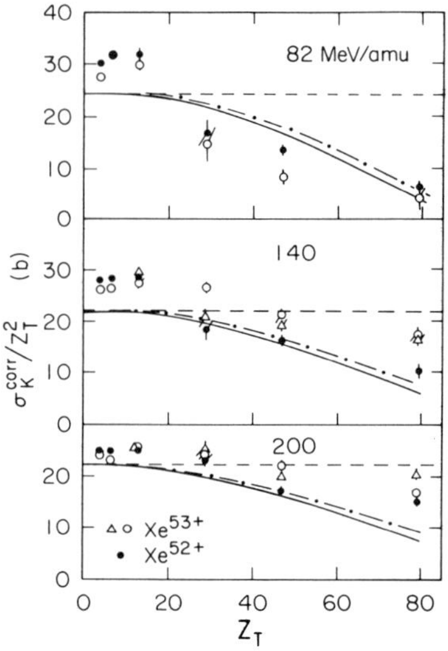


**Figure 7.5.** Single K-electron loss cross section, corrected for target-electron contributions according to Eq. (7.33), for one- and two-electron Xe ions colliding with various targets at the energies indicated, as a function of the target atomic number. The cross section is divided by $Z_T^2$. In addition to the statistical errors shown, the results have possible systematic errors up to 20%. The solid curves have been calculated with the Glauber model. From [AnM87].

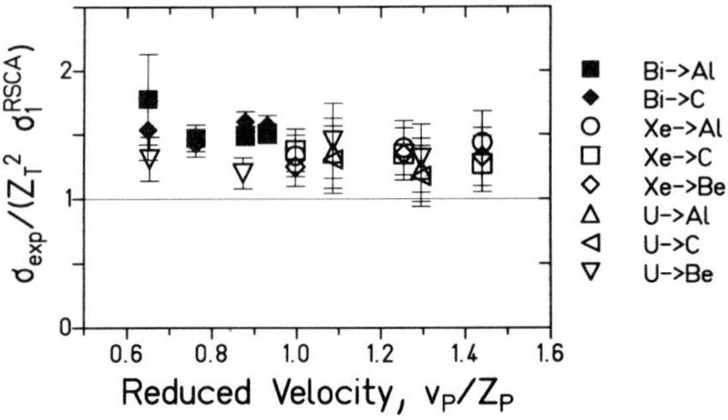


**Figure 7.6.** Ratio between experimental projectile loss cross section and theoretical semiclassical calculations for protons multiplied by $Z_T^2$ [TrR80]. Above the reduced velocity $v_P/Z_P = 1$ (in atomic units), see Eq. (7.3) for the case of target ionization, the cross sections have been corrected for the double-inelastic process according to Eq. (7.33). Coupled-channel calculations [MoG90] lie within 10% of the semiclassical calculations used here. Adapted from [RyS93].

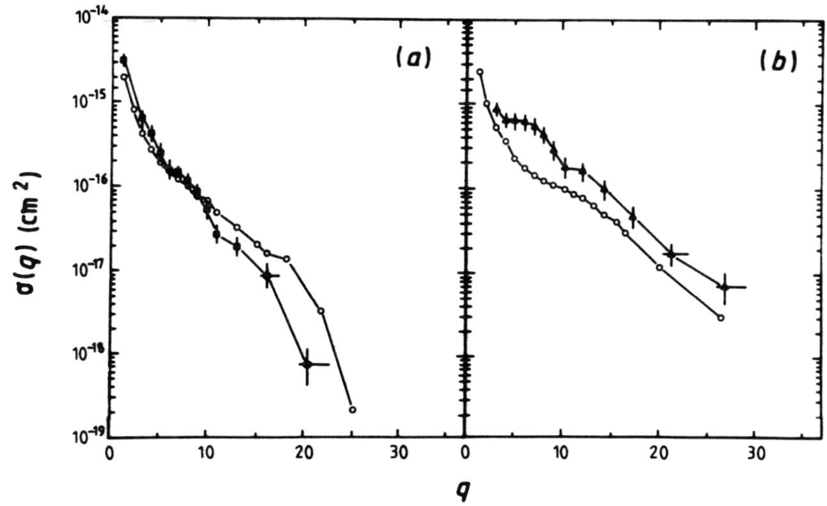


**Figure 7.7.** Measured cross sections for the production of slow recoil ions by 420-MeV/u $U^{91+}$ projectiles impinging on various gas targets as a function of the charge state $q$ of the recoil ions. Data points (solid circles) for (a) Kr and (b) I targets are compared to calculations by [Hor86, HoD86] (open circles). From [KeU86].

to determine the phase-space distribution function at position $\mathbf{r}$ and momentum $p$. This equation implies that each target electron sees the projectile field and an effective screened target field which is time dependent. The value of the target charge parameter $Z_{\mathrm{T}}^{\mathrm{eff}}(t)$ at $t = 0$ is adjusted to give the best fit to the experimental cross sections. Unfortunately, the dependence of the calculated cross sections on $Z_{\mathrm{T}}^{\mathrm{eff}}(0)$ is very sensitive. The calculations shown in Fig. 7.7 use $Z_{\mathrm{T}}^{\mathrm{eff}}(0) = 3.5$.

The calculation assumes that the target electrons can be treated as independent particles. In that case, the overall probability of ionizing $q$ electrons of a target having initially $N(= Z_{\mathrm{T}})$ electrons is given by the binomial distribution

$$P(q, N, b) = \frac{N!}{q!(N-q)!}\,[P_1(b)]^q\,[1 - P_1(b)]^{N-q}\,. \tag{7.35}$$

This equation has to be integrated over $2\pi b\, db$ to obtain the $q$-fold ionization cross section. At relativistic velocities, the effect of capture can be neglected, but at lower velocities it must be included. Figure 7.7 shows that the overall magnitude and trend of the experimental cross section are reproduced. One can also introduce refinements in the theory, such as struc-

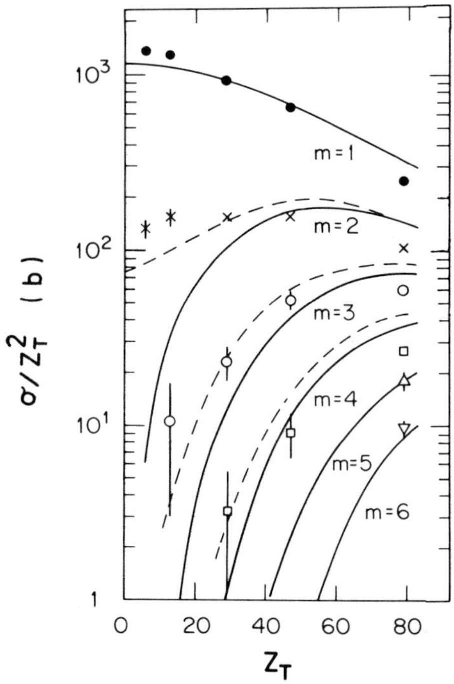


**Figure 7.8.** Single- and multiple-electron loss cross sections for 955-MeV/u $U^{68+}$ ions passing through various targets as a function of the target atomic number. On each curve, $m$ indicates the multiplicity of the loss. The solid curves have been calculated on the basis of the independent-particle model. The dashed curves include corrections for the LMM Auger effect. From [AnM87].

tured projectiles, shell structure of the target, and quantum-theoretical approaches beyond the independent-particle model for the electrons [Hor91].

### 7.5.2 Multiple projectile loss

Multiple electron loss in a single collision of relativistic Xe and U projectiles by a variety of targets has been observed up to sixfold. Figure 7.8 shows experimental and theoretical results for 955-MeV/u $U^{68+}$.

The calculations are again based on the independent-particle model. But here, contrary to the case of target ionization by high-$Z_P$ projectiles, perturbation theory can be applied to compute the one-electron ionization probability $P^{(S)}(b)$ in each shell or subshell S. This procedure is approximate for the U projectiles used, even though the ratio $Z_T/Z_P^{(S)}$

($Z_{\mathrm{P}}^{(\mathrm{S})}$=screened projectile charge) in all the subshells S considered never exceeds unity and the projectile velocity is high enough so that $P_1^{(\mathrm{S})}(b)$ never exceeds unity, even at small values of $b$.

For highly screened projectiles, relativistic effects on the impact parameter dependence remain small for $\beta \leq 0.9$, so that SCA tables of [HaJ75] can be used to estimate $P_1^{(\mathrm{S})}(b)$ for each relevant K, L, and M subshell. Equation (7.35) can then be applied to each subshell, with $N$ equal to the original number of electrons $N_{(\mathrm{S})}$ in the subshell and $q$ equal to the number $n_{(\mathrm{S})}$ of electrons stripped from the subshell S. If electrons are stripped from more than one subshell, e.g., three, the cross section for stripping $m$ electrons from the projectile is given by:

$$\begin{aligned}\sigma_m &= \sum_{n_I+n_{II}+n_{III}=m} \int_0^\infty P^{(I)}(n_I, N_I, b)\, P^{(II)}(n_{II}, N_{II}, b) \\ &\quad \times P^{(III)}(n_{III}, N_{III}, b)\, 2\pi b\, db\,, \qquad (7.36)\end{aligned}$$

where the subscripts and superscripts I, II, and III refer to the three subshells considered.

Figure 7.8 shows that the magnitude and the trend of the cross sections are quite well reproduced by the theory, but only if a side effect, the ejection of electrons by the LMM Auger effect, is included. Since $P_1^{(\mathrm{S})}(b)$ scales approximately as $Z_{\mathrm{T}}^2$, one expects that for one-electron stripping the ratio $\sigma_1/Z_{\mathrm{T}}^2$, as shown in Fig. 7.8, should be independent of $Z_{\mathrm{T}}$. At this high projectile velocity, the perturbation theory should be valid. In contrast to the data in Fig. 7.5, the drop-off with increasing $Z_{\mathrm{T}}$ is due here to the influence of multiple-ionization effects, in particular, due to the factor $[1 - P_1^{(\mathrm{S})}(b)]^{N_{(\mathrm{S})} - n_{(\mathrm{S})}}$ in expression (7.35). As $Z_{\mathrm{T}}$ increases, and hence $P_1^{(\mathrm{S})}$ increases, this factor becomes appreciably less than unity.

One other effect can influence $\sigma_m$. At small values of $Z_{\mathrm{T}}$, the target electrons can enhance the loss probability $P_1$ (Secs. 7.2 and 7.3). This effect had not been taken into account in the calculation of [AnM87] because the relevant expressions were not available at the time [MoM91b, MoM92a].

# Chapter 8

# Charge Exchange

## 8.1 Introduction

The transfer of an electron between target and projectile is one of the basic processes in atomic collision physics. In this chapter, we consider nonradiative or Coulomb capture, while the discussion of another process, namely electron transfer with the simultaneous emission of a photon or "radiative electron capture," is deferred to Chap. 9.

In general, charge changing (nonradiative and radiative) reactions can be studied experimentally by passing an ion beam of known charge-state composition through a target and measuring the fraction of ions that have either captured or lost an electron. Such a measurement yields total cross sections. For testing details of the theory, it would be desirable to specify experimentally the states between which capture occurs. This has been done in a few cases, see Sec. 8.7.2.

From nonrelativistic capture theory it is known (see, e.g., [BrM92]) that for large projectile velocities $v$ the nonradiative charge changing cross section decreases asymptotically as $v^{-12}$ or $E^{-6}$ in first-order and as $v^{-11}$ or $E^{-11/2}$ in second-order perturbation theory. This dramatic velocity dependence occurs because the overlap in momentum space between the initial and final electron wave functions quickly diminishes as the relative velocity increases. The question then arises whether there is any measurable cross section left at *relativistic* velocities. There is a simple qualitative reason why the relativistic cross section decreases less rapidly with the collision energy than the nonrelativistic estimate would indicate: owing to the Lorentz contraction of the electron space wave function in the beam direction, the momentum wave functions acquire (Lorentz-) extended tails

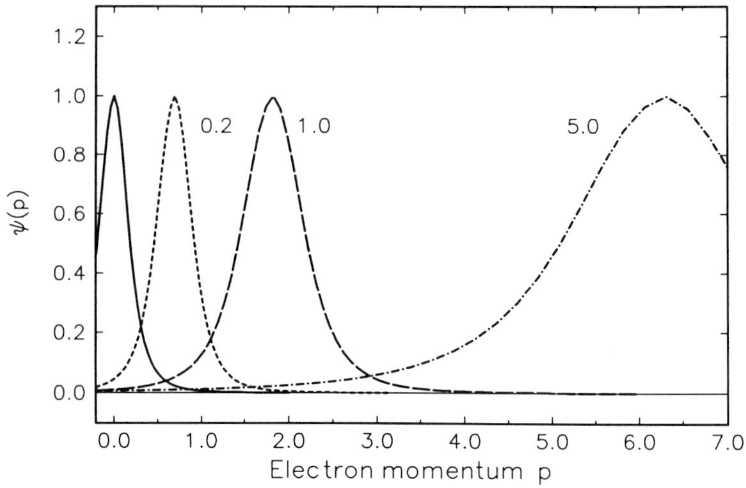


**Figure 8.1.** Overlap of target and projectile 1s momentum wave functions along the beam axis. The atomic wave functions for $Z_T = Z_P = 40$ are taken to be nonrelativistic, the momenta are given in units of $m_e c$. Solid line: momentum wave function of the target; dotted, short-dashed, and long-dashed lines represent the projectile momentum wave function Lorentz transformed to the laboratory system for projectile energies of 0.2, 1.0, and 5.0 GeV/u, respectively. The maximum amplitudes are normalized to unity.

in this direction and thus give rise to a more slowly decreasing overlap in momentum space.

This is illustrated in Fig. 8.1, in which we have used Eq. (2.28) to calculate the projectile momentum distribution as seen from the target system for various projectile energies. With increasing energy, the momentum wave function of the projectile along the beam axis is stretched with the Lorentz factor $\gamma$. Consequently, the overlap between target and projectile momentum wave functions decreases less rapidly than according to nonrelativistic kinematics. In the latter case, the shape of the momentum wave function would be retained and simply shifted by the projectile velocity. Actual calculations show, almost independently of the approximations involved, that the relativistic capture cross section decreases only with about $E^{-1}$ rather than with $E^{-6}$.

We start the presentation in Sec. 8.2 with a discussion of the Born approximation (or the Oppenheimer-Brinkman-Kramers approximation) which is the simplest and historically first approach to relativistic electron capture. It already yields the correct asymptotic $E^{-1}$ energy dependence, but cross sections that are by far too large. This approximation ignores the distortion of the atomic states by the long-range Coulomb interaction. By imposing proper Coulomb boundary conditions in Sec. 8.3, one can obtain a short-range effective interaction and realistic cross sections. In Sec. 8.4 we treat the eikonal approximation which is comparatively easy to implement and also yields reasonable cross sections. In Sec. 8.5, we compare the results of the theories derived so far with one another and with existing measurements. In Sec. 8.6 we discuss attempts to improve the theoretical cross sections by using various distorted-wave methods. In Sec. 8.7, two-center coupled-channel calculations are presented and compared with other theoretical results and with measured cross sections.

We use full units in the major part of this chapter, except for the more technical sections 8.2.2, 8.4.1, and 8.4.3, where we adopt atomic units $\hbar = m_\mathrm{e} = e = 1$, see Appendix.

## 8.2 The Born approximation

### 8.2.1 The cross section in first order

As shown in Chaps. 6 and 7, for excitation and ionization, first-order perturbation theory is a valid approximation, except for very small impact parameters where the interaction may become extremely strong. For rearrangement collisions, the situation is more complicated. We start our discussion with the simplest and historically earliest approach to relativistic electron capture, namely the relativistic version [Mit64, Sha79, MoS80] of the OBK approximation [Opp28, BrK30]. It also serves to introduce some of the concepts and the notation needed for any other treatment. In this method, the initial and final electronic states are assumed to be undistorted atomic eigenstates $\varphi_\mathrm{i}$ and $\varphi_\mathrm{f}$ of target and projectile, respectively. Strictly speaking, this assumption is not valid since a Coulomb field leads to distortions even at infinite distances. The problem of asymptotic distortions is discussed in Secs. 6.1.2 and 6.5.4 for excitation and ionization, and its relevance for electron capture is studied in Sec. 8.3.

For the time being, we may just adopt the amplitude (5.78) given by perturbation theory and rewrite it in the prior form (5.83), with the perturbation (6.56), as

$$A_\mathrm{fi} = i\frac{e^2}{\hbar}\int dt \int d^3r_\mathrm{T}\, \varphi_\mathrm{f}'^{\dagger}(\mathbf{r}_\mathrm{P}')\, e^{iE_\mathrm{f}t'/\hbar}\, S\, \frac{Z_\mathrm{P}}{r_\mathrm{P}'}\, \varphi_\mathrm{i}(\mathbf{r}_\mathrm{T})\, e^{-iE_\mathrm{i}t/\hbar}, \tag{8.1}$$

where $E_\mathrm{i}$ and $E_\mathrm{f}$ are the atomic eigenenergies including the electron rest mass, and $\varphi_\mathrm{i}$ and $\varphi_\mathrm{f}'$ are initial and final bound-state spinor wave functions in the target and projectile frame, respectively. At this point, we may convince ourselves that in the nonrelativistic limit (energies $\epsilon_{\rm i}, \epsilon_{\rm f}$) the phase describing the time oscillation in the projectile frame

$$E_{\rm f}t' = \gamma E_{\rm f}\left(t - \frac{vz_{\rm T}}{c^2}\right) = m_{\rm e}c^2 t + \epsilon_{\rm f} t + \tfrac{1}{2}m_{\rm e}v^2 t - m_{\rm e}(\mathbf{v}\cdot\mathbf{r}_{\rm T}) + \cdots \quad (8.2)$$

consists of an immaterial contribution from the electron mass (which is cancelled by the corresponding term from $E_{\rm i}$), the nonrelativistic eigenoscillation, and the translation factor, see Eqs. (1.6) and (1.9). Thus the nonrelativistic limit of the transition amplitude (8.1) can be obtained as a special case.

By introducing the Fourier transforms $\tilde{\varphi}_{\rm i}^{\mu}$ and $\tilde{g}_{\rm f}^{\mu'}$ through

$$\varphi_{\rm i}^{\mu}(\mathbf{r}_{\rm T}) = (2\pi\hbar)^{-\frac{3}{2}} \int \tilde{\varphi}_{\rm i}^{\mu}(\mathbf{q}) e^{i\mathbf{q}\cdot\mathbf{r}_{\rm T}/\hbar}\, d^3 q \quad (8.3)$$

and

$$\varphi_{\rm f}^{\mu'}(\mathbf{r}'_{\rm P})\,\frac{1}{r'_{\rm P}} = (2\pi\hbar)^{-\frac{3}{2}} \int \tilde{g}_{\rm f}^{\mu'}(\mathbf{p}) e^{i\mathbf{p}\cdot\mathbf{r}'_{\rm P}/\hbar}\, d^3 p \quad (8.4)$$

where $\mu, \mu'$ denote the angular momentum projections (suppressing the label $j$) of the four-spinors $\varphi_{\rm i}^{\mu}, \tilde{\varphi}_{\rm i}^{\mu}, \varphi_{\rm f}^{\mu'}$, and $\tilde{g}_{\rm f}^{\mu'}$ whose spin-angular part is constructed according to Eq. (4.87) with appropriate space or momentum angles. Furthermore, by decomposing $\mathbf{q} = (\mathbf{q}_b, q_z)$ as well as $\mathbf{p} = (\mathbf{p}_b, p_z)$, we are able to move all space-time variables to the exponent and to perform the integrations over $\mathbf{r}_{\rm T}$ and $t$, which yield Dirac delta functions in the momentum vector and in the energy. As a result, we obtain the transition amplitude

$$A_{\rm fi}^{\mu'\mu} = i2\pi\frac{1}{v\gamma}e^2 Z_{\rm P}\int \tilde{g}_{\rm f}^{\mu'\dagger}(\mathbf{q}_b, q_-)\, S\, \tilde{\varphi}_{\rm i}^{\mu}(\mathbf{q}_b, q_+)\, e^{i\mathbf{q}_b\cdot\mathbf{b}/\hbar}\, d^2 q_b, \quad (8.5)$$

where the $z$-components of the momenta $\mathbf{q}$ and $\mathbf{p}$, respectively, are

$$\begin{aligned} q_+ &= (E_{\rm i} - E_{\rm f}/\gamma)/v \\ q_- &= (E_{\rm i}/\gamma - E_{\rm f})/v \end{aligned} \quad (8.6)$$

with the nonrelativistic limits

$$q_{\pm} \approx (\epsilon_{\rm i} - \epsilon_{\rm f})/v \pm \tfrac{1}{2}m_{\rm e}v. \quad (8.7)$$

Here, the + or − sign indicates that the relevant longitudinal momentum points forward or backward, with respect to the beam direction.

In an experiment, it will normally not be possible to specify the initial and final magnetic substates. We are therefore interested in the cross section

$$\sigma_{\rm fi} = \frac{1}{2j+1} \sum_{\mu,\mu'} \int |A_{\rm fi}^{\mu'\mu}|^2 \, d^2b \tag{8.8}$$

averaged over the initial and summed over the final projections $\mu$, $\mu'$, respectively, of the electron angular momenta $j$, $j'$. In evaluating Eq. (8.8) with the amplitude (8.5), it is convenient to construct density matrices from the initial and final four-spinors $\tilde{\varphi}_{\rm i}$ and $\tilde{g}_{\rm f}$ and to perform the sum over the (currently unobservable) angular momentum projections *within* each of the density matrices [Eic85]. Hence we define the matrices

$$P_{\rm i} = \sum_{\mu} P_{\rm i}^{\mu} = \sum_{\mu} \tilde{\varphi}_{\rm i}^{\mu}(\mathbf{q}_b, q_+) \, \tilde{\varphi}_{\rm i}^{\mu\dagger}(\mathbf{q}_b, q_+) \tag{8.9}$$

and

$$P_{\rm f} = \sum_{\mu'} P_{\rm f}^{\mu'} = \sum_{\mu'} \tilde{g}_{\rm f}^{\mu'}(\mathbf{q}_b, q_-) \, \tilde{g}_{\rm f}^{\mu'\dagger}(\mathbf{q}_b, q_-). \tag{8.10}$$

These expressions are diagonal in the momenta $(\mathbf{q}_b, q_\pm)$ because the **b**-integration in Eq. (8.8) produces a Dirac delta function in $\mathbf{q}_b, \mathbf{q}_b'$ owing to the exponential $\exp(i\mathbf{q}_b \cdot \mathbf{b}/\hbar)$ in Eq. (8.5). The total cross section is now written in the compact form

$$\sigma_{\rm fi} = \frac{(2\pi)^4}{2j+1} \frac{\hbar^4}{\beta^2\gamma^2} \left(\frac{e^2}{\hbar c}\right)^2 Z_{\rm P}^2 \int {\rm Tr}\{S \, P_{\rm f} \, S \, P_{\rm i}\} \, d^2 q_b, \tag{8.11}$$

where the integrand is expressed as the trace of a product of $4 \times 4$ matrices. The task is reduced to constructing the matrices $P_{\rm i}$ and $P_{\rm f}$ for each individual case. Owing to axial symmetry of the collision, the right-hand side of Eq. (8.11) is a one-dimensional integral. It turns out that in this case, the post and prior forms of the cross section as defined in Eqs. (5.82) and (5.83) are identical. From Eq. (8.11) and the $\gamma$ dependence of the transformation matrix $S$ it follows that the cross section decreases asymptotically as $\gamma^{-1}$ or $E^{-1}$ as the collision energy $E$ increases. The general dependence on the parameters characterizing the collision can be best seen in the simple analytical expression (8.52).

### 8.2.2 Capture from initial 1s-states into final 1s-states

The presence of high-momentum components in initial and final states renders 1s-1s capture the most important case at collision energies high above the matching energy, see Fig. 5.1.

From the spinor wave functions Eq. (4.86) for the hydrogenic 1s-state [1] (we use atomic units in this subsection)

$$\varphi^{\mu_s}(\mathbf{r}) = (4\pi)^{-\frac{1}{2}} \begin{pmatrix} g(r)\chi_{\mu_s} \\ -\, i f(r)(\boldsymbol{\sigma}\cdot\hat{\mathbf{r}})\chi_{\mu_s} \end{pmatrix} \tag{8.12}$$

with $\hat{\mathbf{r}} = \mathbf{r}/r$, one obtains the Fourier transforms $\tilde{\varphi}_{\mathrm{i}}^{\mu_s}$ and $\tilde{g}_{\mathrm{f}}^{\mu_s'}$ by inverting the defining equations (8.3) and (8.4), respectively, and by inserting the partial-wave expansion (6.42) of the plane wave. Only the $L = 0$ and $L = 1$ terms in the expansion contribute to the integrals over $g$ and $f$, respectively, and leave us with radial integrals whose evaluation yields the radial momentum functions [Eic85]

$$\begin{aligned} a_{\mathrm{i}}(q) &= \frac{1}{q}\frac{N_g\Gamma(s_{\mathrm{T}}+1)}{(Z_{\mathrm{T}}^2+q^2)^{(s_{\mathrm{T}}+1)/2}} \sin\left[(s_{\mathrm{T}}+1)\arctan(\frac{q}{Z_{\mathrm{T}}})\right] \\ b_{\mathrm{i}}(q) &= \frac{1}{q^2}\frac{N_{\mathrm{f}}\Gamma(s_{\mathrm{T}})}{(Z_{\mathrm{T}}^2+q^2)^{s_{\mathrm{T}}/2}} \sin\left[s_{\mathrm{T}}\arctan(\frac{q}{Z_{\mathrm{T}}})\right] \\ &\quad -\frac{1}{q}\frac{N_{\mathrm{f}}\Gamma(s_{\mathrm{T}}+1)}{(Z_{\mathrm{T}}^2+q^2)^{(s_{\mathrm{T}}+1)/2}} \cos\left[(s_{\mathrm{T}}+1)\arctan(\frac{q}{Z_{\mathrm{T}}})\right], \end{aligned} \tag{8.13}$$

where $\mathbf{q} = (\mathbf{q}_b, q_+)$, $s_{\mathrm{T}}$ is the parameter $s$ defined in Eq. (4.94) for $Z = Z_{\mathrm{T}}$, and the normalization factors $N_g$ and $N_{\mathrm{f}}$ are defined in Eq. (4.102). As a result, we get the Fourier-transformed spinor

$$\tilde{\varphi}_{\mathrm{i}}^{\mu_s}(\mathbf{q}) = \frac{1}{\pi\sqrt{2}} \begin{pmatrix} a_{\mathrm{i}}(q)\,\chi_{\mu_s} \\ -\, b_{\mathrm{i}}(q)(\boldsymbol{\sigma}\cdot\hat{\mathbf{q}})\chi_{\mu_s} \end{pmatrix} \tag{8.14}$$

in a form that closely resembles Eq. (8.12). When inserting into Eq. (8.9) and performing the spin summation we obtain the density matrix

$$P_{\mathrm{i}} = \frac{1}{2\pi^2} \begin{pmatrix} a^2 & -ab(\boldsymbol{\sigma}\cdot\hat{\mathbf{q}}) \\ -\, ab(\boldsymbol{\sigma}\cdot\hat{\mathbf{q}}) & b^2 \end{pmatrix} \tag{8.15}$$

which is built from blocks of $2\times 2$ matrices. Here and in the following, we do not explicitly display the unit matrix $I$ multiplying the diagonal quantities. In deriving Eq. (8.15), use is made of the relation $(\boldsymbol{\sigma}\cdot\mathbf{u})(\boldsymbol{\sigma}\cdot\mathbf{v}) =$

[1] We use the notation $\chi_{\mu_s}$ for the Pauli spinor with $\mu_s = \pm\frac{1}{2}$.

$\mathbf{u}\cdot\mathbf{v}+i\boldsymbol{\sigma}\cdot(\mathbf{u}\times\mathbf{v})$. The spin summation within the density matrix leads to a substantial simplification of the subsequent development. In an analogous fashion, we may construct $\tilde{g}_{\mathrm{f}}^{\mu_s'}$ and $P_{\mathrm{f}}$. One obtains the quantities $a_{\mathrm{f}}$ and $b_{\mathrm{f}}$ by replacing $Z_{\mathrm{T}} \to Z_{\mathrm{P}}$, $s_{\mathrm{T}} \to s_{\mathrm{P}} - 1$, and $\mathbf{q} = (\mathbf{q}_b, q_+) \to \mathbf{p} = (\mathbf{q}_b, q_-)$ in Eq. (8.13). From $P_{\mathrm{f}}$ we then assemble the matrix product

$$S\,P_{\mathrm{f}}\,S = \frac{\gamma+1}{8\pi^4} \times \begin{pmatrix} m_{11} & -m_{12}(\boldsymbol{\sigma}\cdot\hat{\mathbf{p}}) + m_{12}'(\boldsymbol{\sigma}\cdot\hat{\mathbf{e}}_z) \\ -m_{12}(\boldsymbol{\sigma}\cdot\hat{\mathbf{p}}) + m_{12}'(\boldsymbol{\sigma}\cdot\hat{\mathbf{e}}_z) & m_{22} \end{pmatrix} \tag{8.16}$$

where $\hat{\mathbf{e}}_z$ is a unit vector in the beam direction, and the coefficients are

$$\begin{aligned} m_{11} &= a_{\mathrm{f}}^2 + 2\delta a_{\mathrm{f}} b_{\mathrm{f}} q_-/p + \delta^2 b_{\mathrm{f}}^2 \\ m_{12} &= a_{\mathrm{f}} b_{\mathrm{f}}(1-\delta^2) \\ m_{12}' &= -\delta(a_{\mathrm{f}}^2 + 2\delta a_{\mathrm{f}} b_{\mathrm{f}} q_-/p + b_{\mathrm{f}}^2) \\ m_{22} &= \delta^2 a_{\mathrm{f}}^2 + 2\delta a_{\mathrm{f}} b_{\mathrm{f}} q_-/p + b_{\mathrm{f}}^2. \end{aligned} \tag{8.17}$$

Here, $\delta = [(\gamma-1)/(\gamma+1)]^{1/2}$ and $p = (q_b^2 + q_-^2)^{1/2}$. In the final matrix multiplication according to Eq. (8.11), we only need consider those terms that contribute to the trace.

*Total 1s – 1s cross sections*

The OBK cross section for capture of an (unpolarized) 1s-electron into both possible 1s-states of the projectile is then obtained (in atomic units) as

$$\sigma_{1s1s}^{\mathrm{OBK}} = 2\pi Z_{\mathrm{P}}^2 \frac{\gamma+1}{v^2\gamma^2} \int_0^\infty Q(q_b^2)\, d(q_b^2) \tag{8.18}$$

where the integrand is

$$Q = m_{11} a_{\mathrm{i}}^2 + \frac{2m_{12}}{pq}\, a_{\mathrm{i}} b_{\mathrm{i}} (q_b^2 + q_+ q_-) - \frac{2m_{12}'}{q} a_{\mathrm{i}} b_{\mathrm{i}} q_+ + m_{22} b_{\mathrm{i}}^2 \tag{8.19}$$

and $q = (q_b^2 + q_+^2)^{1/2}$. We have used the relation $Z_{\mathrm{P}}^2 + q_-^2 = Z_{\mathrm{T}}^2 + q_+^2$ which is rigorously valid for 1s-1s transitions.

It is possible to derive a simple analytical expression (8.52) from Eq. (8.18) which is valid for $\alpha Z \ll 1$. Since the expression so obtained [MoS80] is a special case of a formula derived for the eikonal approximation [Eic85], we defer the discussion to Sec. 8.4.

*Spin-flip cross sections*

It is instructive to use the explicit formulas given in [MoS80] to compare the cross sections $\sigma^{\text{flip}}$ for spin-flip and $\sigma^{\text{non}}$ for non-spin-flip transitions in 1s-1s capture. For $\alpha Z \ll 1$ the ratio is given by

$$\frac{\sigma^{\text{flip}}}{\sigma^{\text{non}}} = \frac{1}{16} \frac{\delta^4}{(1 - \frac{1}{2}\delta^2)[1 - \frac{1}{2}\delta^2 + \frac{5}{18}\pi\alpha(Z_{\text{T}} + Z_{\text{P}})\delta]}. \tag{8.20}$$

In the limit of low velocities, this ratio quickly approaches zero as one would expect since the magnetic field generated by the projectile motion in the target frame decreases with decreasing velocity. While some correction terms entering in Eq. (8.20) are treated in the high-velocity limit, the equation should nevertheless give a useful estimate for low velocities. For example, in p+H collisions at the energy of 100 keV, Eq. (8.20) yields a ratio of $1.8 \times 10^{-10}$ whereas an exact numerical evaluation gives the ratio $3.6 \times 10^{-10}$ [ToE90a]. On the other hand, in the extreme-relativistic regime, $\gamma \to \infty$ or $\delta \to 1$, the ratio tends to 1/4 in the limit of small charges and decreases for higher charges. The approximate validity of Eq. (8.20) has been confirmed [Dec90] for U + U collisions by exact numerical evaluation of the OBK cross sections. For U + U collisions at 500 MeV/u, the ratios following from various other approaches are collected in Table 8.2, see Sec. 8.7.

It should be remarked that spin-flip capture occurs for very high velocities even when target and projectile charges tend to zero. This is due to the Dirac magnetic moment of the free electron. Hence, if the wave functions (8.12) are replaced by some distorted wave functions, this will only affect the terms of the order $\alpha Z_T$ and $\alpha Z_P$ ignored in Eq. (8.20). In other words, we will expect Eq. (8.20) to be approximately valid for more general distorted-wave approaches provided $\alpha Z_T \ll 1$ and $\alpha Z_P \ll 1$.

### 8.2.3 Second-order Born approximation

Capture cross sections calculated in the relativistic first-order Oppenheimer-Brinkman-Kramers approximation (denoted as OBK1 when the second-order term is also considered), Eq. (8.18), are larger than the experimental values, in many cases by an order of magnitude, see, e.g., Fig. 8.8. The question arises whether this defect will be cured in the second order. The first calculations in the relativistic second-order OBK or OBK2 approximation have been performed by Humphries and Moiseiwitsch [HuM84, HuM85] and later, on a larger scale, by Moiseiwitsch [Moi88, Moi89]. Since a second-order OBK theory involves intermediate plane-wave states, multidimensional integrals occur.

The transition amplitude in the post form within the second-order OBK can be written [DeE91a] as

$$\begin{aligned} A^{(+)}_{\text{OBK2}} &= -\frac{i}{\hbar}\int dt \int d^3r_{\text{T}}[\psi'_{\text{f}}(\mathbf{r}'_{\text{P}},t')]^{\dagger}S^{-1}\left(-\frac{Z_{\text{T}}e^2}{r_{\text{T}}}\right)\psi_{\text{i}}(\mathbf{r}_{\text{T}},t) \\ & -\frac{i}{\hbar}\int dt \int d^3r_{\text{T}} \int dt_1 \int d^3r_{T1}[\psi'_{\text{f}}(\mathbf{r}'_{\text{P}},t')]^{\dagger}S^{-1}\left(-\frac{Z_{\text{T}}e^2}{r_{\text{T}}}\right) \\ & \times G_0^{+}(\mathbf{r}_{\text{T}},\mathbf{r}_{T1};t-t_1)S^2\left(-\frac{Z_{\text{P}}e^2}{r'_{P1}}\right)\psi_{\text{i}}(\mathbf{r}_{T1},t_1). \end{aligned} \tag{8.21}$$

While the first term describes a first-order or single-step transition, the second term comes about by a two-step mechanism. In the first step, mediated by the electron-projectile interaction, the bound target state $\psi_{\text{i}}$ is excited to an intermediate free-particle state with momentum $\mathbf{k}$. In the second step, mediated by the electron-target interaction, the free electron is captured into the bound projectile state $\psi_{\text{f}}$. The electron motion in the intermediate state is described by the free-field Feynman propagator [Sak67, Sch61]

$$G_0^{+}(\mathbf{r}_{\text{T}},\mathbf{r}_{T1};t-t_1) = \frac{1}{2\pi\hbar}\int G_0^{+}(\mathbf{r}_{\text{T}},\mathbf{r}_{T1};\varepsilon)\,e^{-i\varepsilon(t-t_1)/\hbar}d\varepsilon \tag{8.22}$$

where

$$G_0^{+}(\mathbf{r}_{\text{T}},\mathbf{r}_{T1};\varepsilon) = \sum_{\mathbf{k},s}\left\{\frac{\varphi^{(+)}_{\mathbf{k},s}(\mathbf{r}_{\text{T}})\varphi^{(+)\dagger}_{\mathbf{k},s}(\mathbf{r}_{T1})}{\varepsilon-|E^{(+)}_{\mathbf{k}}|+i\eta} + \frac{\varphi^{(-)}_{\mathbf{k},s}(\mathbf{r}_{\text{T}})\varphi^{(-)\dagger}_{\mathbf{k},s}(\mathbf{r}_{T1})}{\varepsilon+|E^{(-)}_{\mathbf{k}}|-i\eta}\right\}. \tag{8.23}$$

Here, one has to take the limit $\eta \to +0$. The spinors $\varphi^{(+)}_{\mathbf{k},s}(\mathbf{r}_{\text{T}})$ and $\varphi^{(-)}_{\mathbf{k},s}(\mathbf{r}_{\text{T}})$ for positive and negative energy, respectively, satisfy the free-field Dirac equation

$$(-ic\hbar\boldsymbol{\alpha}\cdot\nabla_{\text{T}} + m_{\text{e}}c^2\gamma^0)\,\varphi^{(\pm)}_{\mathbf{k},s}(\mathbf{r}_{\text{T}}) = \pm|E^{(\pm)}_{\mathbf{k}}|\,\varphi^{(\pm)}_{\mathbf{k},s}(\mathbf{r}_{\text{T}}). \tag{8.24}$$

The initial and final wave functions $\psi_{\text{i}}$ and $\psi_{\text{f}}$ in Eq. (8.21) are taken as hydrogenic $1s_{1/2}$ Dirac wave functions. The summation in (8.23) extends over all free-particle spinors with intermediate momentum $\mathbf{k}$ and both positive and negative energy.

From the analytical form of Eq. (8.21), one may obtain the asymptotic behavior of the OBK2 cross section for very large energies. Expressed in terms of the Lorentz factor $\gamma$, the asymptotically dominant dependence is [JaA85, DeE91a]

$$\sigma_{\text{OBK2}} \propto \frac{(\ln\gamma)^2}{\gamma}. \tag{8.25}$$

Except for the slowly varying factor $(\ln \gamma)^2$, which is caused by the infinite range of the Coulomb interaction and would disappear for a screened potential, the energy dependence is the same as in first-order perturbation theory.

The integrals occurring in (8.21) have been evaluated by using a "peaking approximation" [HuM84, HuM85], in which the integral is evaluated at the maximum of the integrand, and a related "averaging approximation" [Moi88, Moi89]. While the former approximation gave a moderate reduction of the cross sections with respect to the OBK1 values, the latter approximation brought about a large enough reduction to achieve good agreement with the experimental data [Cra79, Anh85a] in most cases. However, it has been shown [DeE91a] that an *exact* numerical evaluation of the 7-dimensional integrals required for the second-order cross section using a Monte-Carlo technique [Kaw86] points in the opposite direction: the exact OBK2 cross sections are not smaller but larger than the OBK1 cross sections by about one order of magnitude. This is, in fact, analogous to the nonrelativistic case where exact OBK2 cross sections are larger than the corresponding OBK1 cross sections for high but not yet asymptotic collision energies. For a more detailed discussion, see [Eic90, DeE91a,b, Moi92, DeE93].

We conclude that peaking-type approximations, which are widely used for mathematical convenience, are often not adequate except for asymptotically high energies. More importantly, however, we learn from the striking discrepancy between the first-order and second-order OBK cross sections for high-$Z$ target atoms that the OBK expansion is not a suitable perturbation series at the energies considered.

## 8.3 The boundary-corrected Born approximation

Similar to its nonrelativistic counterpart, the relativistic OBK approximation discussed in the preceding section has a serious defect: the wave functions in entrance and exit channels do not satisfy the correct asymptotic wave equations, Eqs. (5.48) and (5.49). Indeed the long-range nature of the Coulomb interaction does not allow one to use unperturbed atomic wave functions even at infinite separations. The vital importance of proper boundary conditions for nonrelativistic charge transfer has first been recognized by Dewangan and Eichler [DeE85, DeE86] when they showed that disregard of boundary conditions may lead to singularities in the transition amplitudes of second-order theories. Subsequently, a boundary-corrected first-Born (B1B) approximation has been worked out for many cases (for a review see, e.g., [DeE94]) and was shown to yield good agreement with experimental total cross sections.

In analogy to the nonrelativistic case, it is necessary to construct a first-order transition amplitude from the solutions of the asymptotic Dirac equations (5.48) and (5.49). A boundary-corrected first Born approximation for relativistic electron capture has been formulated in [Eic87a]. Since we keep the nonrelativistic acronyms throughout this book, we here denote it as B1B approximation.

Adopting a formulation in the target inertial frame, the solutions of the asymptotic equations for the initial target and the final projectile states can be written in the forms (5.61) and (5.62) as[2]

$$\Phi_{\mathrm{i}}^{\infty}(\mathbf{r}_{\mathrm{T}}, t) = e^{-i\nu_{\mathrm{P}} \ln(R' - vt')} \psi_{\mathrm{i}}(\mathbf{r}_{\mathrm{T}}, t) \qquad \text{for } t \to -\infty \tag{8.26}$$

and

$$\Phi_{\mathrm{f}}^{\infty}(\mathbf{r}_{\mathrm{T}}, t) = e^{i\nu_{\mathrm{T}} \ln(R + vt)}\, S^{-1}\, \psi_{\mathrm{f}}'(\mathbf{r}_{\mathrm{P}}', t') \qquad \text{for } t \to +\infty \tag{8.27}$$

where the Sommerfeld parameters are given by Eq. (5.52) as $\nu_{\mathrm{T}} = Z_{\mathrm{T}} e^2 / v\hbar$ and $\nu_{\mathrm{P}} = Z_{\mathrm{P}} e^2 / v\hbar$, while the internuclear separations $R$ and $R'$ are defined by Eq. (5.47). The phase factors, which are absent in the OBK approximation, serve to eliminate the asymptotic Liénard-Wiechert interactions caused by the projectile in the laboratory frame [Eq. (5.48)] and by the target in the projectile frame [Eq. (5.50)].

In contrast to the nonrelativistic case, the target wave function (8.26) does not depend on the target time variable $t$ alone but also on the projectile time coordinate $t'$. Conversely, the projectile wave function (8.27) also depends on the target time $t$. Now, since $t' = \gamma(t - v z_{\mathrm{T}}/c^2)$ and $t = \gamma(t' + v z_{\mathrm{P}}'/c^2)$, this means that the *electronic space coordinates* $z_{\mathrm{T}}$ and $z_{\mathrm{P}}'$ enter in the phase factors of (8.26) and (8.27), respectively. This is in contrast to the nonrelativistic distorted-wave theory, where only the *internuclear coordinates* enter in the distorting phase factor.

Once the asymptotic form of the channel wave functions[3] has been established, we are ready to derive the exact transition amplitude for electron capture. According to Eq. (5.80) we have for the post form

$$A_{\mathrm{fi}}^{(+)} = \lim_{t \to \infty} \langle \Phi_{\mathrm{f}}^{\infty} | \Psi_{\mathrm{i}}^{+} \rangle, \tag{8.28}$$

[2] A constant phase factor depending on the unit of length is dropped here and in the following phase factors.

[3] By the term "channel" we denote an asymptotic specification of the fragmentation of the total system, including the characterization by a set of quantum numbers.

where $\Psi_\mathrm{i}^+$ is the exact solution of the full Dirac equation (5.40). With the aid of the corresponding exact solution $\Psi_\mathrm{f}^-$ and by applying the exact and asymptotic Dirac equations we may derive the *prior* form (5.81) in a similar way. In summary, the exact amplitudes may be cast into one of the alternative forms

$$A_{\mathrm{fi}}^{(+)} = \frac{ie^2}{\hbar} \int_{-\infty}^{\infty} dt \left\langle \Phi_\mathrm{f}^\infty \left| \frac{Z_\mathrm{T}}{r_\mathrm{T}} - \frac{Z_\mathrm{T}}{R} \right| \Psi_\mathrm{i}^+ \right\rangle \qquad \text{(post form)} \tag{8.29}$$

or

$$A_{\mathrm{fi}}^{(-)} = \frac{ie^2}{\hbar} \int_{-\infty}^{\infty} dt \left\langle \Psi_\mathrm{f}^- \left| S^2 \left( \frac{Z_\mathrm{P}}{r'_\mathrm{P}} - \frac{Z_\mathrm{P}}{R'} \right) \right| \Phi_\mathrm{i}^\infty \right\rangle \qquad \text{(prior form).} \tag{8.30}$$

The perturbing potential in the matrix elements is no longer of the long-range Coulomb form but of a sufficiently short range to render conventional scattering theory applicable. This is indeed the very reason to choose a particularly convenient gauge by introducing the phase-distorted wave functions $\Phi_{\mathrm{f,i}}^\infty$. In order to obtain a first-order approximation with relativistically modified Coulomb boundary conditions, we have to replace $\Psi_\mathrm{i}^+$ with $\Phi_\mathrm{i}^\infty$ and $\Psi_\mathrm{f}^-$ with $\Phi_\mathrm{f}^\infty$. Inserting Eqs. (8.26) and (8.27), we obtain the explicit amplitudes [Eic87a]

$$\begin{aligned} A_{\mathrm{B1B}}^{(+)} &= \frac{ie^2}{\hbar} \int_{-\infty}^{\infty} dt \int d^3 r_\mathrm{T}\, [\psi'_\mathrm{f}(\mathbf{r}'_\mathrm{P}, t')]^\dagger e^{-i\nu_\mathrm{T} \ln(R+vt)} \qquad \text{(post form)} \\ &\quad \times S^{-1} \left( \frac{Z_\mathrm{T}}{r_\mathrm{T}} - \frac{Z_\mathrm{T}}{R} \right) e^{-i\nu_\mathrm{P} \ln(R'-vt')} \psi_\mathrm{i}(\mathbf{r}_\mathrm{T}, t) \end{aligned} \tag{8.31}$$

and

$$\begin{aligned} A_{\mathrm{B1B}}^{(-)} &= \frac{ie^2}{\hbar} \int_{-\infty}^{\infty} dt \int d^3 r_\mathrm{T}\, [\psi'_\mathrm{f}(\mathbf{r}'_\mathrm{P}, t')]^\dagger e^{-i\nu_\mathrm{T} \ln(R+vt)} \qquad \text{(prior form)} \\ &\quad \times S \left( \frac{Z_\mathrm{P}}{r'_\mathrm{P}} - \frac{Z_\mathrm{P}}{R'} \right) e^{-i\nu_\mathrm{P} \ln(R'-vt')} \psi_\mathrm{i}(\mathbf{r}_\mathrm{T}, t). \end{aligned} \tag{8.32}$$

These transition amplitudes are matrix elements taken between exact solutions $\Phi_\mathrm{i}^\infty$ and $\Phi_\mathrm{f}^\infty$ of the asymptotic Dirac equations (5.48) and (5.49), respectively. In this point, the B1B approximation is distinguished from other distorted-wave approximations for which the distorted waves are *not* exact eigenstates of the asymptotic Hamiltonian.

Similarly as for the OBK approximation, one may convince oneself that the post and prior amplitudes are identical, provided $\psi_\mathrm{i}$ and $\psi'_\mathrm{f}$ obey Eqs. (5.53) and (5.54), respectively.

In general, there are infinitely many ways to generate a short-range perturbing potential. For example, in Eq. (8.32), one may replace the residual interaction $-Z_P/R'$ by

$$U(R') = Z_P[(WR' + 1)\exp(-2WR') - 1]/R', \tag{8.33}$$

where $W > 0$ is a suitable parameter. The modified residual interaction $U(R')$ tends to $-Z_P/R'$ as $R' \to \infty$, but becomes a constant for $R' \to 0$. If $U(R')$ is used in Eq. (8.32), the phase factor distorting the initial state has to be modified accordingly; in general, it will be defined by a time integral [ToI87]. The choice among various distorted-wave approaches is guided by physical arguments and by calculational convenience, see also Sec. 8.6.

From Eqs. (8.31) and (8.32) one may calculate capture cross sections. Owing to the oscillatory factors in the integrand, the exact numerical evaluation is difficult. Cross sections have been obtained in the limit $\alpha Z \ll 1$ [Eic87a]. Some results are presented in Section 8.5 and compared with other theoretical and experimental data. Results for $U^{92+} + U^{91+}$ collisions at 500 MeV/u are included in Table 8.3, Sec. 8.7.1, for 1s capture into various final shells. In these cases, an exact numerical evaluation of the cross section has been performed.

The asymptotic energy dependence of the B1B cross sections for large collision energies is the same as that of the OBK approximation: the cross section decreases as $\gamma^{-1}$ or $E^{-1}$ as the projectile energy increases.

## 8.4 The eikonal approximation

It is the aim of a distorted-wave treatment of electron transfer to incorporate the effect of the projectile on the target states and the effect of the target on the projectile states, the rigorous prescription being given by Eqs. (5.82) and (5.83). A special case is provided by the boundary-corrected Born approximation, Sec. 8.3, in which the asymptotic distortion is taken into account by replacing in the interaction term the electron-nucleus separation with the internuclear distance.

A suggestive ansatz for a more accurate description of the mutual distortion is given by initial and final wave functions, each of them being a product of a bound-state wave function with respect to one of the partners times a Coulomb continuum wave function with respect to the other collision partner. This concept is the basis of the "continuum distorted-wave" (CDW) approximation [Che64] which is successful in nonrelativistic collisions, but difficult to implement for relativistic collisions, see Sec. 8.6.

A related but simpler approximation is obtained if the full Coulomb continuum wave function is replaced with its asymptotic form, namely a phase factor depending on the electronic coordinates. In a more general context, this approach is called the eikonal method which has a time-honored history in optics. It was introduced in potential scattering by Molière [Mol47], in high-energy scattering by Glauber [Gla58] and in fast atomic collisions by Wilets and Wallace [WiW68]. The eikonal approach, in its asymmetric form in which the distortion is included in only one of the channels, was applied to nonrelativistic charge transfer in [Dew75, ChE79a, ChE79b, Eic81] and was shown to be a rather accessible distorted-wave theory which yields good overall agreement with experimental data at high but not yet asymptotic velocities. It includes higher-order Born terms describing multiple soft collisions in an approximate fashion [EiN80] while at the same time preserving the simplicity of the OBK approximation, which drastically overestimates the cross sections.

The relativistic eikonal approximation has been introduced and worked out for arbitrary initial and final states in [Eic85]. Since a relativistic symmetrical treatment gives rise to serious problems and misrepresents the experimental data, see Sec. 8.6, it is reasonable to adopt the unsymmetrical treatment, in which only the initial or the final wave function is distorted. This corresponds to replacing the asymptotic states $\Phi_{\mathrm{P}}^{\infty}$ in Eq. (5.82) for the post form and $\Phi_{\mathrm{T}}^{\infty}$ in Eq. (5.83) for the prior form, respectively, by unperturbed bound final and initial states, while for the exact states $\Psi_{\mathrm{i}}^{+}$ and $\Psi_{\mathrm{f}}^{-}$ phase-distorted eikonal wave functions are substituted.

To be specific, let us choose the prior form. Following the presentation of the OBK approximation in Sec. 8.1, the eikonal transition amplitude in the prior form is written as

$$A_{\mathrm{fi}} = i\frac{e^2}{\hbar}\int dt \int d^3r_{\mathrm{T}}\,[\psi_{\mathrm{f}}'(\mathbf{r}_{\mathrm{P}}', t)]^{\dagger}\, S\,\frac{Z_{\mathrm{P}}}{r_{\mathrm{P}}'}\,\psi_{\mathrm{i}}(\mathbf{r}_{\mathrm{T}}, t). \tag{8.34}$$

In the prior form, the initial and final wave functions[4]

$$\psi_{\mathrm{i}}(\mathbf{r}_{\mathrm{T}}, t) = \varphi_{\mathrm{i}}(\mathbf{r}_{\mathrm{T}})\, e^{-iE_{\mathrm{i}}t/\hbar} \tag{8.35}$$

and

$$\psi_{\mathrm{f}}'(\mathbf{r}_{\mathrm{P}}', t') = \varphi_{\mathrm{f}}'(\mathbf{r}_{\mathrm{P}}')\, e^{-iE_{\mathrm{f}}t'/\hbar}\, e^{i\nu_{\mathrm{T}}' \ln(r_{\mathrm{T}}+z_{\mathrm{T}})} \tag{8.36}$$

are expressed by the unperturbed target and projectile wave functions $\varphi_{\mathrm{i}}$ and $\varphi_{\mathrm{f}}'$ in their respective frames. The Sommerfeld parameter $\nu_{\mathrm{T}}' = Z_{\mathrm{T}}'e^2/\hbar v$ and the charge number $Z_{\mathrm{T}}'$ in the eikonal wave function (8.36) are labeled by a prime in order to provide a unique signature for the electron-target interaction in the *eikonal phase.* This allows one to recover the OBK approximation by simply setting $Z_{\mathrm{T}}' = 0$ or $\nu_{\mathrm{T}}' = 0$.

[4] A constant phase factor depending on the unit of length is dropped in the eikonal phase factor.

In the prior version of the eikonal approach, the final wave function is phase-distorted by the electron-target interaction. The distorting phase factor $\exp[i\nu'_{\mathrm{T}} \ln(r_{\mathrm{T}} + z_{\mathrm{T}})]$ is the asymptotic form of the Coulomb wave function (8.57). It may also be represented by an eikonal integral through the relation

$$\exp[i\nu'_{\mathrm{T}} \ln(r_{\mathrm{T}} + z_{\mathrm{T}})] \propto \exp\left[-i\frac{e^2}{\hbar}\int_t^\infty \frac{Z'_{\mathrm{T}}}{r_{\mathrm{T}}}d\tau\right], \tag{8.37}$$

where the proportionality factor is an irrelevant constant phase factor and where $\mathbf{r}'_{\mathrm{P}}$ is kept fixed when evaluating the integral. This is in line with the condition for the validity of the eikonal approach [EiN80] that the collision is "fast" in the sense of Sec. 5.4.1.

Equation (8.37) shows that the electron-target interaction (occurring in the exponent) acting from the time of capture to infinity is treated nonperturbatively, while the electron-projectile interaction in Eq. (8.34) explicitly enters in first order. Although the wave function (8.36) has the correct asymptotic behavior, the treatment as a whole does not satisfy Coulomb boundary conditions and the interaction in Eq. (8.34) is not of short range, because the initial wave function (8.35) remains undistorted.

The eikonal approximation may also be formulated in the *post* version, which is obtained by inverting the role of target and projectile in Eqs. (8.34) to (8.36), that is by replacing $Z_{\mathrm{P}}$ by $Z_{\mathrm{T}}$ and *vice versa* and by interchanging initial and final states. As the theory is not post-prior symmetric, both versions yield different results.

Since in an asymmetric theory like the eikonal approximation, the stronger one of the electron-target and electron-projectile interaction should be treated nonperturbatively, the usual prescription, which we denote as $Z$-criterion, is

$$\begin{array}{llll} \text{if} & Z_{\mathrm{P}} < Z_{\mathrm{T}} & \text{then use the } \textit{prior} \text{ form} \\ \text{if} & Z_{\mathrm{P}} > Z_{\mathrm{T}} & \text{then use the } \textit{post} \text{ form.} \end{array} \tag{8.38}$$

An alternative prescription, Eq. (8.49), which deviates from (8.38) for higher principal shells has also been proposed [MeA85]; however, in this section, we adopt the standard form (8.38).

## 8.4.1 Capture from initial into final $1s_{1/2}$-states

Similarly as for the OBK approximation, Sec. 8.2.1, we explicitly consider the simplest case which is also the most important one, namely $1s_{1/2}$–$1s_{1/2}$ transitions. We adopt atomic units throughout this subsection.

The analytical treatment of the eikonal approximation starts from the relation [GaM74]

$$e^{-i\nu_T' \ln(r_T+z_T)} = [\Gamma(i\nu_T')]^{-1} \int_0^\infty \lambda^{i\nu_T'-1} e^{-\lambda(r_T+z_T)}\, d\lambda \tag{8.39}$$

and the corresponding Fourier transform $\tilde{h}_i(\mathbf{q}, \lambda)$ defined by

$$\varphi_i^{\mu_s}(\mathbf{r}_T) e^{-\lambda(r_T+z_T)} = (2\pi)^{-\frac{3}{2}} \Gamma(i\nu_T') \int \tilde{h}_i^{\mu_s}(\mathbf{q}, \lambda) e^{i\mathbf{q}\cdot\mathbf{r}_T}\, d^3q \tag{8.40}$$

which replaces $\tilde{\varphi}_i(\mathbf{q})$ in Eqs. (8.3) and (8.5). The prior form of the eikonal transition amplitude is therefore given by

$$A_{fi}^{\mu_s'\mu_s} = i\frac{2\pi}{v\gamma} Z_P \int d^2q_b \int_0^\infty d\lambda\, \tilde{g}_f^{\mu_s'\dagger}(\mathbf{q}_b, q_-)\, S\, \tilde{h}_i^{\mu_s}(\mathbf{q}_b, q_+; \lambda) \lambda^{i\nu_T'-1} e^{i\mathbf{q}_b\cdot\mathbf{b}}, \tag{8.41}$$

where $q_+$ and $q_-$ are defined in Eq. (8.6).

The expression (8.41) is valid for arbitrary initial and final states. The cross section is again conveniently written in terms of density matrices in the form of Eq. (8.11). The density matrix for the final state is the same as for the OBK approximation, Eq. (8.10), and only the initial state density matrix has to be modified compared to Eq. (8.9). The construction of $P_f$ is complicated by the presence of the factor $e^{-\lambda(r_T+z_T)}$ in Eq. (8.40) and by the integration over $\lambda$. While the radial part of the exponent can be combined with that of the wave function, the term $e^{-\lambda z_T}$ is handled conveniently by absorbing it into the plane wave through the definition of a complex wave vector

$$\mathbf{k}_\lambda = \mathbf{q} - i\lambda\hat{\mathbf{e}}_z = (\mathbf{q}_b, q_+ - i\lambda) \tag{8.42}$$

with $k_\lambda^2 = q_b^2 + q_+^2 - \lambda^2 - 2i\lambda q_+$. As a result, the quantities $a_i$, $b_i$, and $\hat{\mathbf{q}}$ in Eqs. (8.14) and (8.15) are replaced with complex integrals. With the abbreviation $\lambda_\pm = \lambda + Z_T \pm ik_\lambda$ and with $s = s_T$ [see Eq. (4.94) for $Z = Z_T$] and $\nu = \nu_T'$, the integrals for 1s-1s transitions take the form

$$A_0(\mathbf{q}) = N_g \frac{\Gamma(s+1)}{\Gamma(i\nu)} \int_0^\infty \frac{d\lambda}{2ik_\lambda} (\lambda_-^{-(s+1)} - \lambda_+^{-(s+1)})\, \lambda^{i\nu-1}$$

$$\begin{aligned} B_m(\mathbf{q}) = {} & N_f \frac{\Gamma(s)}{\Gamma(i\nu)} \int_0^\infty \frac{d\lambda}{2ik_\lambda^3} (\lambda_-^{-s} - \lambda_+^{-s})\, \lambda^{i\nu-1+m} \\ & -N_f \frac{\Gamma(s+1)}{\Gamma(i\nu)} \int_0^\infty \frac{d\lambda}{2k_\lambda^2} (\lambda_-^{-(s+1)} + \lambda_+^{-(s+1)})\, \lambda^{i\nu-1+m}. \end{aligned} \tag{8.43}$$

If the integrals $B_m$ with $m = 0, 1$ are combined to build a vector

$$\mathbf{K} = (B_0 \mathbf{q}_b,\ B_0 q_+ - iB_1), \tag{8.44}$$

the $\lambda$-integral over the spinor function $\tilde{h}_{\mathrm{i}}^{\mu_s}$ assumes a structure analogous to Eq. (8.14), namely

$$\int_0^\infty \tilde{h}_{\mathrm{i}}^{\mu_s}(\mathbf{q}, \lambda)\, \lambda^{i\nu - 1}\, d\lambda = \frac{1}{\pi\sqrt{2}} \begin{pmatrix} A_0 \chi_{\mu_s} \\ -(\boldsymbol{\sigma} \cdot \mathbf{K}) \chi_{\mu_s} \end{pmatrix}. \tag{8.45}$$

The density matrix for the initial state is then obtained as

$$P_{\mathrm{i}} = \frac{1}{2\pi^2} \begin{pmatrix} |A_0|^2 & -A_0(\boldsymbol{\sigma} \cdot \mathbf{K}^*) \\ -A_0^*(\boldsymbol{\sigma} \cdot \mathbf{K}) & (\boldsymbol{\sigma} \cdot \mathbf{K})(\boldsymbol{\sigma} \cdot \mathbf{K}^*) \end{pmatrix} \tag{8.46}$$

in place of Eq. (8.15). The matrix multiplication is now carried out using Eqs. (8.16) and (8.17) in the expression (8.11) for the cross section. The eikonal cross section for the capture of an unpolarized 1s-electron into both possible projectile 1s-states then reads [Eic85]

$$\sigma_{1s-1s}^{\mathrm{eik}} = 2\pi Z_{\mathrm{P}}^2 \frac{\gamma + 1}{v^2 \gamma^2} \int_0^\infty Q^{\mathrm{eik}}(q_b^2)\, d(q_b^2) \tag{8.47}$$

with the integrand

$$\begin{aligned} Q^{\mathrm{eik}} \;\; = \;\; & m_{11}|A_0|^2 + (2m_{12}/q)[\mathrm{Re}(A_0 B_0^*)(q_b^2 + q_+ q_-) - \mathrm{Im}(A_0 B_1^*) q_-] \\ & + 2m'_{12}[-\mathrm{Re}(A_0 B_0^*) q_+ + \mathrm{Im}(A_0 B_1^*)] \\ & + m_{22}[|B_0|^2 (q_b^2 + q_+^2) - 2\,\mathrm{Im}(B_0 B_1^*) q_+ + |B_1|^2]. \end{aligned} \tag{8.48}$$

The momenta $q_\pm$ are defined in Eq. (8.6), the parameters $m_{ik}$ in Eq. (8.17), and the matrix elements $A_0$, $B_k$ in Eq. (8.43).

The calculation of the cross section requires the numerical evaluation of the $\lambda$-integrals in $A_0$, $B_0$, $B_1$ and the additional one-dimensional integration over $q_b^2$. Explicit expressions including initial and final L-shells have also been given [Eic85]. Numerical results have been published in [AnE85] and [MeA85] and are presented in Sec. 8.5.1.

## 8.4.2 Transfer between higher initial and final shells

The method outlined in Section 8.4.1 can be generalized to arbitrary initial and final states. In fact, the eikonal approximation, owing to its intrinsic simplicity, is particularly suited for this extension, and no other perturbative approach has been carried so far. For details of the formulation and of the resulting expression, we refer to [Eic85] and [Eic90].

We just mention here an alternative prescription for choosing between the prior and the post form of the eikonal approximation. If the initial and final shells are different, one may think of replacing the prescription (8.38) by another one in which the weighting of the interaction with the Coulomb wave functions is taken into account. Since the expectation value of the Coulomb potential is equal to $\langle Z/r \rangle_n = (Z/n)^2$ irrespective of the subshell, the effective strength of the potential should be measured by $Z/n$ rather than by $Z$. This leads to an alternative prescription [MeA85], which we denote as $Z/n$-criterion:

$$\begin{array}{llcll} \text{if} & Z_\mathrm{P}/n_\mathrm{P} & < & Z_\mathrm{T}/n_\mathrm{T} & \text{then use the } \textit{prior} \text{ form,} \\ \text{if} & Z_\mathrm{P}/n_\mathrm{P} & > & Z_\mathrm{T}/n_\mathrm{T} & \text{then use the } \textit{post} \text{ form.} \end{array} \tag{8.49}$$

This rule may entail different choices for different combinations of initial and final states within the same projectile-target combination. In Figs. 8.9– 8.11, we compare the prediction of the usual $Z$-criterion (8.38) with that of the $Z/n$-criterion (8.49).

The eikonal approach has been compared with other theoretical calculations as well as with experimental data. For electron capture by 1.05 GeV/u $\mathrm{Ne}^{10+}$ ions from target atoms with charges $Z_\mathrm{T}$ =13, 30, 47, 73, and 92, the eikonal calculations are in rather close agreement with relativistic boundary-corrected Born calculations, see Fig. 8.8, which are post-prior symmetric, in contrast to the eikonal theory. Furthermore, the approach is supported by the reasonable agreement with a variety of experimental data [Eic90]. In some cases, detailed two-center coupled-channel calculations [ToE88a,b], see Sec. 8.7.2, yield cross sections of a similar, but not superior quality.

We now present a survey of calculated cross sections in their dependence on the various parameters entering into the theory. Since capture cross sections vary over many orders of magnitude, even approximate results will help to assess the parameter dependence and the qualitative behavior of the cross section.

In Fig. 8.2, we show the dependence on target and projectile charges for the leading transition ($1\mathrm{s}_{1/2}$-$1\mathrm{s}_{1/2}$) at a fixed projectile energy of 10 GeV/u. As can be seen from the subsequent figures, the other cross sections follow a similar pattern. The set of curves reflects the approximate dependence as $Z_\mathrm{T}^5 Z_\mathrm{P}^5$ of the capture cross section, which is characteristic for all capture theories, see, e.g., [DeE94].

In Figs. 8.3–8.7, we display the capture cross sections for completely

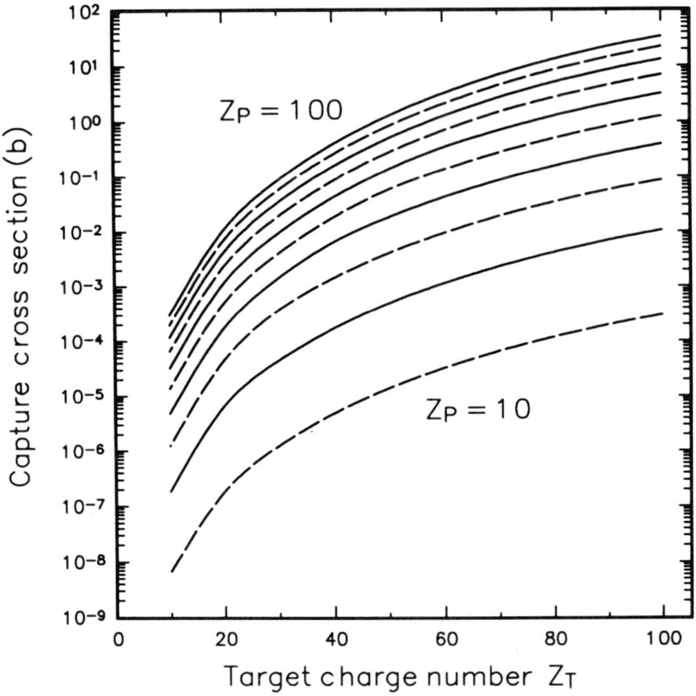


**Figure 8.2.** Cross sections (in barn) for $1s_{1/2}$-$1s_{1/2}$ electron capture at a projectile energy of 10 GeV/u as a function of the target and projectile charges $Z_T$ and $Z_P$ in steps of 10 units. The cross sections are calculated according to the prescription (8.38) and refer to a single target electron captured into a vacant projectile K-shell. They are averaged over the initial and summed over the final angular momentum projections. From [IcS93].

stripped Au ions impinging on hydrogen-like C, Al, Cu, Ag, and Au targets as a function of energy. The data are averaged over the initial and summed over the final angular momentum projections. For other heavy projectiles, one can scale these curves approximately as $Z_P^5$. In the state-to-state cross sections presented, either the initial or the final state is a $1s_{1/2}$ state (in order to reduce the number of curves). As a function of the projectile energy, the curves clearly exhibit the transition from an essentially nonrelativistic to a relativistic behavior: at rather low collision energies around 1 GeV/u, the cross section curves still reflect the rapid decrease as $E^{-6}$ known from nonrelativistic collisions, while in the extreme relativistic energy range above 10 GeV/u, the curves bend over to the $E^{-1}$ dependence.

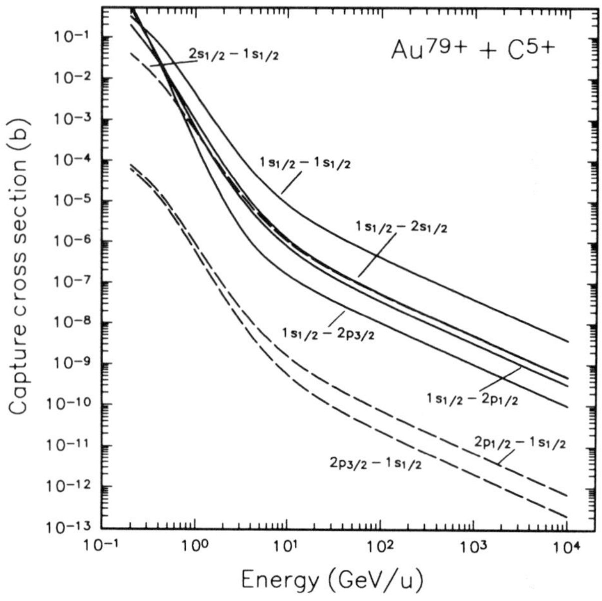


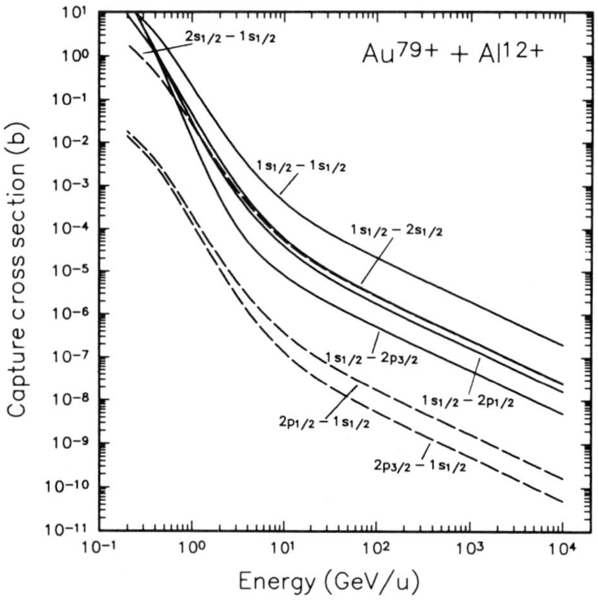


**Figures 8.3 (top) and 8.4 (bottom).** Cross sections for electron capture by $Au^{79+}$ ions as a function of the collision energy. See caption of Fig. 8.2.

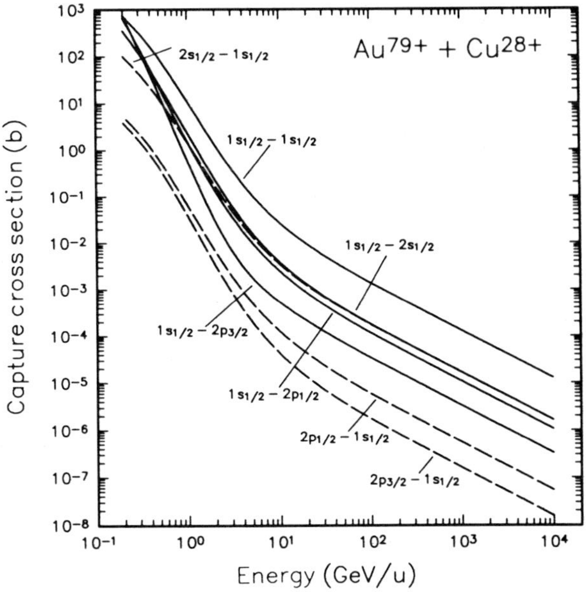


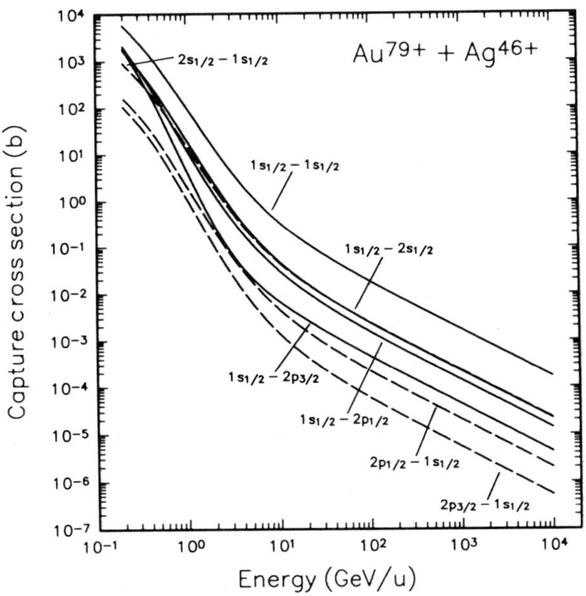


**Figures 8.5 (top) and 8.6 (bottom).** Cross sections for electron capture by $Au^{79+}$ ions as a function of the collision energy. See caption of Fig. 8.2.

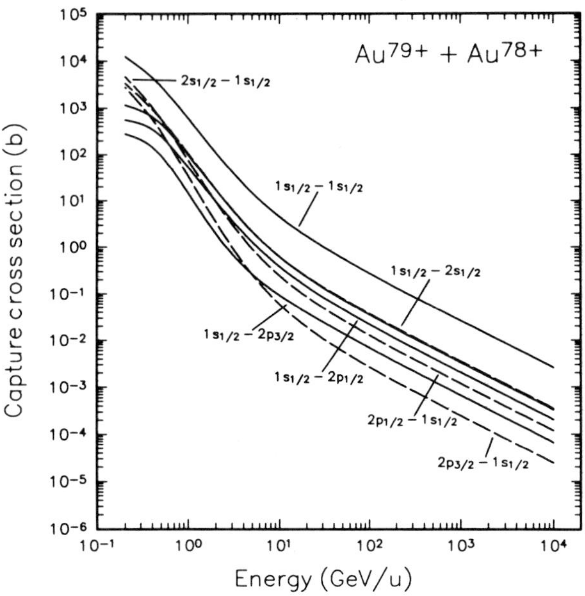


**Figure 8.7.** Cross sections for electron capture by $Au^{79+}$ ions as a function of the collision energy. See caption of Fig. 8.2.

For the more asymmetric systems, the spread of curves is much larger than for symmetric collisions. One also observes the effect of increased spin-orbit splitting with increasing charge of the collision partner.

The eikonal cross sections presented here are available in tabular form for more initial and final states and also for uranium projectiles [IcS93]. We believe that these cross sections are useful global estimates. While it is difficult to quantitatively assess the accuracy of the results, it is expected to increase with the collision energy.

### 8.4.3 An approximate formula for 1s-1s transitions

Although the eikonal cross section Eq. (8.47) can be evaluated numerically, it is useful to give an approximate analytical expression. Any relativistic treatment includes two kinds of relativistic effects, namely the relativistic projectile motion, characterized by the parameter $\delta = [(\gamma - 1)/(\gamma + 1)]^{1/2}$, and the relativistic motion of the electron, characterized by $\alpha Z$, in its initial and final atomic orbits. Compared to the nonrelativistic case, the relativistic electron motion has two important consequences. (1) The electron orbitals and their binding energies are modified. (2) The electron acquires a Dirac magnetic moment which, in turn, interacts with the induced magnetic field produced by the projectile motion. The occurrence of a Dirac magnetic moment is, of course, independent of the charge $Z$.

In order to derive an approximate cross section formula we expand the electronic wave functions in powers of $\alpha Z_{\mathrm{P,T}} \ll 1$ keeping only the leading terms and at the same time adopting a high-energy approximation for the correction terms in the cross section. This means that in Eq. (8.13) we take $|q_\pm| \approx 137\,\delta \gg Z_{\mathrm{P,T}}$ and $\arctan(q/Z_{\mathrm{T}}) \approx \pi/2$. Correspondingly, in (8.43) we assume that contributions to the integrals arise mainly from values $\lambda \ll 137$. Writing the Sommerfeld parameter $\nu_{\mathrm{T}} = Z_{\mathrm{T}}/v$ (in atomic units) explicitly, Eichler [Eic85] has derived an analytical formula both for the eikonal cross section (setting $Z'_{\mathrm{T}} = Z_{\mathrm{T}}$) and for the OBK cross section (setting $Z'_{\mathrm{T}} = 0$). For convenience, we state the two expressions separately. The 1s-1s eikonal cross section per initial electron summed over the two final spin states is given by

$$
\begin{aligned}
\sigma^{\mathrm{eik}}_{1s-1s} &= \frac{2^8 \pi Z_{\mathrm{P}}^5 Z_{\mathrm{T}}^5}{5 v^2 (Z_{\mathrm{T}}^2 + q_+^2)^5} \, \frac{\gamma+1}{2\gamma^2} \, \frac{\pi \nu_{\mathrm{T}}}{\sinh(\pi \nu_{\mathrm{T}})} \\
&\quad \times e^{-2\nu_{\mathrm{T}} \arctan(q_+/Z_{\mathrm{T}})} \, (S_{\mathrm{eik}} + S_{\mathrm{mag}} + S_{\mathrm{orb}})
\end{aligned}
\tag{8.50}
$$

with

$$
\begin{aligned}
S_{\mathrm{eik}} &= 1 - \tfrac{5}{4} v^{-1} q_+ + \tfrac{5}{12} v^{-2} q_+^2 + \tfrac{1}{6} v^{-2} Z_{\mathrm{T}}^2 \\
S_{\mathrm{mag}} &= -\delta^2 + \tfrac{5}{16}\delta^4 + \tfrac{5}{8}\delta^2 \frac{\gamma}{\gamma+1} + \tfrac{1}{4}\delta^2 v^{-2} Z_{\mathrm{T}}^2 + \tfrac{5}{48}\,\delta^4 v^{-2} Z_{\mathrm{T}}^2 \\
S_{\mathrm{orb}} &= \tfrac{5}{18}\pi\delta\alpha(Z_{\mathrm{P}} + Z_{\mathrm{T}}) - \tfrac{5}{36}\pi\delta^3\alpha(Z_{\mathrm{P}} + Z_{\mathrm{T}}) - \tfrac{5}{8}\delta\alpha v^{-1} Z_{\mathrm{T}}^2 (1 - \tfrac{1}{2}\delta^2) \\
&\quad - \tfrac{5}{18}\pi\,\delta \frac{\gamma}{\gamma+1}\,\alpha Z_{\mathrm{P}} + \tfrac{5}{28}\,\pi\delta \left(\frac{\gamma}{\gamma+1}\right)^2 \alpha Z_{\mathrm{P}} \\
&\quad - \tfrac{5}{28}\,\pi\delta \frac{\gamma}{\gamma+1}\,\alpha(Z_{\mathrm{P}} + Z_{\mathrm{T}} - \delta^2 Z_{\mathrm{P}}),
\end{aligned}
\tag{8.51}
$$

where $q_+$ is defined in Eq. (8.6). This expression can be used to estimate capture cross sections and allows us to discuss a few limiting cases. (a) If we only keep $S_{\mathrm{eik}}$ in Eq. (8.50), we obtain the eikonal cross section for relativistic projectiles but nonrelativistic spinless electrons. (b) If, furthermore, we let $\gamma \to 1$ and use the nonrelativistic limit $q_+^{\mathrm{nonrel}} = (\epsilon_{\mathrm{i}} - \epsilon_{\mathrm{f}})/v + \frac{1}{2}v$ of Eq. (8.7), we recover the nonrelativistic eikonal cross section [ChE79a,b, EiC79]. (c) The term $S_{\mathrm{mag}}$ survives for vanishing target charges [since the last two terms originally arise from $Z'_{\mathrm{T}}$ or $\nu'_{\mathrm{T}}$ in (8.34) and hence represent a final-state interaction] and tends to zero with decreasing projectile velocity.

Therefore, it is interpreted as a magnetic contribution to capture mediated by the interaction between the relativistically induced magnetic field of the projectile and the Dirac magnetic moment of the electron. (d) The quantity $S_{\rm orb}$ is composed of terms that explicitly include $\alpha Z_{\rm P}$ or $\alpha Z_{\rm T}$ and hence are interpreted as correction terms arising from a relativistic modification of the electronic orbitals.

Setting $Z'_{\rm T} = 0$, the approximate OBK cross section [MoS80] is simultaneously derived as

$$\sigma^{\rm OBK}_{1s-1s} = \frac{2^8 \pi Z_{\rm P}^5 Z_{\rm T}^5}{5v^2(Z_{\rm T}^2 + q_+^2)^5} \frac{\gamma+1}{2\gamma^2} \left(1 + S^{\rm OBK}_{\rm mag} + S^{\rm OBK}_{\rm orb}\right) \tag{8.52}$$

with

$$\begin{aligned} S^{\rm OBK}_{\rm mag} &= -\delta^2 + \tfrac{5}{16}\delta^4 \\ S^{\rm OBK}_{\rm orb} &= \tfrac{5}{18}\pi\delta\alpha(Z_{\rm P} + Z_{\rm T}) - \tfrac{5}{36}\pi\delta^3\alpha(Z_{\rm P} + Z_{\rm T}). \end{aligned} \tag{8.53}$$

Both in the relativistic and in the nonrelativistic case [DeE94], the OBK cross sections considerably overestimate the experimental results. For a rough estimate of total cross sections involving higher principal shells, one may use Eq. (8.50) by applying a scaling rule. One starts from the observation that in the nonrelativistic OBK approximation and in the nonrelativistic eikonal approximation up to second order in the expansion of the eikonal phase factor, the cross section scales with $Z/n$ where $n$ is the principal quantum number of the shell in question. Thus one may formulate an approximate scaling rule [MeA85].

Approximate relativistic capture cross sections averaged over subshells within arbitrary initial and final principal shells specified by their principal quantum numbers $n_{\rm T}$ and $n_{\rm P}$, respectively, can be obtained from Eq. (8.50) by replacing

$$\begin{aligned} Z_{\rm T} &\rightarrow \frac{Z_{\rm T}}{n_{\rm T}} \\ Z_{\rm P} &\rightarrow \frac{Z_{\rm P}}{n_{\rm P}} \end{aligned} \tag{8.54}$$

in the prior form. To obtain the post form, the target and the initial and final states are interchanged.

If this scaling rule is adopted, the decision between the prior and the post form will be automatically based on the $Z/n$-criterion (8.49). A numerical comparison between the exact and the approximate form of the eikonal cross section is given in Table 8.1 and, more systematically, in [Eic90].

**Table 8.1.** Theoretical cross sections per electron (in barn) for 1s-1s electron capture by 1050 MeV/u $Ne^{10+}$ ions. The number in square brackets gives the power of 10 multiplying the preceding number. The OBK, exact, and approximate eikonal results are calculated from Eqs. (8.18), (8.47), and (8.50), respectively, the B1B results are taken from [Eic87a].

| $Z_T$ | OBK | exact eikonal | approx. eikonal | B1B |
|---|---|---|---|---|
| 13 | 2.00[−4] | 3.93[−5] | 3.85[−5] | 5.53[−5] |
| 30 | 1.19[−2] | 1.81[−3] | 1.76[−3] | 2.31[−3] |
| 47 | 8.47[−2] | 1.10[−2] | 1.10[−2] | 1.31[−2] |
| 73 | 3.35[−1] | 4.10[−2] | 4.81[−2] | 4.86[−2] |
| 92 | 4.15[−1] | 5.60[−2] | 8.34[−2] | 7.24[−2] |

## 8.5 Theoretical and experimental cross sections

### 8.5.1 Theoretical total cross sections

We now have at our disposal several theories which are all based on a perturbative treatment. It is worthwhile to compare their predictions for capture cross sections among each other and with existing experimental data.

In order to compare various theoretical approximations, we present in Table 8.1 cross sections for $1s_{1/2}$-$1s_{1/2}$ electron capture by 1050 MeV/u bare Ne ions from various single-electron (idealized) targets. A comparison with experimental data is given below. Besides the OBK approximation, which yields considerably too large cross sections, we show results for the eikonal approximation in an exact [Eq. (8.47)] and an approximate [Eq. (8.50)] evaluation, and the B1B approximation, see Sec. 8.2. It is seen that the eikonal and the B1B approximations yield similar results. Of course the numbers for the highest charges $Z_T$ have to be taken with some caution, since one gets close to the energy matching, see Sec. 5.5.1, between electron and projectile, and also the $\alpha Z$ expansion used for the approximate eikonal and for the B1B approximation will lead to inaccuracies. Nevertheless, a comparison between the exact and the approximate eikonal results shows that the deficiencies of the $\alpha Z$ expansion are not too dramatic in this case.

For a comparison of various theories for the (because of the smallness of the cross sections, somewhat academic) case of p + H collisions up to 1

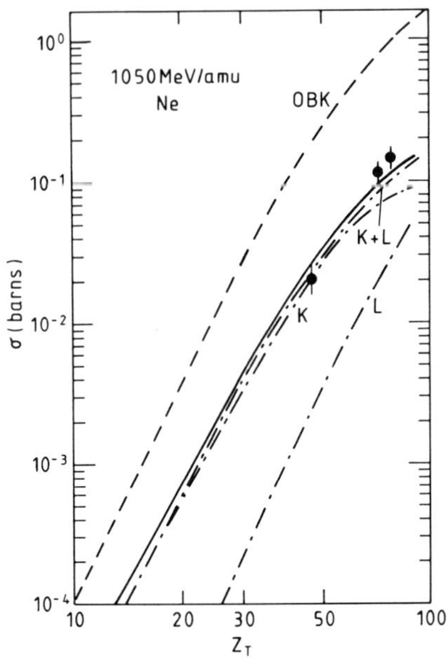


**Figure 8.8.** Experimental and theoretical capture cross sections for 1050 MeV/u $Ne^{10+}$ ions versus target charge number. Data from [Cra79, Anh85a]; dashed curve: OBK approximation for K-K capture; dash-dot curve: eikonal calculations (with shielding) for target K, L, and the sum of K and L capture [AnE85, Eic90]; solid curve: K-K capture from the B1B approximation [Eic87a], see Section 8.2.

GeV, we refer to [Eic90]. In that review, a systematic comparison (including higher principal shells) is made between the exact [Eq. (8.47)] and the approximate [Eq. (8.50)] evaluation of the eikonal cross section.

## 8.5.2 Comparison with experimental data

There is a qualitative difference between charge exchange involving low-$Z$ and high-$Z$ ions. In the former case, $1s_{1/2}$-$1s_{1/2}$ capture is clearly dominant whereas in the latter case charge exchange between excited target and projectile states is of equal importance in the energy range below 2 GeV/u. For both cases, we present a comparison between experimental and theoretical total cross sections.

Figure 8.8 shows a comparison between experimental capture cross sections and theoretical results for 1050 MeV/u bare Ne ions impinging on various targets. The dashed line represents the OBK approximation for K-

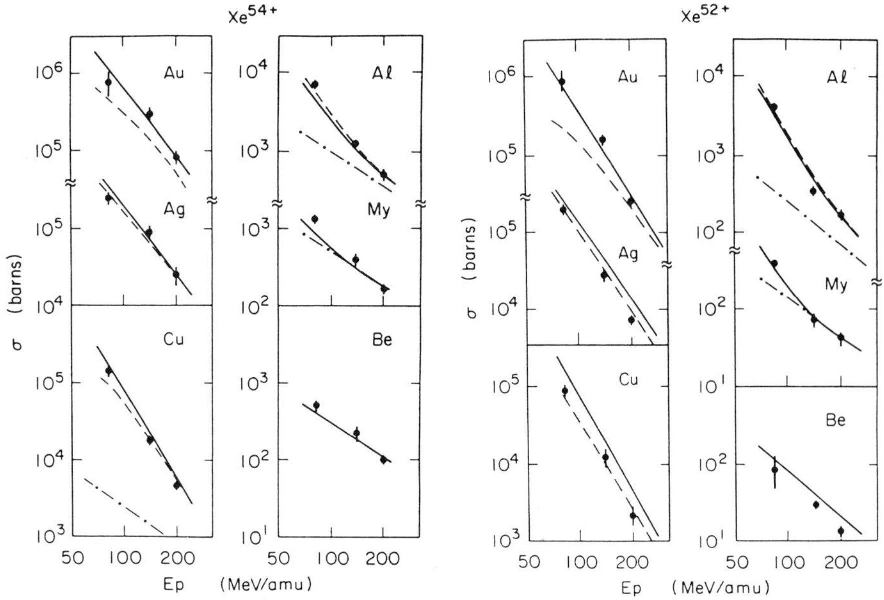


**Figures 8.9 (left) and 8.10 (right).** Total (nonradiative plus radiative) capture cross section [MeA85] for $Xe^{54+}$ and $Xe^{52+}$, respectively, impinging on various targets (My denotes mylar foils). Solid lines denote eikonal calculations using the $Z/n$ criterion Eq. (8.49), dashed lines the conventional $Z$ criterion (8.38) for selecting between the post and prior form. For low-$Z$ targets such as Be and mylar, the radiative electron capture (dash-dot lines) is dominant.

K capture which overestimates the data by about one order of magnitude. The dash-dot lines indicate eikonal calculations using Slater-screened effective charges for K-K and L-K capture and also show the sum of both. These results agree reasonably well with experimental data. The solid curve represents K-K capture cross sections calculated from the relativistic boundary-corrected Born approximation B1B, see Section 8.3. This curve (based on Slater-screened effective charges) lies slightly above the eikonal K-K curve and represents the data quite well considering that a contribution from L-K capture has to be added.

Figures 8.9 and 8.10 show a comparison between measured and theoretical eikonal cross sections for $Xe^{54+}$ and $Xe^{52+}$ projectiles [MeA85]. The difference between the two sets of curves can be ascribed to capture into $1s_{1/2}$ states as discussed in Sec. 8.7.2. It is seen that in most cases the shell-size $Z/n$ criterion, Eq. (8.49), (solid line) for choosing the post or prior form is in better accord with the data than the conventional $Z$-criterion (dashed line). In some cases, though, the $Z$-criterion seems to yield a better agreement with the data. In all calculations, the full nuclear charge was used for the projectile. For the target atoms, effective nuclear charges $Z_T^\star = Z_T - \Delta Z$ based on the Slater rules were adopted, with $\Delta Z = 0.3$ for the K-shell, 4.15 for the L-shell, 11.0 for the 3s and 3p shells, and 21.0 for the 3d shell. For the L- and M-shells, nonrelativistic target wave functions were used, which was found to have a very small effect on the total cross section summed over subshells. Experimental and theoretical results in Figures 8.9 and 8.10 include radiative electron capture (REC), see Chap. 9, indicated by a dash-dot line. For Be and mylar targets, REC is dominant, whereas for higher target charges, REC is negligible.

### 8.5.3 Theoretical angle-differential cross sections

For most theories, angle-differential cross sections have not been calculated because, at present, measurements with sufficient angular resolution are not feasible. However, when experiments have developed to this point, it is no major problem to calculate differential cross sections using Eqs. (5.84) or (5.87) within the theories described here.

For purely theoretical reasons, differential cross sections are of interest in order to study the Thomas peak [Tho27], which can be interpreted as arising from a double-scattering mechanism at asymptotically high velocities. In the first step of the classical Thomas model, the electron scatters off the projectile by $60^o$ with a speed equal to the speed of the projectile. In the second step, the electron is elastically rescattered off the target nucleus by again $60^o$, so that it travels along with the projectile. The connection of this classical model with the nonrelativistic second-order OBK approximation has been discussed in great detail in the literature , see e.g. [ShS79, Dri55, BrM92].

Since these discussions invariably assume asymptotically high nonrelativistic velocities, it is necessary to make a proper relativistic treatment. According to classical mechanics [Sha79, MoS79], a peak of the differential cross section in the laboratory frame should occur at the Thomas angle

$$\Theta_{\mathrm{Th}} \approx \frac{\sqrt{2\gamma+1}}{\gamma+1}\,\frac{m_{\mathrm{e}}}{M_{\mathrm{P}}}, \tag{8.55}$$

where $m_e$ is the electron mass and $M_P$ the projectile mass. It is interesting to note that the classical total cross section for the double-scattering mechanism decreases for large values of $\gamma$ as $\gamma^{-3}$ [Sha79]. Hence this mechanism is not asymptotically dominant as it is in nonrelativistic charge transfer.

Exact quantum mechanical calculations of differential cross sections within a second-order OBK approximation (OBK2) yield the following results [DeE91a]: (a) For p + H collisions at 1 GeV, the differential OBK2 cross section has a sharp spike at the Thomas angle $\Theta_{\mathrm{Th}}$. This confirms earlier results [HuM84, Moi85] which are based on peaking approximations [Als88, DeE91a]. (b) On the other hand, for U + U collisions at energies of 0.5, 5, and 100 GeV/u, the differential OBK2 cross sections do not show any indication of a structure that might suggest a Thomas peak [DeE91b].

## 8.6 Other distorted-wave theories

While the eikonal approximation discussed above is reasonably successful in accounting for experimental cross sections, it is not quite satisfactory theoretically. It does not satisfy Coulomb boundary conditions (8.26) and (8.27) and is not post-prior symmetric. We discuss briefly some other approaches in which multiple electron-target or electron-projectile interactions are taken into account in an approximate way. We use atomic units in this section.

*Impulse approximation*

One of the earliest theories for relativistic electron capture, due to Jakubassa-Amundsen and Amundsen [JaA80], is based upon the impulse approximation which includes an infinite number of electron-target interactions. In this approximation, the exact projectile eigenfunctions $\Psi_{\mathrm{f}}$ is replaced with a superposition of continuum target Coulomb eigenfunctions $\varphi_{\omega,k}(\mathbf{r}_{\mathrm{T}}, t)$ with energy $E = \hbar\omega$ and momentum $\hbar\mathbf{k}$ weighted with the momentum distribution of the bound projectile state $\psi'_{\mathrm{f}}(\mathbf{r}'_{\mathrm{P}}, t')$. This is implemented in two steps. First, the projectile state $\psi'_{\mathrm{f}}$ is expanded in terms of free projectile eigenstates $v'_{\omega',k'}(\mathbf{r}', t') = u'_{E',p'} \exp(i\mathbf{k}' \cdot \mathbf{r}'_{\mathrm{P}} - i\omega' t')$ where $u'_{E',p'}$ is a free-particle spinor. Subsequently, the free-particle wave functions $v'_{\omega',k'}$ are transformed into the target system and expanded into free target eigenstates $v_{\omega,k} = u_{E,p} \exp(i\mathbf{k} \cdot \mathbf{r}_{\mathrm{T}} - i\omega t)$. Finally, each momentum component in the target system is weighted with the target Coulomb eigenfunction in momentum space, so that one arrives at an ionization matrix element leading from a bound target state $\psi_{\mathrm{i}}$ through the projectile perturbation to a continuum target state. Denoting space-time integrations by angular brackets and introducing the spinor labels $s, s'$ one gets the transition amplitude in the impulse approximation as

$$A_{\rm fi}^{IA} = -\frac{i}{\hbar}\sum_{s.s'}\int d\omega'\, d^3k' \int d\omega\, d^3k \langle\psi_{\rm f}'|v_{\omega',k'}^{\prime(s')}\rangle\langle v_{\omega',k'}^{\prime(s')}|S^{-1}|v_{\omega,k}^{(s)}\rangle \times \left\langle \varphi_{\omega,k}^{(s)} \left| S^2\left(-\frac{Z_{\rm P}e^2}{r_{\rm P}'}\right)\right| \psi_{\rm i}\right\rangle . \qquad (8.56)$$

This equation is studied at asymptotically high collision energies within a peaking approximation. It is found that the cross section decreases asymptotically as $(\ln\gamma)^2/\gamma$ in agreement with the result (8.25) of the second-order OBK approximation and in contrast to the OBK1, the B1B, and the eikonal approximations which all give a decrease as $\gamma^{-1}$. Owing to the complicated integrals appearing in the expression (8.56), the impulse approximation has not yet been evaluated at finite energies.

*Target continuum distorted-wave approximation*

A "target continuum distorted wave" (TCDW) approximation derived from Eq. (5.82) is proposed by McCann [Mcc85]. In this approximation, the eikonal phase factor multiplying the final-state wave function of Eq. (8.36) is replaced by

$$G_1 = e^{\frac{1}{2}\pi\nu_{\rm T}}\Gamma(1+i\nu_{\rm T})\,{}_1F_1[-i\nu_{\rm T}, 1, -i\gamma(vr_{\rm T}+vz_{\rm T})m_{\rm e}/\hbar] \qquad (8.57)$$

which for large values of $r_{\rm T}$ has the same behavior as the eikonal phase factor and for finite values reflects the Coulomb distortion. Initial and final wave functions are approximated by the Darwin form (4.109). Everything else is the same as in Eqs. (8.34), (8.35), and (8.36). As a consequence, in the nonrelativistic limit, spin-flip transitions are forbidden, as expected, since the magnetic field produced by the projectile motion tends to zero in this limit. For 500 MeV/u p + H collisions, the ratio $\sigma^{\rm flip}/\sigma^{\rm non} = 3\times10^{-3}$ which is in reasonable agreement with the OBK value (8.20). Numerical calculations have been performed for p + H and p + $Be^{4+}$ collisions between 1 and $10^5$ MeV.

*Symmetric eikonal approximation*

The TCDW treatment as well as the eikonal treatment, see Sec. 8.4, are *asymmetric* in target and projectile which certainly is an undesirable feature. Moiseiwitsch [Moi86, Moi87] and Deco and Rivarola [DeR87b] have studied a *symmetrized* eikonal approximation (SE) and Deco and Rivarola a symmetric "continuum-distorted wave" model [DeR86]. Both approaches use distorted initial and final wave functions in the general form

$$\begin{aligned}\chi_{\rm i} &= U_{\rm i}\psi_{\rm i}(\mathbf{r}_{\rm T}, t) \\ \chi_{\rm f} &= U_{\rm f}\psi_{\rm f}(\mathbf{r}'_{\rm P}, t'),\end{aligned} \tag{8.58}$$

where $\psi_{\rm i}$ and $\psi_{\rm f}$ are the initial and final Dirac bound-state wave functions. Considering first the SE approximation, the functions[5]

$$\begin{aligned}U_{\rm i} &= e^{-i\nu_{\rm P}\ln(r'_{\rm P}+z'_{\rm P})} \\ U_{\rm f} &= e^{i\nu_{\rm T}\ln(r_{\rm T}+z_{\rm T})}\end{aligned} \tag{8.59}$$

are (phase) distortion factors. The impact-parameter dependent transition amplitude in the *prior* form then is given by [Moi86, Moi87, DeR87b]

$$\begin{aligned}A^{\rm SE}_{\rm fi}(b) &= -\frac{i}{\hbar}\int_{-\infty}^{\infty} dt \\ &\left\langle S^{-1}\,U_{\rm f}\psi'_{\rm f}\left| -\frac{Z_{\rm P}e^2}{r'_{\rm P}}\,S\left(1+\frac{1}{\beta}\,\frac{\boldsymbol{\alpha}\cdot\mathbf{r}'_{\rm P}+\alpha_z r'_{\rm P}}{r'_{\rm P}+z'_{\rm P}}\right)S\right| U_{\rm i}\psi_{\rm i}\right\rangle .\end{aligned} \tag{8.60}$$

Remembering that the Dirac matrices $\alpha_x$ and $\alpha_y$ mediate spin-flip transitions, the explicit form (8.60) with the factor $\beta = v/c$ in the denominator of the second term suggests the existence of large spin-flip contributions at small velocities. This is unrealistic since spin-flip is caused by the motion-induced magnetic field of the projectile interacting with the magnetic moment of the electron. These features appear as an artifact of the SE approximation, which also significantly underestimates the experimental data.

The spurious contributions in the expression (8.60) are analyzed in [ToE90a], who show that they vanish if either $U_{\rm i} = 1$ or $U_{\rm f} = 1$, a choice which corresponds to the usual (nonsymmetric) eikonal approximation. The conclusions regarding the nonrelativistic limit of the SE approach are confirmed by exact numerical evaluations of capture cross sections [ToE90a]. While the OBK, eikonal, and B1B results show that the spin-flip contribution is exceedingly small for p + H collision at 100 keV, the SE approximation gives a spin-flip cross section that is by many orders of magnitude too large and is of the same order as the non-spin-flip cross section.

In order to avoid the spurious spin-flip contributions in the nonrelativistic limit, a modified symmetric eikonal approximation has been introduced

[5]Constant phase factors depending on the unit of length are dropped in the eikonal phase factors.

[GlM92] which, however, does not possess the correct relativistic behavior.

*Continuum distorted-wave approximation*

Similar arguments as for the symmetric eikonal approximation hold for the CDW approach [DeR86]. The only difference is that the functions $U_{\mathrm{i,f}}$ of Eq. (8.59) are replaced by functions of the type (8.57). Only, if one of the channels is left undistorted as in the TCDW approximation, Sec.8.4.2, spurious terms do not arise.

In [DeR87a], a "matrix continuum-distorted wave" model has been adopted. In this version, $U_{\mathrm{i,f}}$ is taken as a matrix operator that produces an approximate Sommerfeld-Maue continuum wave function for $\alpha Z \ll 1$ when applied to nonrelativistic spinors. This procedure leads to the nonrelativistic CDW limit, but important terms in the residual interaction are neglected. Post-prior symmetry is not achieved. Moreover, the resulting expressions for the transition amplitudes are very complicated, so that peaking and various additional approximations had to be used in order to arrive at expressions that were amenable to numerical evaluation. A "relativistic continuum-distorted wave approximation" is proposed in [GlM94], which removes some of these defects, but still does not have the correct relativistic behavior and is not applicable to high-$Z$ systems, see [ToE95].

At this level of complication with improved distorted-wave formulations, and in view of the persisting limitations, it may be more appropriate to resort to sophisticated numerical methods, like two-center coupled-channel methods discussed in the next section.

## 8.7 Two-center coupled-channel methods

As discussed in Sec. 6.5 for excitation and ionization, one may obtain an approximate solution of the time-dependent Dirac equation (5.40) in the laboratory frame with the two-center coupled-channel method. In this approach, the problem is treated rigorously within a limited space of basis functions. For this purpose, the wave function $\Psi(\mathbf{r}_{\mathrm{T}}, t)$ is expanded in terms of a suitable set of basis states $\psi_k(\mathbf{r}_{\mathrm{T}}, t)$. The degree of approximation then is essentially determined by the choice and the truncation of the basis set. The method is suitable if the projectile energy is not too high, otherwise the rapid oscillations of the phase factors in time and space prohibit a numerical evaluation of the matrix elements (6.60) and (6.61). See Sec. 5.4.1 for a definition of "fast" and "slow" collisions.

The time-dependent Dirac equation (5.40) for relativistic ion-atom collisions is solved in Sec. 6.5.1 using the expansion (6.55) of the time-dependent wave function $\Psi(\mathbf{r}_\mathrm{T}, t)$ in terms of target and projectile eigenstates $\psi_k(\mathbf{r}_\mathrm{T}, t)$ and $\psi_{k'}(\mathbf{r}'_\mathrm{P}, t')$, respectively. These methods apply equally for charge transfer.

However, according to the discussion in Sec. 6.5.4, it is more appropriate to use a set of basis states that satisfy Coulomb boundary conditions (8.26) and (8.27). We therefore replace Eq. (6.55) with the basis expansion (6.69)

$$\begin{aligned}\Psi(\mathbf{r}_\mathrm{T}, t) &= \sum_k a_k(t) e^{-i\nu_\mathrm{P}\ln(R'-vt')} \psi_k(\mathbf{r}_\mathrm{T}, t) \\ &+ \sum_{k'} a_{k'}(t) e^{i\nu_\mathrm{T}\ln(R+vt)}\, S^{-1}\psi_{k'}(\mathbf{r}'_\mathrm{P}, t'). \end{aligned} \tag{8.61}$$

The first term contains target states asymptotically phase-distorted by the projectile, while the second sum contains projectile states asymptotically phase-distorted by the target.

When we insert (8.61) into the time-dependent Dirac equation (5.39) and write down the equations corresponding to (6.56) and (6.57), we find that on the right-hand side, the time differentiation of the logarithmic phase factors yields additional terms so that for the perturbing potential we get the replacement

$$\begin{aligned}\frac{Z_\mathrm{P}}{r'_\mathrm{P}} &\rightarrow \frac{Z_\mathrm{P}}{r'_\mathrm{P}} - \frac{Z_\mathrm{P}}{R'} \\ \frac{Z_\mathrm{T}}{r_\mathrm{T}} &\rightarrow \frac{Z_\mathrm{T}}{r_\mathrm{T}} - \frac{Z_\mathrm{T}}{R}. \end{aligned} \tag{8.62}$$

The phase factors themselves drop out in the one-center integrals since they are state-independent, but they remain in the two-center integrals. We denote the expansion (8.61) as "boundary corrected" and in contrast (6.55) as "unperturbed" since the asymptotic Coulomb distortion is not accounted for in the basis set.

If the expansion coefficients $a_k(t)$ and $a_{k'}(t)$ in Eq. (8.61) are combined into a column vector $\mathbf{a} = (a_k, \cdots; a_{k'}, \cdots)$, the coupled equations can be written in a compact matrix form (6.59) as

$$\dot{\mathbf{a}} = -\frac{i}{\hbar}\mathbf{N}^{-1}\,\mathbf{V}\,\mathbf{a}. \tag{8.63}$$

The overlap matrix $\mathbf{N}$ and the interaction matrix $\mathbf{V}$ are each built from four submatrices corresponding to target-target, target-projectile, projectile-target, and projectile-projectile transitions. The matrix elements are defined in Eqs. (6.60) and (6.61) [with the replacements (8.62) for the expansion (8.61)] and the cross section are given by Eq. (6.65).

**Table 8.2.** Comparison of theoretical cross sections per electron (in barn) for 1s capture in collisions of $U^{92+} + U^{91+}$ at 500 MeV/u. CC: 36-state coupled-channel calculations [ToE88b]; B1B: relativistic boundary-corrected Born approximation [Eic87a], see Sec. 8.3 (cross sections obtained by exact numerical integration); EA prior and EA post: prior and post form eikonal approximation [Eic85]; OBK: relativistic OBK approximation. The numbers in square brackets give the powers of 10 multiplying the preceding number. From [ToE88b]. The table also includes the ratio of spin-flip to non-spin-flip cross sections [ToE90a].

| final shell | CC | B1B | EA prior | EA post | OBK |
|---|---|---|---|---|---|
| $1s_{1/2}$ | 4.12[3] | 3.39[3] | 4.14[3] | 4.14[3] | 4.91[4] |
| $2s_{1/2}$ | 1.06[3] | 5.13[2] | 7.62[2] | 5.67[2] | 5.28[3] |
| $2p_{1/2}$ | 3.91[2] | 5.95[2] | 1.35[3] | 4.44[2] | 9.02[3] |
| $2p_{3/2}$ | 2.31[2] | 2.74[2] | 1.24[3] | 1.07[2] | 7.61[3] |
| $3s_{1/2}$ | 3.28[2] | 1.51[2] | 2.30[2] | 1.58[2] | 1.41[3] |
| $3p_{1/2}$ | 1.03[2] | 1.75[2] | 4.37[2] | 1.30[2] | 2.60[3] |
| $3p_{3/2}$ | 7.25[1] | 9.15[1] | 4.62[2] | 3.61[1] | 2.56[3] |
| sum | 6.31[3] | 5.19[3] | 8.62[3] | 5.58[3] | 7.76[4] |
| $\sigma^{\rm flip}/\sigma^{\rm non}$ | 5.73[−2] | 3.41[−2] | 5.84[−2] | 5.83[−2] | 7.30[−3] |

### 8.7.1 Application to $U^{92+} + U^{91+}$ collisions

The time evolution of the transfer probabilities in uranium-uranium collisions at 500 MeV/u is included in Fig. 6.12 and the impact parameter dependence of the final transfer probability in Fig. 6.13. The 36-state fully relativistic coupled-channel treatment [ToE88b] for 500-MeV/u $U^{92+}$ + $U^{91+}$ collisions may serve as a standard calculation for the basis set (6.55) of unperturbed states.

The detailed results for the $U^{92+} + U^{91+}$ test system invite a comparison with the results of other theories. For charge transfer, a comparison is presented in Table 8.2. In addition to B1B (Sec. 8.3) and OBK (Sec. 8.2) cross section (calculated numerically without assuming the $\alpha Z$ expansion), the prior and post versions of the relativistic eikonal approximation (Sec. 8.4) are considered. We observe that in most cases the B1B cross sections have a magnitude intermediate between the prior and post form eikonal values. For some final subshells, in particular for $1s_{1/2}$, the B1B results are close to

**Table 8.3.** Excitation and transfer cross sections (in barn) between target $1s_{1/2}(\frac{1}{2})$ and projectile $nl_j(\mu_j)$ states in $U^{92+}$ + $U^{91+}$ collisions at 1.0 GeV/u laboratory energy. The 36-state close-coupling calculations with unperturbed [Eq. (6.55)] or boundary-corrected [Eq. (8.61)] basis states include target and projectile K-, L-, and M-shells. The number in square brackets gives the power of 10 multiplying the preceding number. From [ToE90b].

| | excitation | | | transfer | |
|---|---|---|---|---|---|
| final state | unpert. | corrected | Born appr. | unpert. | corrected |
| $1s_{1/2}(\frac{1}{2})$ | | | | 6.28[2] | 7.46[2] |
| $1s_{1/2}(-\frac{1}{2})$ | | | | 4.61[1] | 5.02[1] |
| $2s_{1/2}(\frac{1}{2})$ | 4.95[3] | 1.66[3] | 2.13[3] | 2.22[2] | 2.45[2] |
| $2s_{1/2}(-\frac{1}{2})$ | 3.94[2] | 1.14[2] | 4.60[1] | 1.21[1] | 1.28[1] |
| $2p_{1/2}(\frac{1}{2})$ | 3.12[3] | 6.48[2] | 4.23[2] | 8.80[1] | 7.26[1] |
| $2p_{1/2}(-\frac{1}{2})$ | 5.83[3] | 4.71[3] | 5.64[3] | 2.30[1] | 2.31[1] |
| $2p_{3/2}(\frac{3}{2})$ | 5.51[3] | 4.60[3] | 6.11[3] | 6.60[0] | 6.63[0] |
| $2p_{3/2}(\frac{1}{2})$ | 7.54[3] | 1.48[3] | 9.02[2] | 3.44[1] | 3.24[1] |
| $2p_{3/2}(-\frac{1}{2})$ | 1.16[3] | 1.01[3] | 1.36[3] | 1.09[1] | 4.75[0] |
| $2p_{3/2}(-\frac{3}{2})$ | 1.06[2] | 9.94[1] | 1.17[2] | 9.15[-1] | 4.92[-1] |

the coupled-channel results, for others they deviate by as much as a factor of two. The total K-shell capture cross sections are all in fair agreement with one another except for those of the OBK theory, which systematically yields cross sections that are by one order of magnitude too high. As has been discussed in Section 8.2.2, the inclusion of second-order terms in the OBK expansion further increases the cross sections by one order of magnitude. Similarly, the ratio of spin-flip to non-spin-flip cross sections are in close agreement with one another except again for the OBK approximation.

In Sec. 6.5.4, the influence of boundary conditions in coupled-channel calculations for excitation and ionization is discussed. We observe in Fig. 6.12 that with a given basis size, the excitation probabilities are appreciably changed when proper Coulomb boundary conditions are imposed on the basis functions by adopting the expansion (8.61). We may ask whether the same effect becomes visible for charge exchange as well.

The effect of Coulomb boundary conditions is examined in Table 8.3

showing excitation and transfer cross sections for $U^{92+} + U^{91+}$ collisions at 1 GeV/u, resulting from 36-state two-center coupled-channel calculations. Atomic states up to the $3p_{3/2}$ shell at both centers are included. In all cases, the initial state is assumed to be $1s_{1/2}(\frac{1}{2})$. We see that for excitation, the inclusion of boundary corrections in coupled-channel calculations decreases the cross section by a factor of about two, while for charge exchange the corrections are of the order of 10% or less in most cases. The Born cross section, which for excitation is also shown, does not have a systematic correlation to the coupled-channel cross sections with or without boundary conditions. In fact, at 1 GeV/u, the projectile energy may still be too low for the Born approximation to be valid.

### 8.7.2 Comparison with experiment: Xe on Ag and Au

It is desirable to compare calculated cross sections with experimental values. Total cross sections have been measured for a variety of projectile-target combinations [Cra79, MeA85, AnG86]. Coupled-channel calculations are available for $Xe^{54+}$ on Ag and Au at collision energies of 82, 140, and 197 MeV/u. Since the coupled-channel theory is constructed for a single-electron target, additional assumptions have to be made for multielectron targets. Transitions between occupied target shells cannot occur, therefore the contribution of each target subshell has been calculated separately [ToE88b] by including only this specific target state in the two-center basis and by blocking all other occupied target states. For the bare projectile, all 28 states of the K-, L-, and M-shells are coupled. One after the other, all target subshells up to $2p_{3/2}$ for Ag and up to $3p_{3/2}$ for Au have been individually considered using effective charges $Z^\star = Z_T$ for the K-shell (the value of $Z_T - 0.3$ gives almost identical results), $Z^\star = Z_T - 4.15$ for the L-shell, and $Z^\star = Z_T - 11.0$ for the $3s$ and $3p$ shells.

In Figure 8.11, experimental and theoretical total experimental cross sections for $Xe^{54+}$ + Ag and $Xe^{54+}$ + Au are plotted and compared with calculated eikonal and coupled-channel cross sections. While the latter are in reasonable agreement with the data for Au, they overestimate the cross section for Ag by 20-40%.

An independent piece of experimental information can be extracted from existing capture data for $Xe^{52+}$ projectiles [MeA85], see Fig. 8.10. Aside from a small screening correction, the difference between the total cross sections for capture into $Xe^{54+}$ and $Xe^{52+}$ may be identified with the cross section for capture into the K-shell of $Xe^{54+}$. This experimentally obtained number can be directly compared with the corresponding theoretical value, so that the contribution to the cross section from higher excited and contin-

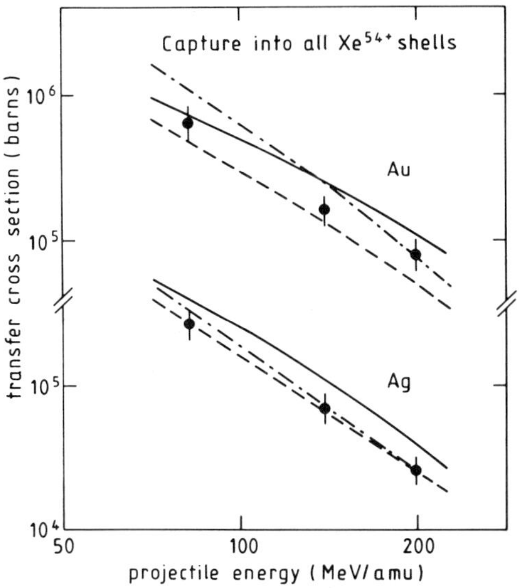


**Figure 8.11.** Cross sections for charge transfer in $Xe^{54+}$ + Ag and Au collisions as a function of the laboratory projectile energy. The error bars include estimated 20% systematic errors. The solid lines represent the results of coupled-channel calculations [ToE88b]. The dashed and dash-dotted lines use the eikonal approximation with the $Z$ [Eq. (8.38)] and $Z/n$ [Eq. (8.54)] criteria, respectively. Adapted from [MeA85].

uum projectile states is eliminated, see Fig. 8.10. The exact eikonal curves are calculated according to [Eic85]. For Au, the results of both prescriptions, see Eqs. (8.38) and (8.49), to choose between the post and prior forms are included, for Ag both versions coincide. The agreement with the more specific experimental data is quite satisfactory, in particular for higher energies.

## 8.8 Many electrons. Multiple capture

As noted in the last subsection, in any comparison with experiment, one necessarily deals with many electrons on the target, whereas the preceding calculations assume a single electron in the collision system. The complexity of the experimental situation can be appreciated by considering Fig. 8.13. This figure shows calculated relative capture cross sections from filled target shells with $n_T = 1, 2, 3$ to vacant projectile shells with $n_P = 1, 2, 3$, and

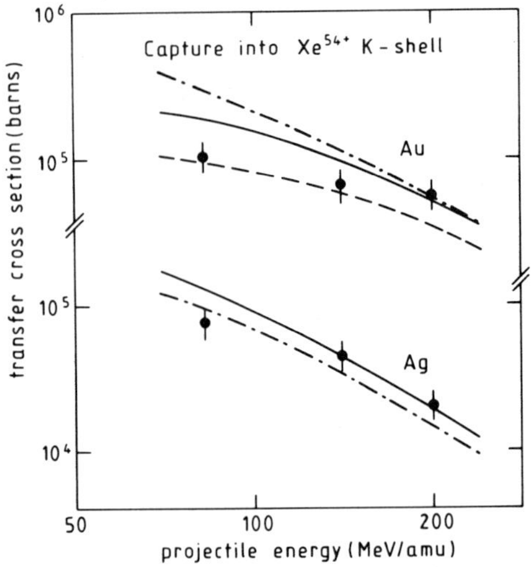


**Figure 8.12.** Cross sections for charge transfer into the projectile K-shell in $Xe^{54+}$ + Ag and $Xe^{54+}$ + Au collisions as a function of the laboratory projectile energy. The data points have been obtained as the difference between total capture cross sections into $Xe^{54+}$ and into $Xe^{52+}$. The notation for the theoretical curves is the same as in Figure 8.11. For Ag, the two eikonal versions coincide. From [ToE88b].

4-10 for several collision systems: 1050-MeV/u $Ne^{10+}$ + Ag, 197-MeV/u $Xe^{54+}$ + Ag, and 82-MeV/u $Xe^{54+}$ + Au. The calculations are based on the eikonal equation (8.50) with the substitutions (8.54), using the $Z/n$ criterion, Eq. (8.49). (The absolute values of the cross sections can be obtained with the help of Table 8.1 and Fig. 8.12.)

One sees from Fig. 8.13 that for low values of $Z_P$ and high projectile velocities, 1s → 1s capture is dominant, but for the opposite case of high $Z_P$ and low velocities, capture from all filled target states into many projectile states has to be considered.

Since absolute cross sections increase as $Z_P$ increases and the velocity decreases, one can expect increasing multiple capture in such collisions. This is indeed the case, as Fig. 8.14 shows. The figure gives measured multiple-capture cross sections for 105- and 220-MeV/u uranium projectiles with various charge states as a function of the multiplicity $m$ of the capture process [AnM87]. One finds an approximately exponential drop of the cross section for a given collision system with increasing multiplicity. The curves are based on a crude model of multiple capture, which

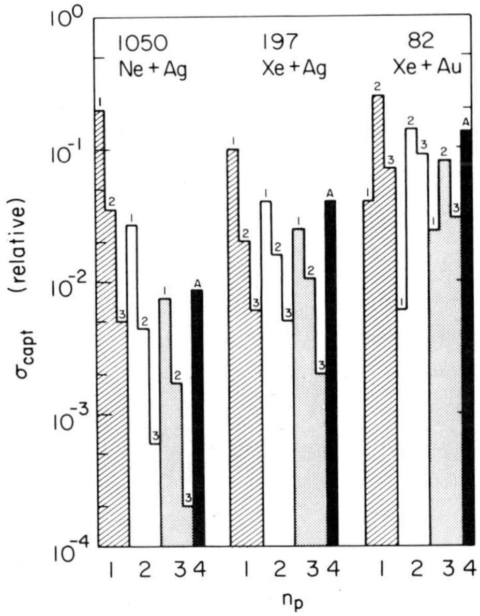


**Figure 8.13.** Relative capture cross sections from filled target states with $n_T = 1, 2, 3$ to vacant projectile shells with $n_P = 1, 2, 3$, and $4 - 10$ for 1050-MeV/u $Ne^{10+}$ + Ag, 197-MeV/u $Xe^{54+}$ + Ag, and 82-MeV/u $Xe^{54+}$ + Au. The numbers above each bar are the values of $n_T$, but for $n_P = 4 - 10$, only the sum over all three target shells (A) is displayed. From [MeA85].

assumes that the (mean) capture probability per electron is independent of the other electrons or vacancies. This assumption is based on the observation that the relative capture cross sections are only little dependent on the projectile charge state (which implies that the capture processes must go predominantly to excited projectile states, in agreement with Fig. 8.13).

If one calls $P_c$ the mean capture probability per electron, the theoretical cross section for $m$-fold capture is

$$\sigma_m^{\mathrm{th,c}} = \int (P_c)^m \, d^2b. \tag{8.64}$$

In the present model, this cross section must be interpreted as the cross section for the simultaneous capture of $m$ target electrons, independently of what happens to the other electrons, i.e., it includes $(m+1, m+2, ...)$-fold capture. But an experimental cross section $\sigma_m^{\mathrm{ex,c}}$, based on a charge-state

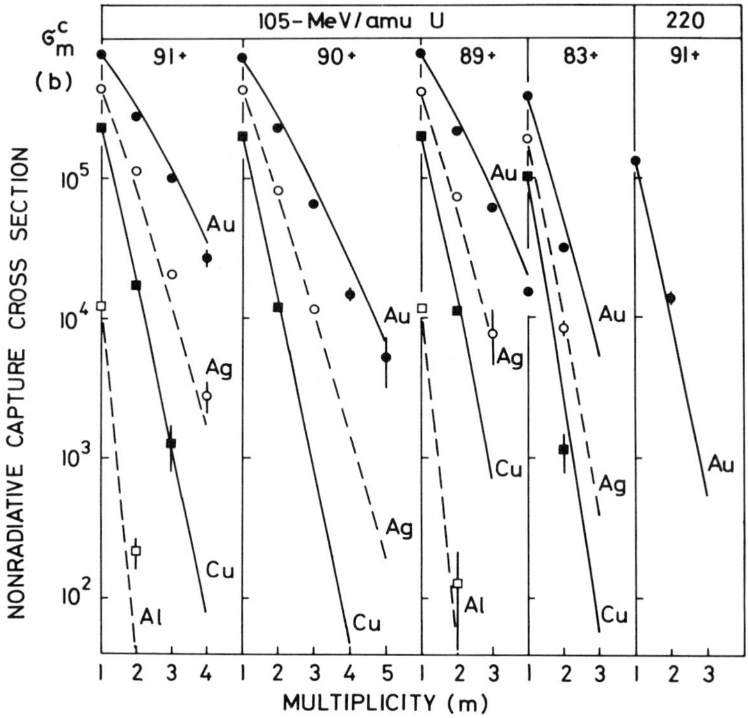


**Figure 8.14.** Multiple-capture cross sections of various charge states of 105- and 220-MeV/u U ions passing through Al, Cu, Ag, and Au target foils, as a function of the multiplicity $m$ of the capture. The curves are normalized to the experimental $m = 1$ cross sections. From [AnM87].

measurement (Chap. 13), determines only the $m$-fold events. Hence,

$$\sigma_m^{\mathrm{th,c}} = \sigma_m^{\mathrm{ex,c}} + \sigma_{m+1}^{\mathrm{ex,c}} + \sigma_{m+2}^{\mathrm{ex,c}} + \ldots \tag{8.65}$$

or

$$\sigma_m^{\mathrm{ex,c}} = \sigma_m^{\mathrm{th,c}} - \sigma_{m+1}^{\mathrm{th,c}}. \tag{8.66}$$

[AnM87] derive a semiempirical expression for $\sigma_m^{\mathrm{th,c}}$, which is normalized to the $\sigma_1^{\mathrm{ex,c}}$ values given in Fig. 8.14. The trend of the data is reproduced.

# Chapter 9

# Radiative Electron Capture

In Chap. 8, we discussed Coulomb or nonradiative electron capture. In the nonrelativistic energy regime, the cross section for this process falls off asymptotically with increasing projectile velocity $v$ as $v^{-12}$ (or as $v^{-11}$ in second order). This rapid decrease is caused mainly by the requirement that a given momentum component in the initial electronic wave function has to find its counterpart in the final momentum wave function displaced by the momentum $m_e v$ of an electron traveling with the speed of the projectile. If, however, the electron transfer is accompanied by the emission of electromagnetic radiation, the emitted photon acts as a third body carrying away energy and momentum released by the formation of the final bound state. Hence, the condition of momentum matching will be relaxed so that the cross section for radiative electron capture (REC) falls off as $v^{-5}$ for high nonrelativistic projectile velocities.

For free electrons, capture cannot take place at all without the emission of photons owing to energy and momentum conservation. This means, qualitatively, that electrons loosely bound in low-$Z$ target atoms or in outer shells are more likely to be captured *with* photon emission than *without*. Hence for low-$Z$ target atoms at high projectile energies (e.g., for $Z_T = 1$ with $E_P \geq 10$ MeV/u, [Spi79]), the REC cross section exceeds the cross section for Coulomb capture. From this point of view, the REC mechanism deserves particular attention.

Starting in Sec. 9.1 with a preliminary pictorial description of REC within the impulse approximation, we turn in Sec. 9.2 to the direct perturbative formulation. An insight into qualitative features, in particular the angular distribution, is obtained from a nonrelativistic treatment in Sec. 9.3. As a consequence of these considerations, we argue that presently the best description of REC is provided by treating it essentially as the inverse of the photoelectric effect. Therefore, in Sec. 9.4, we discuss several ways to obtain photoelectric cross sections so that in Sec. 9.5 we can use these results to derive the desired REC cross sections. Finally, in Sec. 9.6, we present further theoretical and experimental results.

We use full units in the major part of this chapter, except for the more technical sections 9.4.3 and 9.5.3, where we adopt relativistic (natural) units $\hbar = m_e = c = 1$, see Appendix.

## 9.1 General features

A loosely bound target electron may be considered as approximately free in a high-energy collision. In this limit, REC is identical with radiative recombination (RR), in which an electron initially moving with the velocity $-\mathbf{v}$ in the projectile frame is captured into a bound state with the simultaneous emission of a photon of energy $\hbar\omega'$ and wave number $\mathbf{k}'$. Experimentally, the process of REC has been first identified by Raisbeck and Yiou [RaY71], by Schnopper et al. [ScB72], and by Kienle et al. [KiK73], and since that time has been extensively studied both experimentally and theoretically.

In the following, we start with a brief outline of the treatment. We assume that the cross section $\sigma_{\mathrm{RR}}$ for radiative recombination is known since it is essentially the inverse of the atomic photoelectric effect. It is then natural to refer all momenta to the projectile frame. If the target electron has the momentum $\mathbf{q}$ with respect to the target nucleus and the projectile momentum is characterized by the Lorentz factor $\gamma$, its momentum in the projectile frame $\mathbf{q}'$ is obtained from Eq. (2.29). If the momentum $\gamma m_e v$ of an electron traveling with the relative speed of the target towards the projectile is large compared to the electron momentum $q$ in a low-$Z$ target atom, one may use the impulse approximation [KlJ75] to write the double-differential REC cross section in the projectile frame (primed coordinates) as

$$\frac{d^2\sigma_{\mathrm{REC}}}{d\Omega'\, d(\hbar\omega')} = \int d^3q\, \frac{d\sigma_{\mathrm{RR}}(q')}{d\Omega'}\, |\tilde{\varphi}_{\mathrm{i}}(\mathbf{q})|^2 \delta(\hbar\omega' + E'_{\mathrm{f}} - E'_{\mathrm{i}}), \qquad (9.1)$$

where $\tilde{\varphi}_{\mathrm{i}}(\mathbf{q})$ is the Fourier transform of the initial electronic target wave function. The delta function expresses the energy conservation between the final electronic energy $E'_{\mathrm{f}}$ and the photon energy $\hbar\omega'$ in the projectile frame on the one hand and the initial electronic energy $E'_{\mathrm{i}}$ (also in the projectile frame) on the other hand. According to Eq. (2.29), these energies are written as

$$E'_{\rm i} = \gamma(E_{\rm i} - vq_z) = \gamma m_{\rm e}c^2 - \gamma|\epsilon_{\rm i}| - \gamma v q_z \quad (9.2)$$

and

$$E'_{\rm f} = m_{\rm e}c^2 - |\epsilon'_{\rm f}|. \quad (9.3)$$

Here $|\epsilon_{\rm i}|$ and $|\epsilon'_{\rm f}|$ are the binding energies of initial target and final projectile states. Inserting the expressions (9.2) and (9.3) into Eq. (9.1), we have

$$\frac{d^2\sigma_{\rm REC}}{d\Omega'\, d(\hbar\omega')} = \int d^3q\, \frac{d\sigma_{\rm RR}(q')}{d\Omega'}\, |\tilde{\varphi}_{\rm i}(\mathbf{q})|^2 \delta(\hbar\omega' + E'_{\rm f} - \gamma E_{\rm i} + \gamma v q_z) = \int d^3q\, \frac{d\sigma_{\rm RR}(q')}{d\Omega'}\, |\tilde{\varphi}_{\rm i}(\mathbf{q})|^2 \delta(\hbar\omega' - |\epsilon'_{\rm f}| + \gamma|\epsilon_{\rm i}| - T_{\rm e} + \gamma v q_z), \quad (9.4)$$

where $T_{\rm e} = m_{\rm e}c^2(\gamma - 1)$ is the kinetic energy of an electron traveling with the same speed as the projectile. Since the initial momentum distribution of the target electron is peaked around $\mathbf{q} \approx 0$ and since we usually consider $Z_{\rm T} < Z_{\rm P}$, the recombination cross section varies slowly and can be taken outside the integral over the transverse momentum. In this approximation, we get

$$\frac{d\sigma_{\rm REC}}{d\Omega'\, d(\hbar\omega')} = \frac{1}{\gamma v} \left(\frac{d\sigma_{\rm RR}}{d\Omega'}\right)_{q'_z = \gamma(-m_{\rm e}v + q_z)} \mathcal{J}_{\rm i}(q_z), \quad (9.5)$$

where

$$\mathcal{J}_{\rm i}(q_z) = \int d^2q_b |\tilde{\varphi}_{\rm i}(\mathbf{q})|^2 \quad (9.6)$$

is the Compton profile. The integration extends over the transverse momentum $\mathbf{q}_b$. In Eq. (9.5), the longitudinal momentum $q_z$ is fixed by $\omega'$, so that the shape of the photon spectrum

$$\hbar\omega' = |\epsilon'_{\rm f}| - \gamma|\epsilon_{\rm i}| + T_{\rm e} - \gamma v q_z \quad (9.7)$$

is determined by the spread of the longitudinal momentum $q_z$ around its mean value $q_z = 0$, which yields the peak energy

$$\hbar\omega'_0 = \gamma E_{\rm i} - E'_{\rm f} = |\epsilon'_{\rm f}| - \gamma|\epsilon_{\rm i}| + T_{\rm e}. \quad (9.8)$$

The term $\gamma v q_z$ in Eq. (9.7) describes the Doppler broadening through the Compton profile.

However, since the cross section $\sigma_{\rm RR}$ decreases with increasing x-ray energy, see Sec. 9.5, the actually measured peak is usually shifted to lower energy.

In the laboratory system, the photon energy $\hbar\omega$ depends on the emission angle $\theta$ with respect to the beam direction via

$$\hbar\omega = \frac{|\epsilon'_{\mathrm{f}}| - \gamma|\epsilon_{\mathrm{i}}| + T_{\mathrm{e}} - \gamma v q_z}{\gamma(1-\beta\cos\theta)}, \tag{9.9}$$

which follows from Eq. (2.51). Similarly, if Eqs. (9.4) or (9.5) lead to an angular distribution in the projectile system as a function of the emission angle $\theta'$ with respect to the beam axis, we obtain the angular distribution in the laboratory system by the transformations

$$\cos\theta' = \frac{\cos\theta - \beta}{1-\beta\cos\theta} \qquad \text{and} \qquad \frac{d\Omega'}{d\Omega} = \frac{1}{\gamma^2\,(1-\beta\cos\theta)^2}, \tag{9.10}$$

given by Eqs. (2.53) and (2.54), respectively.

## 9.2 The transition amplitude for REC

While Sec. 9.1 describes an intuitively appealing approach based upon the impulse approximation, a complete theoretical description should start from the general treatment of Sec. 5.4.2. We set out from the formulation of the perturbative transition amplitude. In contrast to the cases treated there, the perturbation giving rise to REC is the electromagnetic interaction of the active electron with the photon field. After analyzing this method in Sec. 9.3, we turn to a different description in Sec. 9.4.

As in the preceding chapters, we use a straight-line version of the impact parameter method to describe the projectile motion, see Fig. 3.2. It is convenient to describe the radiating system in its own rest system; therefore we refer all coordinates to the projectile frame. In order to ensure the transverse polarization of the photons, we work in the radiation gauge (also called Coulomb gauge) for the electromagnetic vector potential $\mathbf{A}'$ with $\nabla' \cdot \mathbf{A}' = 0$. We then can write the interaction operator as $-e\gamma^\mu A'_\mu = e\boldsymbol{\gamma}\cdot\mathbf{A}' = e\gamma^0\boldsymbol{\alpha}\cdot\mathbf{A}'$, where the vector potential appropriate for the emission of a photon with a wave vector $\mathbf{k}'$, energy $\hbar\omega' = \hbar c k'$ and unit polarization vector $\hat{\mathbf{u}}'_\lambda$ with $(\lambda = 1, 2)$ is given by [Sak67]

$$\mathbf{A}'_\lambda = \left(\frac{2\pi\hbar c^2}{V\omega'}\right)^{\frac{1}{2}} \hat{\mathbf{u}}'_\lambda e^{-i(\mathbf{k}'\cdot\mathbf{r}' - \omega' t')}. \tag{9.11}$$

Here, $V$ is the normalization volume, for which periodic boundary conditions are required. If all integrations are carried out in the projectile system, the impact-parameter-dependent transition amplitude between specific initial and final atomic states is written as

$$A_{\rm fi}^{(\lambda)} = -\frac{i}{\hbar}\int dt' \int d^3r'_{\rm P}\, \overline{\Psi}'_{\rm f}(\mathbf{r}'_{\rm P}, t')(e\gamma^0\boldsymbol{\alpha}\cdot\mathbf{A}'_\lambda)\, S\,\psi_{\rm i}(\mathbf{r}_{\rm T}, t). \tag{9.12}$$

As long as exact wave functions are used for the motion of a single electron in the field of the residual target atom and the projectile, this expression is exact within the one-photon approximation, i.e., within first-order perturbation theory in the photon field.

Since, in practice, it is not possible to evaluate Eq. (9.12) with exact solutions of an effective three-body problem, approximate wave functions have to be introduced. For the time being, it is instructive to adopt the Born approximation, that is, to use undistorted eigenstates $\varphi_{\rm i}$ and $\varphi'_{\rm f}$ of target and projectile with eigenenergies (in their rest system) $E_{\rm i}$ and $E'_{\rm f}$. The implications of this drastic simplification are discussed following Eq. (9.19). Inserting explicitly the time dependence of the wave functions (4.74) and writing $E_{\rm i} = \hbar\omega_{\rm i}$ and $E'_{\rm f} = \hbar\omega'_{\rm f}$, we get

$$\begin{aligned} A_{\rm fi}^{(\lambda)}(b) &= -\frac{ie}{\hbar}\left(\frac{2\pi\hbar c^2}{V\omega'}\right)^{\frac{1}{2}}\int dt' \int d^3r'_{\rm P}\,\varphi_{\rm f}'^{\dagger}(\mathbf{r}'_{\rm P})e^{i\omega'_{\rm f}t'} \\ &\quad \times \boldsymbol{\alpha}\cdot\hat{\mathbf{u}}'_\lambda e^{-i(\mathbf{k}'\cdot\mathbf{r}'_{\rm P}-\omega' t')} S\varphi_{\rm i}(\mathbf{r}_{\rm T})e^{-i\omega_{\rm i}t}. \end{aligned} \tag{9.13}$$

After introducing the Fourier transforms $\tilde{\varphi}_{\rm i}(\mathbf{q}) = \tilde{\varphi}_{\rm i}(\mathbf{q}_b, q_z)$ and $\tilde{\varphi}'_{\rm f}(\mathbf{p}) = \tilde{\varphi}'_{\rm f}(\mathbf{p}_b, p_z)$ of the initial and final states, respectively, and transforming the target-frame coordinates $\mathbf{r}_{\rm T}, t$ to the projectile system by the Lorentz transformation (3.31), we are able to perform the space and time integrations to obtain

$$\begin{aligned} A_{\rm fi}^{(\lambda)}(b) &= -2\pi i\frac{e}{\gamma}\left(\frac{2\pi\hbar c^2}{V\omega'}\right)^{\frac{1}{2}}\int d^3q\;\tilde{\varphi}_{\rm f}'^{\dagger}(\mathbf{q}_b - \hbar\mathbf{k}'_b, q_- - \hbar k'_z) \\ &\quad \times \boldsymbol{\alpha}\cdot\hat{\mathbf{u}}'_\lambda\, S\,\tilde{\varphi}_{\rm i}(\mathbf{q})\,\delta(\hbar\omega' + E'_{\rm f} - \gamma E_{\rm i} + \gamma v q_z)e^{i\mathbf{q}_b\cdot\mathbf{b}/\hbar}, \end{aligned} \tag{9.14}$$

where $q_- = \gamma q_z - \gamma(v/c^2)E_{\rm i}$. The delta function establishes the relation (9.8) for the photon energy and allows one to eliminate $q_z$ so that one is left with the two-dimensional integral

$$\begin{aligned} A_{\rm fi}^{(\lambda)}(b) &= -2\pi i\frac{e}{v\gamma}\left(\frac{2\pi\hbar c^2}{V\omega'}\right)^{\frac{1}{2}}\int d^2q_b\;\tilde{\varphi}_{\rm f}'^{\dagger}(\mathbf{q}_b - \hbar\mathbf{k}'_b, q_- - \hbar k'_z) \\ &\quad \times \boldsymbol{\alpha}\cdot\hat{\mathbf{u}}'_\lambda\, S\,\tilde{\varphi}_{\rm i}(\mathbf{q}_b, q_+)e^{i\mathbf{q}_b\cdot\mathbf{b}/\hbar}, \end{aligned} \tag{9.15}$$

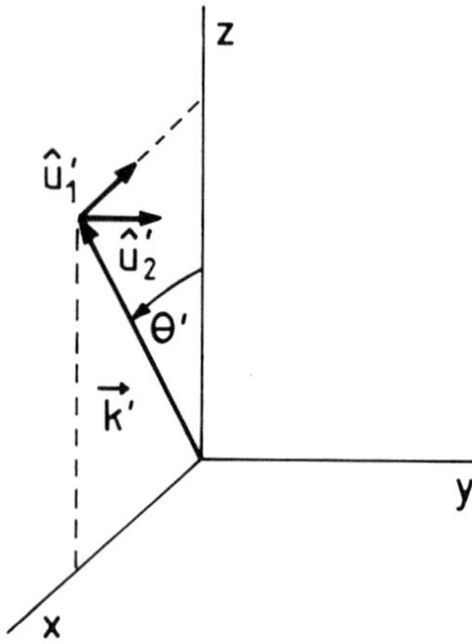


**Figure 9.1.** Orientation of the wave vector $\mathbf{k}'$ and the polarization vectors $\hat{\mathbf{u}}_1'$ and $\hat{\mathbf{u}}_2'$.

with

$$\begin{aligned} q_+ &= [E_\mathrm{i} - (E_\mathrm{f}' + \hbar\omega')/\gamma]/v \\ q_- &= [E_\mathrm{i}/\gamma - (E_\mathrm{f}' + \hbar\omega')]/v. \end{aligned} \tag{9.16}$$

For Coulomb capture, the definitions (9.16) reduce to the definitions (8.6). With photon emission, the resonance condition is $q_+ = 0$. From the transition amplitude (9.15), the state-to-state cross section is obtained as

$$\frac{d^2\sigma}{d\Omega'\, d(\hbar\omega')} = \frac{V(\hbar\omega')^2}{(2\pi\hbar c)^3} \sum_{\lambda=1,2} \int d^2b\, |A_\mathrm{fi}^{(\lambda)}(b)|^2. \tag{9.17}$$

When inserting Eq. (9.15) into (9.17), the integration over the impact parameter plane can be performed to yield the cross section in the projectile frame as

$$\begin{aligned} \frac{d^2\sigma}{d\Omega'\, d(\hbar\omega')} &= \frac{e^2}{\hbar c}(2\pi\hbar)^2\hbar\omega' \frac{1}{v^2\gamma^2} \sum_{\lambda=1,2} \int d^2q_b |\tilde{\varphi}_\mathrm{f}'^\dagger(\mathbf{q}_b - \hbar\mathbf{k}_b', q_- - \hbar k_z') \\ &\times \boldsymbol{\alpha}\cdot\hat{\mathbf{u}}_\lambda'\, S\, \tilde{\varphi}_\mathrm{i}(\mathbf{q}_b, q_+)|^2. \end{aligned} \tag{9.18}$$

If the initial and final electron polarizations are not detected, one has to average over initial and to sum over final electron substates in order to get the cross section per electron. Let $\theta'$ be the emission angle of the photon with respect to the beam direction in the projectile system. After the $\mathbf{b}$-integration has been carried out, it is possible and convenient to choose a coordinate system in which the wave vector $\mathbf{k}'$ of the photon and the linear polarization vector $\hat{\mathbf{u}}_1'$ lie in the $x-z$ plane, and the other polarization vector $\hat{\mathbf{u}}_2'$ perpendicular to it, that is in the $y$ direction, see Fig. 9.1. Explicitly written in components, we then have

$$\begin{aligned} \mathbf{k}' &= (k' \sin\theta', 0, k' \cos\theta') \\ \hat{\mathbf{u}}_1' &= (\cos\theta', 0, -\sin\theta') \\ \hat{\mathbf{u}}_2' &= (0, 1, 0). \end{aligned} \tag{9.19}$$

The angular distribution in the laboratory system is obtained from Eq. (9.18) with the aid of Eqs. (9.9) and (9.10).

We now turn to a discussion of the limitations introduced by the use of approximate wave functions. The first point is the role played by the current operator $-ec\boldsymbol{\alpha}$ entering in the formulation given above. In general, in a charge transfer collision, one will try to use approximations in which the initial state is an eigenstate of the channel Hamiltonian $H_{\mathrm{i}}$, see Eq. (5.53), while the final state is an eigenstate of another Hamiltonian $H_{\mathrm{f}} \neq H_{\mathrm{i}}$, see Eq. (5.54). If this is indeed the case, the operator $-ec\boldsymbol{\alpha}$ will no longer satisfy a continuity equation like Eq. (4.17) together with the charge density and hence will not qualify as a current operator $\mathbf{j}$ to be used with the electromagnetic interaction $\mathbf{j} \cdot \mathbf{A}/c$. In fact, one may convince oneself that a source term resulting from $(H_{\mathrm{i}} - H_{\mathrm{f}})$ appears in the continuity equation. Since in suggestive approximations, e.g., those successfully used in Coulomb capture, see Chap. 8, the Hamiltonians $H_{\mathrm{i}}$ and $H_{\mathrm{f}}$ belong to different centers and hence are drastically different, there is no current operator consistent with such an approach. As a consequence, it is not possible to define a Born approximation or distorted-wave Born approximation for REC, see, e.g., [HiW87, HiW89], consistent with the usual definition $\mathbf{j} = -ec\boldsymbol{\alpha}$ for the current operator.

In the absence of an exact three-body wave function, a consistent description is obtained if both initial and final states are constructed as eigenstates of *either* $H_{\mathrm{i}}$ *or* $H_{\mathrm{f}}$. Since usually in REC one considers systems with $Z_{\mathrm{T}} \ll Z_{\mathrm{P}}$, the only practical way to treat the process is to consider initial and final states as continuum and bound states of the *projectile*. The target atom is taken into account only insofar as it generates an initial electronic momentum distribution. This is the impulse approximation in the projectile frame discussed in Section 9.1.

## 9.3 A nonrelativistic treatment

For qualitative purposes, it is instructive to specialize the transition amplitude (9.15) to a nonrelativistic collision. In view of the foregoing discussion we choose the projectile system as a reference frame. Since, furthermore, it is also known [KlJ75, BrD74] that the plane-wave impulse approximation or, equivalently, the Born approximation leads to acceptable results in the projectile frame, we use unperturbed nonrelativistic hydrogenic-bound states as initial and final states. Assuming spinless particles for the moment, we replace $\boldsymbol{\alpha}$ by $\mathbf{v}_{\mathrm{el}}/c = \mathbf{p}_{\mathrm{el}}/(m_{\mathrm{e}}c)$.

For high projectile velocities, REC into projectile 1s states will be dominant, and, although this limitation is easily removed, we also assume 1s target states. With the 1s hydrogenic momentum wave function being given by [BeS57]

$$\tilde{\varphi}_{1\mathrm{s}}(q) = \frac{2\sqrt{2}\, Q_Z^{5/2}}{\pi(q^2 + Q_Z^2)^2}, \tag{9.20}$$

where $Q_Z = m_{\mathrm{e}}e^2 Z/\hbar$, the 1s-1s REC cross section per electron (9.18) in the projectile frame can be written as

$$\frac{d^2\sigma}{d\Omega'\, d(\hbar\omega')} = \frac{e^2}{\hbar c}\left(\frac{2\pi\hbar}{c}\right)^2 \frac{\hbar\omega'}{(m_{\mathrm{e}}v)^2} \sum_{\lambda} \int d^2 q_b\, |\tilde{\varphi}_{\mathrm{f}}(p)(\mathbf{p}\cdot\hat{\mathbf{u}}'_{\lambda})\tilde{\varphi}_{\mathrm{i}}(q)|^2, \tag{9.21}$$

where

$$\begin{aligned} \mathbf{q} &= (\mathbf{q}_b, q_+) \\ \mathbf{p} &= (\mathbf{q}_b, q_-) - \hbar\mathbf{k}' \end{aligned} \tag{9.22}$$

and, in analogy to Eq. (8.7),

$$q_{\pm} = [\epsilon_{\mathrm{i}} - (\epsilon'_{\mathrm{f}} + \hbar\omega')]/v \pm \tfrac{1}{2} m_{\mathrm{e}} v. \tag{9.23}$$

The energy conservation expressed by Eq. (9.23) is a result of retaining the full space-time dependence $\exp(-i\mathbf{k}'\cdot\mathbf{r}' + i\omega' t')$, sometimes denoted as retardation, of the photon wave function.

Kleber and Jakubassa [KlJ75] show that the Born approximation Eq. (9.21) in the impact parameter picture is identical to the plane-wave impulse approximation Eq. (9.1).

*Angular distribution*

Let us now concentrate on the angular distribution of the emitted photons, assuming that the photon polarization is not detected. While the momentum wave function $\tilde{\varphi}_{\mathrm{i}}$ of the initial state enters only insofar as it cuts off the

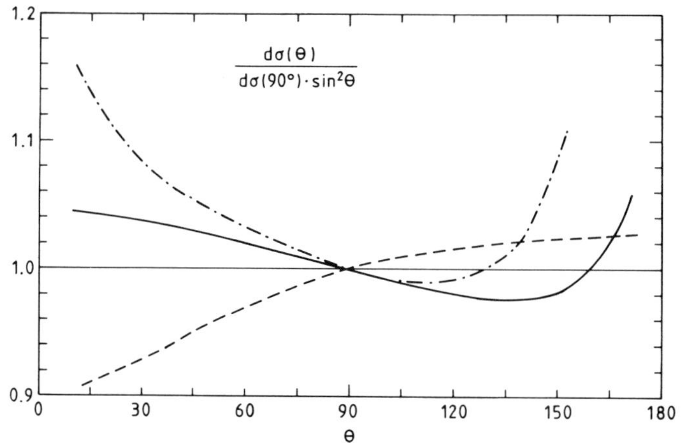


**Figure 9.2.** Calculated angular distributions of K REC photons in 197-MeV/u Xe + Be collisions divided by $\sin^2\theta$ and normalized at $90^0$. Dashed curve: Sauter formula (9.32), see Sec. 9.4.2; solid curve: nonrelativistic Born approximation calculation, Eq. (9.24); dash-dot curve: relativistic approximation calculation for non-spin-flip transitions only, Eq. (9.18). See also Fig. 9.8.

integration over $\mathbf{q}_b$ at large values of $q_b$, the emission angle $\theta'$ enters directly into $\tilde{\varphi}_\mathrm{f}$ when $\mathbf{p}$ is inserted from Eq. (9.22). If $\chi$ denotes the azimuthal angle of $\mathbf{q}_b$, i.e., the angle between $\mathbf{q}_b$ and $\mathbf{k}'_b$ in the impact parameter plane, we may write

$$\sum_\lambda |\tilde{\varphi}_\mathrm{f}(p)(\mathbf{p}\cdot\hat{\mathbf{u}}'_\lambda)\tilde{\varphi}_\mathrm{i}(q)|^2 \propto$$

$$\frac{(q_b\cos\chi\cos\theta' - q_-\sin\theta')^2 + q_b^2\sin^2\chi}{(q_b^2 - 2\hbar q_b k'\cos\chi\sin\theta' + \hbar^2 k'^2 + q_-^2 - 2\hbar q_- k'\cos\theta' + Q_{Z_\mathrm{P}}^2)^4}$$
$$\times\frac{1}{(q_b^2 + q_+^2 + Q_{Z_\mathrm{T}}^2)^4}. \tag{9.24}$$

In order to illustrate the main features of this expression, it is useful to introduce some approximations. The main contribution to the integral in Eq. (9.21) arises for $q_+ = 0$, in agreement with the resonance condition (9.8). The width $\Delta(\hbar\omega') \approx vQ_{Z_\mathrm{T}}$ of the resonance is determined by the Doppler spread in $q_+$ permitted by the target charge $Z_\mathrm{T}$. In the limit of a vanishing target charge $Z_\mathrm{T}$ or of $\epsilon_\mathrm{i} \approx 0$, the contributions of transverse momenta $q_b$ are strongly peaked around $q_b = 0$. At the resonance, we then have $\epsilon'_\mathrm{f} + \hbar\omega' = \frac{1}{2}m_\mathrm{e}v^2$ and $q_- \approx -m_\mathrm{e}v$. Since furthermore $\hbar\omega' \ll 2m_\mathrm{e}c^2$, we can write the photon angular distribution in the projectile system as

$$\frac{d\sigma}{d\Omega'} \propto \frac{\sin^2\theta'}{(1+\beta\cos\theta')^4}. \tag{9.25}$$

Using Eq. (9.10) to transform Eq. (9.25) into the laboratory system, one finds at the resonance frequency

$$\frac{d\sigma}{d\Omega} \propto \sin^2\theta. \tag{9.26}$$

Hence the cross section vanishes in the forward and backward directions. Indeed, since a transverse photon carries the angular momentum $\pm 1$ in its direction of propagation, transitions into atomic s-states of a spinless electron moving along the $z$-axis are forbidden if the photon is emitted along the positive or negative $z$-axis [FaM59]. In a relativistic description (Sec. 9.5.3) where the electron spin is taken into account, forward and backward emission is consistent with angular momentum conservation.

In an exact nonrelativistic calculation, if the integration over $\mathbf{q}_b$ is actually carried out, one obtains an angular distribution very close to Eq. (9.26), see Fig. 9.2. In order to derive the result (9.25), it is essential to take into account the retardation of the photon momentum, i.e., *not* to use the dipole approximation replacing $\exp(-i\mathbf{k}'\cdot\mathbf{r}') \to 1$. This would lead to a photon angular distribution in the *projectile frame* proportional to $\sin^2\theta'$. As Eq. (9.26) shows, the forward-backward asymmetry (9.25) in the projectile frame caused by the retardation is fully compensated by the Lorentz transformation to the laboratory system. This was first pointed out by Spindler [Spi79, SpB79].

The angular distribution (9.26) is in good agreement with experimental data for intermediate projectile charges, see, e.g., Fig. 9.8, even at relativistic velocities when the approximations leading to this result are no longer justified.

*The role of electron spin*

So far, in Eq. (9.21), we have disregarded the electron spin. If one takes into account the interaction of the electron magnetic moment with the magnetic field of the photon, the situation changes. The conservation of the angular momentum along the polar axis no longer prohibits photon emission in the forward or backward direction, because a spin-flip transition can compensate for the angular momentum carried away by the photon. This means that in a relativistic description *REC into l=0 states at forward or backward angles can occur only by spin-flip processes mediated by magnetic interactions*. Therefore, angular distributions may change significantly, see, e.g., Fig. 9.10.

In leading order, spin-flip transitions are mediated by the magnetic dipole term in the multipole expansion [Mos66] of the electromagnetic interaction, which does not act on the space part of the wave functions. Therefore, if the magnetic dipole interaction is evaluated between *exact* bound and continuum radial wave functions, its matrix element vanishes owing to orthogonality so that only the next order in the expansion of the photon plane wave can give a finite contribution. On the other hand, if the magnetic dipole interaction is evaluated with *approximate* wave functions, as provided by the Born or distorted-wave Born approximation, there will be, in general, a finite overlap giving rise to spurious contributions. The argument shows that spin-flip transitions are extremely sensitive to the use of accurate wave functions.

We may immediately draw a conclusion regarding relativistic collisions. Since the starting equation (9.12) for relativistic REC is based on the Dirac $\boldsymbol{\alpha}$ operator which embraces both non-spin-flip as well as spin-flip transitions, any Born type approximation will lead to incorrect spin-flip contributions to the angular distribution.

On the other hand, if one confines oneself to non-spin-flip transitions, a relativistic Born treatment leads to results that are similar to those of a nonrelativistic theory, disregarding the spin degree of freedom. Figure 9.2 shows the deviation of various angular distributions from a pure $\sin^2\theta$ dependence in the laboratory system. Besides relativistic and nonrelativistic cross sections integrated over the REC resonance, the figure also includes the differential cross section derived from the Sauter theory [Sau31] of the relativistic photoeffect, see Sec. 9.4.2. In all cases, the departure from a $\sin^2\theta$ behavior around $90^0$ is very small at this energy. The cross sections vanish at forward and backward angles as is required by angular momentum conservation.

Much more than for non-spin-flip transitions, is it crucial for spin-flip transitions and hence for spin-averaged relativistic angular distributions to describe the initial electronic continuum state (with respect to the projectile) by an *exact* Coulomb wave function or at least by a wave function that is orthogonal to the final bound state.

## 9.4 The photoelectric cross section

In order to avoid the problems discussed in Sec. 9.3, it is reasonable to describe radiative electron capture as the modified inverse of the atomic photoelectric effect: a continuum electron is captured by the projectile with the simultaneous emission of a photon. As is discussed preliminarily in Sec. 9.1, the original three-body problem is then reduced to a two-body problem, and the target merely enters by providing an initial momentum distribution of the continuum electrons as seen from the projectile nucleus. The initial momentum spread of the electrons bound in the target atom leads to a broadening of the photon resonance.

As a first step in deriving REC cross sections, we discuss the photoelectric cross section. For the photoelectric effect, approximate formulas for the differential cross section are available [PrL64a,b, PrR73], mostly for K-electron emission. When a photon with energy $\hbar\omega$ interacts with a hydrogenic target atom of charge $Z$, an electron with total energy $E$ (including the rest mass) may be emitted. If $|\epsilon_b|$ is the binding energy of the initially bound electron, the conservation of energy requires that

$$E = \hbar\omega + m_e c^2 - |\epsilon_b|, \tag{9.27}$$

so that there exists a one-to-one relation between $E$ and $\hbar\omega$. Before discussing the general procedure for calculating the exact photoelectric cross section in Sec. 9.4.3, we first consider a nonrelativistic approximation in Sec. 9.4.1 and in Sec. 9.4.2 a relativistic approach valid for low-$Z$ atoms.

### 9.4.1 The Stobbe formula

Assuming $\hbar\omega \ll m_e c^2$ and $\alpha Z \ll 1$, it is appropriate to use a nonrelativistic description. Let $E_{\text{kin}} = \frac{1}{2} m_e v^2$ be the kinetic energy of an electron ejected from the K-shell. Then Eq. (9.27) is replaced with

$$E_{\text{kin}} = \hbar\omega - |\epsilon_b|, \tag{9.28}$$

where

$$\frac{\hbar\omega}{E_{\text{kin}}} = 1 + \frac{|\epsilon_b|}{E_{\text{kin}}} = 1 + \nu^2 \tag{9.29}$$

is expressed by the Sommerfeld parameter $\nu = Ze^2/\hbar v$, see Eq. (5.52). Using exact nonrelativistic hydrogenic initial and final states, disregarding retardation effects, and adopting the dipole approximation, Stobbe [Sto30] has obtained the photoelectric cross section per K-shell electron in the form [BeS57, Som39]

$$\sigma_{\text{ph}} = \frac{2^8 \pi^2 \alpha}{3} \frac{m_e c^2}{\hbar\omega} \lambda\!\!\!^{-}_c{}^2 \left( \frac{\nu^2}{1+\nu^2} \right)^3 \frac{e^{-4\nu \arctan(1/\nu)}}{1 - e^{-2\pi\nu}}, \tag{9.30}$$

where $\lambda\!\!\!^{-}_c = \hbar/m_e c$ is the Compton wavelength of the electron.

The corresponding differential cross section is

$$\frac{d\sigma_{\text{ph}}}{d\Omega_{\text{ph}}} = \sigma_{\text{ph}} \frac{3}{8\pi} \sin^2\theta. \tag{9.31}$$

The Stobbe cross section proves to be quite useful in REC calculations up to projectile energies of a few hundred MeV/u, corresponding to electron kinetic energies $(\gamma - 1)m_{\text{e}}c^2$, well below the electron rest energy.

### 9.4.2 The Sauter formula

Sauter [Sau31] has derived an analytic expression for the cross section for the relativistic photoeffect assuming $\beta \approx 1$ and disregarding terms of the order $\alpha^2 Z^2$ in an $\alpha Z$ expansion of the matrix element. For a discussion of the approximations, see Fano et al. [FaM59] and Pratt et al. [PrR73]. The Sauter formula has been used to derive the corresponding REC cross sections and to compare them with experimental data [RaY71, AnM84]. According to the Sauter formula [Sau31, Nag60, PrR73], the differential photoionization cross section per K-shell electron for unpolarized photons and electrons is

$$\begin{aligned}\frac{d\sigma_{\text{ph}}}{d\Omega_{\text{ph}}} &= \alpha^6 Z^5 \left(\frac{m_{\text{e}}c^2}{\hbar\omega}\right)^5 ƛ_{\text{c}}^2 \frac{\beta^3}{\gamma} \\ &\times \frac{\sin^2\theta}{(1-\beta\cos\theta)^4}\left[1 + \frac{\gamma(\gamma-1)(\gamma-2)}{2}(1-\beta\cos\theta)\right]. \end{aligned} \tag{9.32}$$

By integrating over the solid angle, one obtains the total cross section per K-electron as

$$\begin{aligned}\sigma_{\text{ph}} &= 2\pi\alpha^6 Z^5 \left(\frac{m_{\text{e}}c^2}{\hbar\omega}\right)^5 ƛ_{\text{c}}^2 \\ &\times \beta^3\gamma^3 \left\{\frac{4}{3} + \frac{\gamma(\gamma-2)}{\gamma+1}\left[1 - \frac{1}{2\beta\gamma^2}\ln\frac{1+\beta}{1-\beta}\right]\right\}. \end{aligned} \tag{9.33}$$

The Sauter formulas (9.32) and (9.33) have been improved by Gavrila [Gav59, Gav60], by Nagel [Nag60, HuN67], and by Pratt [PrR73], who included higher-order terms in $\alpha Z$.

### 9.4.3 Exact relativistic calculations

The exact evaluation of the relativistic photoelectric cross section requires a partial-wave expansion of the Coulomb continuum function. This means that closed-form expressions can no longer be derived, and one has to resort to numerical methods. Detailed formulations and tabulations of cross sections can be found in [Nag60, AlJ65, HuN67, PrR73, Sco73]. However, the existing tabulations are not always sufficient for use in connection with REC, which requires differential cross section on a sufficiently dense mesh. Since for this purpose independent computer codes have to be developed [IcS94], we here present an outline of the formulation, assuming unpolarized photons and a polarization-insensitive detector for the emitted electrons.

For a given photon energy $\hbar\omega$ and binding energy $|\epsilon_b|$ of the electron in the bound initial state $b$, the final electron energy $E$ (including the rest mass) is determined by Eq. (9.27). In treating the electron spin, one may either adopt a quantization in a fixed space direction or, alternatively, in the direction of the electron momentum. Each of these approaches has certain advantages.

*Quantization of the electron spin in the z-direction*

We first consider the formally simpler representation, in which the electron spin is quantized with respect to an arbitrary $z$-direction. This is possible if the electron spin is not measured and sometimes allows for a numerically faster evaluation.

The differential cross section for a single electron is [PrR73][1]

$$\frac{d\sigma_{\text{ph}}}{d\Omega} = \frac{\alpha\, m_{\text{e}}c^2}{4\,\hbar\omega}\,\frac{\lambdabar_{\text{c}}^2}{2j_b+1}\sum_{\mu_b}\sum_{m_s}\left|M_{\mathbf{p},b}(m_s,\lambda_+,\mu_b)\right|^2, \tag{9.34}$$

where we have averaged over the $(2j_b+1)$ angular momentum projections $\mu_b$ in the bound state and have summed over the spin components $m_s = \pm\frac{1}{2}$ of the emitted electron. Furthermore, we have averaged over the circular polarizations $\lambda_+ = 1$ and $\lambda_- = -1$ of the incoming photon.[2] Because of the summation over all other angular momentum projections, $\mu_b, m_s$, this is equivalent to taking one photon polarization, e.g., $\lambda_+ = 1$ only. The quantity $\alpha$ denotes the fine-structure constant.

In analogy to Eq. (9.12), the transition matrix element (adopting relativistic units $\hbar = c = m_{\text{e}} = 1$ in this subsection from here on) is

$$M_{\mathbf{p},b}(m_s,\lambda_+,\mu_b) = \int \psi^{\dagger}_{\mathbf{p},m_s}(\mathbf{r})\,\boldsymbol{\alpha}\cdot\hat{\mathbf{u}}_+\,e^{i\mathbf{k}\cdot\mathbf{r}}\,\psi_{j_b,\mu_b}(\mathbf{r})\,d^3r, \tag{9.35}$$

---

[1] Equation (4.2.22) of [PrR73] for the normalization of the radial wave functions contains a misprint. The expressions $[(E\pm1)^{1/2}/2E]$ should be replaced with $[(E\pm1)/2E]^{1/2}$. In this case, agreement with [HuN67, AlJ65] is obtained. The expression (9.34) is consistent with the corrected normalization (9.37).

[2] In conjunction with an angular momentum representation, it is convenient to sum over the circular rather than over the linear polarizations $\lambda_{1,2}$.

where, according to Eq. (4.123), but with a different normalization,

$$\begin{aligned}\psi_{\mathbf{p},m_s}(\mathbf{r}) &= 4\pi \sum_{\kappa,\mu} i^l e^{-i\Delta_\kappa} \left( \begin{array}{cc|c} l & \frac{1}{2} & j \\ \mu - m_s & m_s & \mu \end{array} \right) Y^*_{l,\mu-m_s}(\theta,0) \\ &\times \left( \begin{array}{c} g_\kappa(r)\,\chi^\mu_\kappa \\ i f_\kappa(r)\,\chi^\mu_{-\kappa} \end{array} \right) \end{aligned} \tag{9.36}$$

is the partial-wave expansion of the wave function describing the relativistic electron emitted with asymptotic momentum $\mathbf{p}$ ($p^2 = E^2 - 1$, $p_z = p\cos\theta$) and spin projection $m_s$ with respect to the $z$-axis. The summation in Eq. (9.36) extends over the relativistic quantum numbers $\kappa = \pm(j + \frac{1}{2})$ and the angular momentum projections $\mu$. Associated with $\kappa$ are the orbital angular momenta $l = l_A$ and $l' = l_B$ of the upper and lower component, respectively, given in Eq. (4.85). The $\chi^\mu_\kappa$ are the normalized two-component spin-angular functions (4.87), and the $g_\kappa(r)$ and $f_\kappa(r)$ are the (real) radial continuum wave functions (4.114).

The Coulomb wave functions are normalized on the energy scale and their asymptotic behavior is given by Eq. (4.118), that is

$$\begin{aligned} g_\kappa &\simeq \frac{1}{r}\sqrt{\frac{E+1}{\pi p}}\cos(pr + \sigma_\kappa), \\ f_\kappa &\simeq -\frac{1}{r}\sqrt{\frac{E-1}{\pi p}}\sin(pr + \sigma_\kappa). \end{aligned} \tag{9.37}$$

In the limit of a point nucleus, the Coulomb phase shift (4.120) is

$$\sigma_\kappa = \Delta_\kappa + \eta \ln(2pr), \tag{9.38}$$

where

$$\begin{aligned} \Delta_\kappa &= \tfrac{1}{2}\arg\left(\frac{-\kappa + i\eta/E}{s + i\eta}\right) - \arg\Gamma(s + i\eta) - \tfrac{1}{2}\pi s \\ \eta &= \frac{\alpha Z E}{p}, \qquad s = \sqrt{\kappa^2 - (\alpha Z)^2}. \end{aligned} \tag{9.39}$$

Since angular distributions are sensitive to phases, one has to make sure that the phase conventions Eqs. (9.37) and (9.38) (or another set of consistent phase conventions) are simultaneously satisfied.[3]

[3] In [IcS94] the generalization for an extended nucleus given by Müller et al. [MuR73] has been used.

Returning to Eq. (9.35), the exact hydrogenic bound-state wave function is identified with Eq. (4.86). If we now insert the expansion (6.42) of the photon plane wave and perform the spinor algebra, we arrive at integrals over three spherical harmonics which can be expressed by Clebsch-Gordan coefficients (4.88). Collecting the results, we obtain

$$M_{\mathbf{p},b}(m_s,\lambda_+,\mu_b) = 4\pi\sqrt{2}\sum_{L=0}^{\infty}\sum_{\kappa} i^{L+1-l}\, e^{i\Delta_\kappa} \times \left(\begin{array}{cc|c} l & \frac{1}{2} & j \\ \mu_b+1-m_s & m_s & \mu_b+1 \end{array}\right) F^+_{\mu_b}(L,\kappa)\, Y_{l,\mu_b+1-m_s}(\theta,0), \quad (9.40)$$

where

$$F^+_{\mu_b}(L,\kappa) = \left[A_L(j_b\mu_b j; l_b' l)\int_0^\infty g_\kappa(r)\, j_L(kr)\, f_b(r)\, r^2\, dr - A_L(j_b\mu_b j; l_b l')\int_0^\infty f_\kappa(r)\, j_L(kr)\, g_b(r)\, r^2\, dr\right], \quad (9.41)$$

and

$$A_L(j_b\mu_b j; l_1 l_2) = (2L+1)\sqrt{\frac{2l_1+1}{2l_2+1}}\left(\begin{array}{cc|c} l_1 & \frac{1}{2} & j_b \\ \mu_b+\frac{1}{2} & -\frac{1}{2} & \mu_b \end{array}\right) \times \left(\begin{array}{cc|c} l_2 & \frac{1}{2} & j \\ \mu_b+\frac{1}{2} & \frac{1}{2} & \mu_b+1 \end{array}\right)\left(\begin{array}{cc|c} l_1 & L & l_2 \\ \mu_b+\frac{1}{2} & 0 & \mu_b+\frac{1}{2} \end{array}\right) \times \left(\begin{array}{cc|c} l_1 & L & l_2 \\ 0 & 0 & 0 \end{array}\right). \quad (9.42)$$

It is not difficult to carry the analytical reduction further [PrR73,HuN67] by squaring the matrix element (9.40) and by expanding the resulting products of spherical harmonics. In this way, additional summations are introduced. However, for numerical evaluation, it is often convenient to take directly the modulus square of the matrix element (9.40) and to introduce it into Eq. (9.34).

*Helicity representation*

An alternative description, convenient for calculating total cross sections (9.46) and necessary for angular correlations, see, e.g., [Eic94], employs the helicity representation (4.124) of the electron continuum wave function, in which the electron spin is quantized along the asymptotic momentum. Adopting a different normalization than in Eq. (4.124),

$$
\begin{aligned}
\psi_{\mathbf{p},\sigma}(\mathbf{r}) &= \sum_{\kappa,\mu} i^l e^{-i\Delta_\kappa} \sqrt{4\pi(2l+1)} \left( \begin{array}{cc|c} l & \frac{1}{2} & j \\ 0 & \sigma & \sigma \end{array} \right) D^j_{\mu\sigma}(\hat{\mathbf{z}} \to \hat{\mathbf{p}}) \\
&\times \begin{pmatrix} g_\kappa(r)\, \chi^\mu_\kappa \\ i f_\kappa(r)\, \chi^\mu_{-\kappa} \end{pmatrix}
\end{aligned}
\tag{9.43}
$$

is the partial-wave expansion with the spin projection $\sigma$ with respect to $\mathbf{p}$. The rotation matrix $D^j_{\mu\sigma}(\hat{\mathbf{z}} \to \hat{\mathbf{p}})$ carries each partial wave from an original quantization along the $z$-axis (expressed by the Clebsch-Gordan coefficient) into a quantization axis along $\mathbf{p}$.

The differential cross section is then obtained as an expansion in terms of Legendre polynomials $P_\nu(\cos\theta)$ as

$$
\frac{d\sigma_{\mathrm{ph}}}{d\Omega} = \frac{\alpha\, m_e c^2}{4\hbar\omega} \frac{\chi_c^2}{2j_b+1} \sum_\nu B_\nu\, P_\nu(\cos\theta).
\tag{9.44}
$$

with the expansion coefficients

$$
\begin{aligned}
B_\nu &= 8\pi \sum_{\mu_b} (-1)^{\mu_b - 1/2} \sum_{L\kappa} \sum_{\overline{L\kappa}} (-1)^{(L-l-\overline{L}+\overline{l})/2} \cos(\Delta_\kappa - \Delta_{\overline{\kappa}}) \\
&\times \sqrt{(2l+1)(2j+1)(2\overline{l}+1)(2\overline{j}+1)} \left( \begin{array}{cc|c} l & \overline{l} & \nu \\ 0 & 0 & 0 \end{array} \right) \\
&\times \left( \begin{array}{cc|c} j & \overline{j} & \nu \\ \mu_b+1 & -\mu_b-1 & 0 \end{array} \right) \left\{ \begin{array}{ccc} l & j & \frac{1}{2} \\ \overline{j} & \overline{l} & \nu \end{array} \right\} \\
&\times F^+_{\mu_b}(L,\kappa)\, F^+_{\mu_b}(\overline{L},\overline{\kappa}),
\end{aligned}
\tag{9.45}
$$

where $F^+_{\mu_b}(L,\kappa)$ is given by Eq. (9.41). The summations occurring in Eq. (9.45) are limited by the selection rules embodied in the Clebsch-Gordan coefficients and $6j$-symbols $\{\cdots\}$. In particular, if the electron is initially bound in a $1s_{1/2}$ state, a rather small number of terms contributes. Nevertheless, the double sums occurring in Eq. (9.45) usually require longer computing times than needed for squaring the undeveloped matrix element (9.40). For numerically calculating the differential cross section, the latter is therefore a more practical starting point.

Most of the published calculations have been performed for the K-shell photoeffect, taking into account screening corrections which are needed for neutral atoms. However, if one wants to use existing tabulations to derive

REC cross sections for *bare* projectiles, one has to compensate for these undesired screening corrections [MeA85].

*The total cross section*

The total photoelectric cross section is obtained by integrating Eq. (9.44) over the photon angles. Then only the term with $\nu = 0$ contributes in the sum, and the total cross section per electron is

$$\sigma_{\text{ph}} = \frac{\alpha\pi\, m_{\text{e}}c^2}{\hbar\omega} \frac{\lambda\!\!\!^{-}_{\text{c}}^2}{2j_b + 1} B_0 \tag{9.46}$$

with the positive-definite expression

$$B_0 = 8\pi \sum_{\mu_b} \sum_{\kappa} \left| \sum_L i^L F^{+}_{\mu_b}(L,\kappa) \right|^2 . \tag{9.47}$$

*Radiative corrections*

Radiative corrections to the atomic photoeffect [BoG82, McG77] can be expressed by a corrective factor $[1 + (\alpha/\pi)\delta]$ multiplying the basic differential cross section (9.34) or (9.44). The quantity $\delta$, calculated to lowest order in $\alpha$ and $\alpha Z$, increases with the photon energy and the electron ejection angle. The correction term $(\alpha/\pi)\delta$ is always negative and of the order 0.2% for 100-keV, 1.5% for 1-MeV, and 5% for 5-MeV photon energy. For the REC experiments discussed in Sec. 9.6, the radiative correction is well within the experimental uncertainty. Since for large photon energies the correction becomes quite appreciable, it would be interesting to increase the experimental accuracy to a point that allows for a test of these results.

## 9.5 From the atomic photoeffect to REC

### 9.5.1 Radiative recombination

Once the cross section $\sigma_{\text{ph}}(\omega', \theta')$ of the photoelectric effect in the projectile frame (we reintroduce primed quantities) has been computed, it is a simple matter to write down the cross section $\sigma_{\text{RR}}(E', \theta')$ for radiative recombination. This is the cross section for capturing a free electron of energy $E'$ into a bound atomic state (binding energy $|\epsilon_b|$) with the simultaneous emission of a photon of energy $\omega'$. Since according to Eq. (9.27), $\hbar\omega' = E' - m_{\text{e}}c^2 + |\epsilon_b|$, we may use $E'$ as well as $\omega' = k'c$ to characterize the process. We now choose the direction of the incoming electron as the $z$ direction, so that $\theta'$ is the emission angle of the photon.

By the principle of detailed balance, the cross section for radiative recombination into a vacant shell is written as

$$\begin{aligned} \frac{d^2\sigma_{\mathrm{RR}}(E',\theta')}{dE'\,d\Omega'_{\mathrm{ph}}} &= (2j_b+1)\,\frac{(\omega'/c)^2}{p'^2}\,\frac{d^2\sigma_{\mathrm{ph}}(E',\theta')}{dE'\,d\Omega'_{\mathrm{el}}} \\ &= (2j_b+1)\left(\frac{\hbar\omega'}{m_{\mathrm{e}}c^2}\right)^2 \frac{1}{\beta^2\gamma^2}\,\frac{d^2\sigma_{\mathrm{ph}}(E',\theta')}{dE'\,d\Omega'_{\mathrm{el}}}. \end{aligned} \tag{9.48}$$

Here, $\sigma_{\mathrm{ph}}$ is the photoelectric cross section per electron and the factor $(2j_b + 1)$ converts the averaging of Eq. (9.34) over bound-state angular momentum projections into a summation over a filled shell while the factor $(\omega'/c)^2/p'^2$ replaces the phase space factor of the outgoing electron by that of the emitted photon.

In applications to relativistic ion-atom collisions, the $z$-direction is usually defined as the direction of the projectile motion. This is opposite to the direction of the electron momentum as seen from the projectile. Hence for REC, the angle $\theta'$ of the photoelectric effect or of the radiative recombination has to replaced by $\pi - \theta'$, or $\cos\theta'$ is replaced by $-\cos\theta'$.

The total cross section for radiative recombination into an empty projectile subshell $j_b$ is obtained from Eqs. (9.48) and (9.46) in the form

$$\sigma_{\mathrm{RR}}^{\mathrm{exact}} = \frac{\alpha\pi}{\beta^2\gamma^2}\,\frac{\hbar\omega}{m_{\mathrm{e}}c^2}\,\lambda\!\!\!^{-}_{\mathrm{c}}{}^2\,B_0, \tag{9.49}$$

where $B_0$ is given by Eq. (9.47).

*The Stobbe cross section*

In the nonrelativistic limit, the factor in Eq. (9.48) is $(\hbar\omega'/m_{\mathrm{e}}c^2)^2\,(1/\beta^2\gamma^2) = (\hbar\omega/2m_{\mathrm{e}}c^2)(1+\nu^2)$, so that the Stobbe cross section (9.30) for the photoelectric effect can be rewritten for radiative electron capture into an empty K-shell ($j_b = \frac{1}{2}$) as

$$\sigma_{\mathrm{RR}}^{\mathrm{Stobbe}} = \frac{2^8\pi^2\alpha}{3}\,\lambda\!\!\!^{-}_{\mathrm{c}}{}^2\left(\frac{\nu^3}{1+\nu^2}\right)^2 \frac{e^{-4\nu\arctan(1/\nu)}}{1-e^{-2\pi\nu}}, \tag{9.50}$$

where $\lambda\!\!\!^{-}_{\mathrm{c}} = \hbar/m_{\mathrm{e}}c$ is the Compton wavelength of the electron and $\nu = Ze^2/\hbar v$ is the Sommerfeld parameter. The constants in front of the $\nu$-dependent terms make up a factor of 9164.7 barn.

As mentioned above, the Stobbe cross section proves to be quite useful to estimate REC into the K-shell up to projectile energies of a few hundred

MeV/u, corresponding to electron kinetic energies $(\gamma - 1)m_e c^2$ well below the electron rest energy. The angular distribution is discussed in Sec. 9.4.1.

*The Sauter cross section*

For relativistic electron energies, one may apply the approximate Sauter formula (9.33). Inserting into Eq. (9.48), one obtains the cross section for radiative recombination into an empty K-shell in the form

$$\begin{aligned} \sigma_{\mathrm{RR}}^{\mathrm{Sauter}} &= 4\pi\alpha^6 Z^5 \left(\frac{m_e c^2}{\hbar\omega}\right)^3 \lambda\!\!\!^{-}{}_c^2 \\ &\times \beta\gamma \left\{ \frac{4}{3} + \frac{\gamma(\gamma-2)}{\gamma+1} \left[ 1 - \frac{1}{2\beta\gamma^2} \ln\frac{1+\beta}{1-\beta} \right] \right\}. \end{aligned} \tag{9.51}$$

The angular distribution is given by Eq. (9.32). We notice that for moderately relativistic collisions ($\gamma \leq 2$), the leading term in Eq. (9.32) has the same $\sin^2\theta$ angular dependence in the laboratory system as the nonrelativistic REC expression of Eq. (9.25). However, with increasing values of $\gamma$, the second term in (9.32) becomes more important and leads to a shift of the maximum in the angular distribution to forward angles. In the extreme relativistic energy regime, the Sauter angular distribution (9.32) behaves as $\cos^2\frac{\theta}{2}$ except in the very forward direction. This dependence, in its effect on REC, is illustrated in Figure 9.3 for various values of $\gamma$. In the exact forward direction $\theta = 0$ as well as in the backward direction $\theta = \pi$, the Sauter cross section vanishes, although, with the inclusion of the electron spin, the conservation of angular momentum does not enforce such a behavior. It has been conjectured [Nag60] that this curious feature of the differential cross section calculated in first order of $\alpha Z$, observed already by Fano et al. [FaM59], may be accidental for the Coulomb potential. Higher-order terms in $\alpha Z$ yield a finite contribution at $\theta = 0$ and at $\theta = \pi$ [SaW55, Nag60]. Exactly calculated differential cross sections are discussed in Sec. 9.6.3.

For the exact relativistic cross section, Sec. 9.4.3, no closed formula can be given. Numerically evaluated total cross sections are presented in Sec. 9.6.2 and are compared with the Stobbe cross sections.

## 9.5.2 The momentum distribution in the target

In relativistic electron capture, the velocity of the captured electron is essentially determined by the velocity $-\mathbf{v}$ of the target atom with respect to the projectile. However, an electron bound in the target (which is assumed to have a low charge $Z_T \ll Z_P$) has a momentum distribution $\rho(\mathbf{q})$, so that there is a momentum spread around the momentum $-\mathbf{p}_0 = -\gamma m_e \mathbf{v}$.

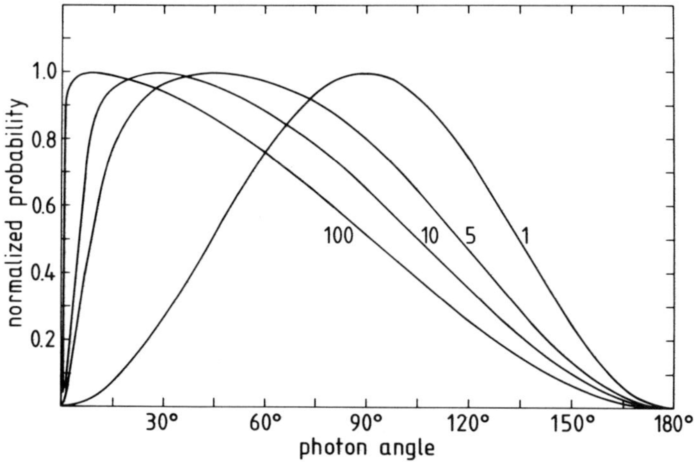


**Figure 9.3.** Normalized Sauter angular distribution (9.32) for the emission of REC photons integrated over the resonance and transformed to the laboratory system. The curves are labeled by the projectile energy $T_{\mathrm{P}}$ in GeV/u. The angular distributions for $T_{\mathrm{P}} < 1$ GeV/u are almost indistinguishable from the curve for 1 GeV/u.

*Hydrogenic wave functions*

If we assume a spherically averaged distribution for a single electron (averaging over a complete principal shell $n$), the nonrelativistic hydrogenic momentum distribution is given by

$$\rho_n(q^2) = \frac{1}{n^2} \sum_l \sum_m |\tilde{\varphi}_{nlm}(\mathbf{q})|^2 = \frac{8\, Q_n^5}{\pi^2\, (q^2 + Q_n^2)^4}, \tag{9.52}$$

where $\tilde{\varphi}_{nlm}(\mathbf{q})$ is the Fourier transform of the electronic wave function in the state $\{nlm\}$ and

$$Q_n = \frac{m_{\mathrm{e}} e^2 Z_{\mathrm{T}}}{n\hbar}. \tag{9.53}$$

In order to approximate the full momentum distribution $\rho(q^2)$ in a multielectron target atom, one has to sum (9.52) over all target electrons, inserting the appropriate effective charges in Eq. (9.53)

The momentum distribution (9.52) has a maximum at $q = 0$ and a width determined by $Q_n$: the distribution becomes more and more peaked around $q = 0$, the lower $Z_{\mathrm{T}}$, and the higher $n$.

Using Eq. (9.52), the Compton profile (9.6) can be written in the form

$$\mathcal{J}_n(q_z) = 2\pi \int_{q_z}^{\infty} \rho_n(q^2) q\, dq = \frac{8\, Q_n^5}{3\pi\, (q^2 + Q_n^2)^3}. \tag{9.54}$$

*Roothaan-Hartree-Fock wave functions*

For multielectron atoms, the hydrogenic density distribution may not be accurate enough. More realistic momentum distributions can be obtained from Roothaan-Hartree-Fock (RHF) wave functions tabulated by Clementi and Roetti [ClR74]. In these calculations, single-electron orbitals $i$ of symmetry $l, m$ are represented as linear combinations of Slater-type orbitals (STO) in the form

$$\varphi_{i,lm}(\mathbf{r}) = \sum_n C_{i,ln} R_{ln}(r) Y_{lm}(\hat{\mathbf{r}}), \tag{9.55}$$

where $n = \{n, p\}$ refers to the integer nodal quantum number $n$ as well as to the label $p$ distinguishing basis functions with the same $n, lm$, while $R_{ln}(r)$ is the STO radial wave function

$$R_{ln}(r) = N_{ln} r^{n-1} e^{-\zeta_n r} \tag{9.56}$$

with the normalization factor $N_{ln}$, the expansion coefficients $C_{i,ln}$, and the optimized values $\zeta_n$ being tabulated in [ClR74] for all atoms and specified principal quantum numbers $i$.

From Eqs. (9.55) and (6.42), we obtain the Fourier transform of the electron wave function as

$$\tilde{\varphi}_{i,lm}(\mathbf{q}) = (2\pi\hbar)^{-3/2}\, 4\pi\, i^l\, Y_{lm}(\hat{\mathbf{q}}) \sum_n C_{iln} I_{ln}(q), \tag{9.57}$$

where

$$I_{ln}(q) = N_{ln} \int_0^{\infty} r^{n-1} e^{-\zeta_n r} j_l(qr) r^2\, dr \tag{9.58}$$

and $j_l(qr)$ is the spherical Bessel function.

For convenience, we give the integrals (9.58) for s- and p-states, which are derived in the same way as Eq. (8.13).

$$\begin{aligned} I_{0n} &= \frac{N_{0n}}{q} \frac{n!}{(\zeta_n^2 + q^2)^{(n+1)/2}} \sin\left((n+1) \arctan \frac{q}{\zeta_n}\right) \\ I_{1n} &= \frac{N_{1n}}{q^2} \frac{(n-1)!}{(\zeta_n^2 + q^2)^{n/2}} \sin\left(n \arctan \frac{q}{\zeta_n}\right) \\ &\quad - \frac{N_{1n}}{q} \frac{n!}{(\zeta_n^2 + q^2)^{(n+1)/2}} \cos\left((n+1) \arctan \frac{q}{\zeta_n}\right). \end{aligned} \tag{9.59}$$

Integrals for higher angular momenta are similar. The spherically averaged momentum density for a single electron in an $i, l$ shell is

$$\begin{aligned} \rho(q^2) &= \frac{1}{2l+1} \sum_m |\tilde{\varphi}_{i,lm}(\mathbf{q})|^2 \\ &= \frac{2}{\pi} \left( \sum_n C_{i,ln}\, I_{ln} \right)^2 . \end{aligned} \tag{9.60}$$

Using the tabulated values, it is not difficult to calculate the momentum distribution. In analogy to Eq. (9.54), one may also compute the Compton profile by numerical integration.

### 9.5.3 Convolution with the target momentum density

Since, with a probability given by $\rho(\mathbf{q})$, the electron momentum within the target has a component $q_b$ transverse to the beam direction, the momentum seen from the projectile deviates slightly from the negative $z$ direction (which is defined by the beam) and the corresponding angular distribution will be tilted by this amount. For example, for the rather low-energy collision of 197 MeV/u $Xe^{54+}$ on Be, see Fig. 9.5, one estimates a deviation from the negative $z$-direction by $\arctan(Q_1/\beta\gamma) \approx 4.3^0$, while for higher energies the angle will be less and for higher target charges greater. In general, the effect on the angular distribution will be small, but for completeness, we give a formulation which includes this effect.

For a given electron momentum $\mathbf{q}$ in the target, the Lorentz transformed momentum $\mathbf{q}'$ in the projectile forms the axis of an axially symmetric photon angular distribution. If we denote the polar and azimuthal angles of the photon with respect to $\mathbf{q}'$ as $\overline{\theta}'$ and $\overline{\phi}'$, a given photon direction $\theta'$ (we choose the $x$-axis such that the photon azimuthal angle $\phi' = 0$) in the projectile frame will receive contributions from various axes $\hat{\mathbf{q}}' = (\theta'_q, \phi'_q)$ and, associated with them, from corresponding angles $\overline{\theta}'$ and $\overline{\phi}'$. The angles are illustrated in Fig. 9.4. Let us consider the spherical triangle composed of the pole $P_0 = (\theta = 0)$, of the point $P_\gamma = (\theta', \phi' = 0)$ giving the direction of the emitted photon, and of the point $P_q = (\theta'_q, \phi'_q)$ giving the direction of the electron momentum $\mathbf{q}'$. The sides in this triangle are $s_0 = \overline{\theta}'$, $s_\gamma = \theta'_q$, and $s_q = \theta'$. Knowing the angle $\phi'_q$ at the pole, we can express the opposite side $s_0 = \overline{\theta}'$ by the law of cosines as

$$\cos\overline{\theta}' = \cos\theta'_q \cos\theta' + \sin\theta'_q \sin\theta' \cos\phi'_q . \tag{9.61}$$

In this way, we may eliminate the angle $\overline{\theta}'$ by which the photon is emitted with respect to the direction $\hat{\mathbf{q}}'$.

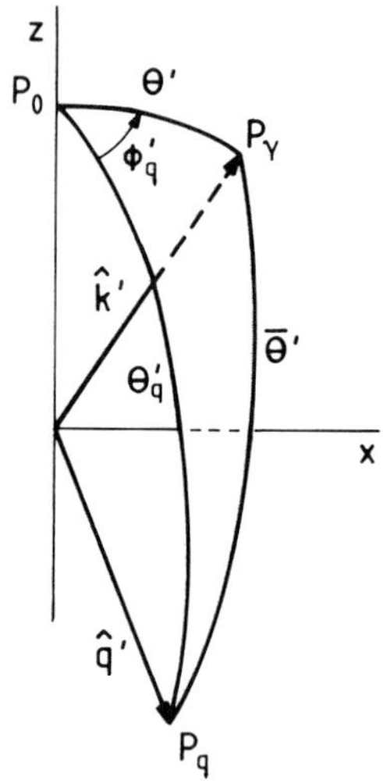


**Figure 9.4.** The angles characterizing the photon direction and the electron momentum in the projectile system.

We thus get a generalization of Eq. (9.1) which in $d\sigma_{\mathrm{RR}}/d\Omega'$ takes into account the deviations of the electron momenta $\mathbf{q}'$ from the negative beam axis and integrates over their contributions. Adopting relativistic units, $\hbar = c = m_{\mathrm{e}} = 1$, in this subsection, the differential REC cross section in the projectile frame is

$$\frac{d^2\sigma_{\mathrm{REC}}(\omega',\theta')}{d\omega'\,d\Omega'} = \frac{1}{2\pi}\,\frac{d^2\sigma_{\mathrm{REC}}(\omega',\theta')}{d\omega'\,d\cos\theta'} = \frac{1}{2\pi}\int d^2q_b \int dq_z \int d\cos\overline{\theta'}$$

$$\times\,\frac{d^2\sigma_{\mathrm{RR}}(\mathbf{q}',\overline{\theta}')}{d\omega' d\cos\overline{\theta}'}\left|\frac{\partial\cos\overline{\theta}'}{\partial\cos\theta'}\right|\rho(q^2)\,\delta(\omega' + E_{\mathrm{f}}' - \gamma E_{\mathrm{i}} + \gamma v q\cos\theta_q)$$

$$\times\,\delta(\cos\overline{\theta}' - \cos\theta_q'\cos\theta' - \sin\theta_q'\sin\theta'\cos\phi_q'). \qquad (9.62)$$

Here, $\phi_q'$ is measured with respect to $\phi' = 0$ chosen to define the $x$ axis.

In order to evaluate Eq. (9.62), we apply Eq. (2.29) to derive the Lorentz- transformed vector $\mathbf{q}'$ from $\mathbf{q}$. For the angles, we get

$$\begin{aligned}\sin\theta_q' &= \frac{q}{q'}\sin\theta_q\\ \cos\theta_q' &= \frac{\gamma}{q'}(q\cos\theta_q - v\sqrt{1+q^2}),\end{aligned} \qquad (9.63)$$

where

$$q \cos\theta_q = q_z = \frac{\omega_0' - \omega'}{\gamma v} \equiv q_0, \tag{9.64}$$

with the resonance photon energy $\omega_0' = \gamma E_i - E_f'$ given by Eq. (9.8). Correspondingly, the kinetic energy of the target electron, which, within the impulse approximation, is assumed to be free, is fully determined by the transverse momentum $q_b$, so that

$$E = \sqrt{q_b^2 + q_0^2 + 1}. \tag{9.65}$$

From the Lorentz transformation (2.29), we obtain

$$q'^2 = E'^2 - 1 = \gamma^2 \left( \sqrt{q_b^2 + q_0^2 + 1} \quad - \frac{\omega_0' - \omega'}{\gamma} \right)^2 - 1. \tag{9.66}$$

With these relations and (9.61), we are now able to evaluate Eq. (9.62). The REC cross section is obtained as a two-dimensional integral over the cross section for radiative recombination

$$\frac{d^2\sigma_{\rm REC}(\omega', \theta')}{d\omega'\, d\Omega'} = \frac{1}{2\pi\gamma v} \int d^2 q_b \frac{d^2\sigma_{\rm RR}(\mathbf{q}', \overline{\theta}')}{d\omega' d\cos\overline{\theta}'} \left| \frac{\partial \cos\overline{\theta}'}{\partial \cos\theta'} \right| \rho(q_b^2 + q_0^2). \tag{9.67}$$

In the limiting case $q_b \ll q' \approx v$, so that the axis of the angular distribution coincides with the $z$-axis, we have $\sin\theta_q' = 0$, $\cos\theta_q' = -(\gamma v/q') = -1$, $\partial \cos\overline{\theta}'/\partial\cos\theta' = -1$, and $\cos\overline{\theta}' = \cos\theta'$. We then get

$$\frac{d^2\sigma_{\rm REC}(\omega', \theta')}{d\omega'\, d\Omega'} = \frac{1}{2\pi\gamma v} \frac{d^2\sigma_{\rm RR}(q', \pi - \theta')}{d\omega'\, d\cos\theta'} \int d^2 q_b\, \rho(q_b^2 + q_0^2), \tag{9.68}$$

which is identical to the usual result, Eqs. (9.5) and (9.6). The cross section has a peak at $q = 0$ or

$$\omega_0' = \gamma E_{\rm i} - E_{\rm f}' \tag{9.69}$$

with a Doppler width given by $Q_n$ as defined in Eq. (9.53). The target momentum density enters the cross section only via the Compton profile.

## 9.6 Theoretical and experimental results

### 9.6.1 Radiative recombination

The general dependence of REC cross sections on the projectile charge and energy is best represented if we consider radiative recombination in which

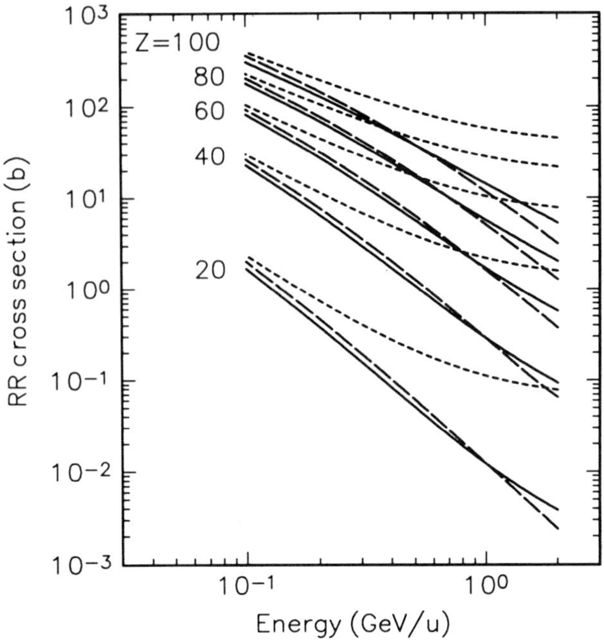


**Figure 9.5.** Energy and charge dependence of the cross section for radiative recombination (in barns) with capture into the projectile K-shell. Solid lines: exact (relativistic) calculated cross sections; short-dashed lines: Stobbe cross sections, Eq. (9.50), assuming the correct relativistic relation between projectile energy and velocity; long-dashed lines: Eq. (9.50), however assuming the nonrelativistic relation $E = \frac{1}{2}Mv^2$ between projectile energy and velocity.

the electron is assumed to be at rest in the initial state. In this case, no target parameters enter. The exact total cross sections are given for the K-shell by the solid lines in Fig. 9.5 and for the L-shells in Fig. 9.6.

A curious feature was noted in [StK92] by comparing tabulated exact photoionization cross sections with the simple result (9.50) of the nonrelativistic dipole approximation. Although these treatments are widely different, the total cross sections for radiative recombination agree rather well for energies below 1 GeV/u, a result that was confirmed in systematic relativistic calculations [IcS94]. Figure 9.5 presents a comparison of the exact cross section calculated from Eq. (9.49) by including all partial waves $|\kappa| \leq 20$ (solid line) with the cross section calculated according to the nonrelativistic Stobbe formula (9.50) for radiative electron capture into an empty K-shell. The short-dashed lines are computed by assuming the correct relation be-

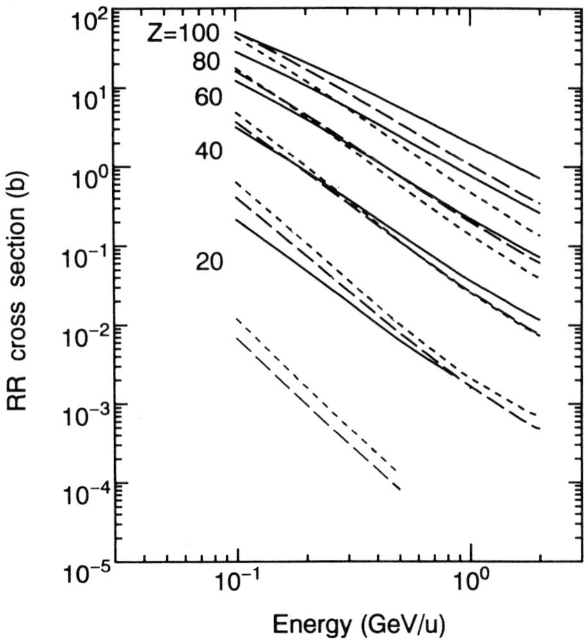


**Figure 9.6.** Energy and charge dependence of the cross section for radiative recombination (in barns) with capture into the projectile L-shell. All results are obtained from exact relativistic calculations. Solid lines: capture into the $L_1$-shell; long-dashed lines: capture into the $L_2$-shell; and short-dashed lines: capture into the $L_3$-shell.

tween the velocity $v$ and the projectile energy, while the long-dashed line uses the relation $E_{\text{kin}} = \frac{1}{2} M_P v^2$, as if the projectile speed were nonrelativistic. For the latter case, the results of the nonrelativistic dipole approximation do not deviate much from those of the exact calculations, even for relativistic energies (up to 1 GeV/u), where the energy-velocity relation is far from the nonrelativistic limit. This astonishing agreement must be regarded as fortuitous.

A similar equivalence cannot hold for the differential cross section, which nonrelativistically would be represented by a strict $\sin^2\theta$ distribution (9.31), while in the exact cross sections, considerable deviations from this form occur. One reason is that spin-flip contributions to the cross section yield nonzero values at forward and/or backward angles, see Fig. 9.10.

In analogy to Fig. 9.5, Fig. 9.6 shows the exact relativistic RR cross sections for capture into the $L_-1$, $L_2-$, and $L_3-$shells corresponding to final

$2s_{1/2}$, $2p_{1/2}$, and $2p_{3/2}$ states. Here, capture into the $L_1$-shell dominates for all charges and energies.

## 9.6.2 Total REC cross sections

In order to present total REC cross sections in a systematic way, it is useful to render different collision systems comparable (a) by normalizing the cross sections *per target electron*, (b) in the case of incompletely stripped projectiles, by considering cross sections *per projectile vacancy* in the shell considered, and (c) by displaying cross sections as a function of the "adiabaticity parameter" $\eta$, which decides whether the collision is fast ($\eta > 1$) or slow ($\eta < 1$) in the sense of Sec. 5.4.1. According to Eq. (5.63), we define

$$\eta = \frac{m_e c^2(\gamma - 1)}{E_{\text{bind}}} \approx 40.31 \, \frac{E_{\text{kin}}(\text{MeV/u})}{Z_P^2}. \tag{9.70}$$

The last approximation assumes a nonrelativistic total binding energy and a nonrelativistic energy-velocity relation.

Experimental cross sections normalized per target electron according to (a) may be compared with theoretical cross sections for radiative recombination, see Eq. (9.49). In this way, target properties do not enter since the electrons are assumed to be at rest with respect to the target nucleus, which is a good approximation for low-$Z$ targets. Figure 9.7 shows that the available experimental total REC cross sections per target electron as a function of the adiabaticity parameter $\eta$ fall approximately on a single universal curve [StK92]. This behavior finds its counterpart in the exact theoretical total cross sections for radiative recombination. It turns out that for calculating the total cross section, it is sufficient to add up the contributions from the K-, L-, and M-shells [IcS93, StK95]. While for the K- and L-shells, the curves for the different projectile charge numbers are well separated as a function of the collision energy, as shown in Figs. 9.5 and 9.6, and similarly for the M-shells, they form a close-lying bundle as a function of the parameter $\eta$ in the representation of Fig. 9.7. The theoretical curves, which all range from 100 MeV/u to 1.0 GeV/u, are in rather good agreement with the experimental data points. With one exception, only experimental data for projectile energies larger than 140 MeV/u are included. However, there exist many more data points at lower energies which follow the bundle of theoretical curves to higher values of $\eta$.

We conclude that total REC cross sections can be reliably estimated by theoretical cross sections for radiative recombination, which are dominated within each principal shell by capture into s-states. Now it is a general observation [RoG94] that nonrelativistic dipole total cross sections (but

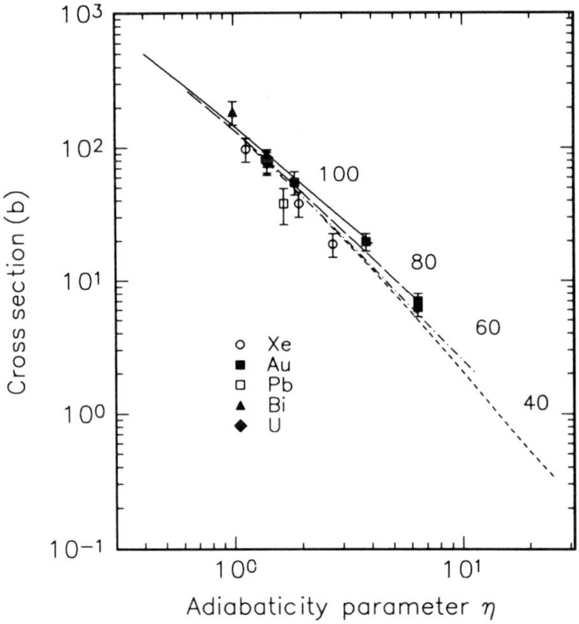


**Figure 9.7.** Experimental total REC cross sections for capture into bare projectiles per target electron in barn as a function of the adiabaticity parameter $\eta$ defined in Eq. (9.70). The projectiles are indicated explicitly; the target atoms range from Be to Al. In deriving total cross sections, realistic theoretical angular distributions are taken into account. From [StK95]. The curves represent exact theoretical cross sections for radiative recombination into the K + L + M shells calculated according to Eq. (9.49) [IcS94]. The numbers give the projectile charges.

not angular distributions) for the photoelectric effect of s-states (and other processes involving s-states) are valid far beyond their anticipated energy limits, see also Fig. 9.5. Therefore it is not surprising that exact theoretical REC cross sections are well approximated by consistently nonrelativistic Stobbe cross sections. If one introduces the latter in Fig. 9.7, the velocities and charge numbers enter only in the combination $\nu \propto 1/\sqrt{\eta}$, see Eq. (9.50), so that the bundle of exact theoretical curves contracts to a single universal curve.

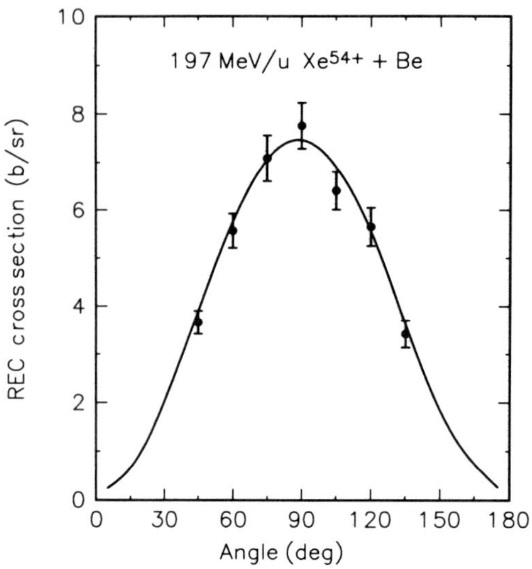


**Figure 9.8.** Differential REC cross section in barn/sr for 197 MeV $Xe^{54+}$ on Be as a function of the laboratory angle. Measured values are from [AnA84]. The solid curve represents calculated results [IcS94] multiplied with a normalization factor of 0.77 in order to fit the data. Hartree-Fock wave functions for the target atom [ClR74] are used.

### 9.6.3 Differential REC cross sections

As discussed in Sec. 9.3, in the theoretical description of nonrelativistic REC, two ingredients combine in such a way that the resulting photon angular distribution is simply proportional to $\sin^2\theta$ in the laboratory system. The two ingredients are (a) the inclusion of the full retardation, thereby keeping the complete exponential of the photon wave function in Eq. (9.11), and (b) the transformation (9.10) of the angles from the projectile system to the laboratory system.

As an example, Fig. 9.8 shows measured and calculated REC cross sections for 197 MeV/u $Xe^{54+}$ projectiles on Be targets. The calculated cross section is normalized to fit the data. The normalization factor is 0.77, which may be within the systematic errors of the experiment. Aside from the overall normalization, the experimental angular dependence is well reproduced by the theory. For the momentum distribution within the target atom, Hartree-Fock wave functions [ClR74] have been used. The calculated total K-REC cross section is 81.85 b, compared to an independently mea-

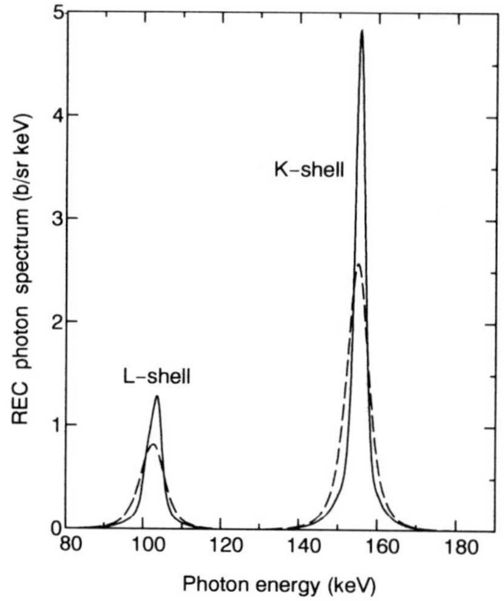


**Figure 9.9.** Calculated REC photon spectrum for 295 MeV/u $U^{92+}$ on N atoms for the laboratory photon angle of $132^0$. Solid line: calculated with approximate Roothaan-Hartree-Fock wave functions for the target atom; dashed line: calculated with a hydrogenic momentum distribution. From [IcS94].

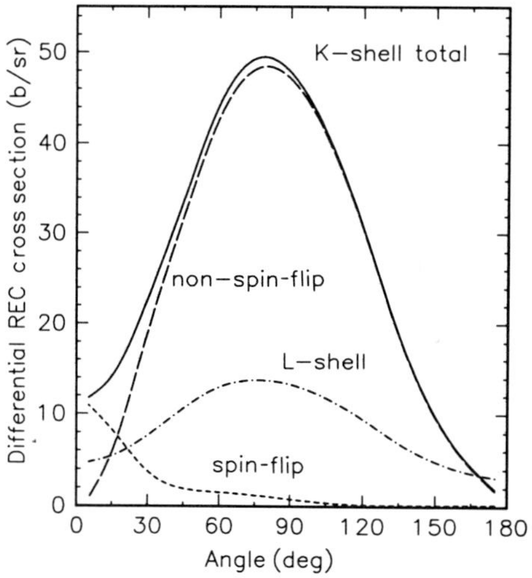


**Figure 9.10.** Calculated differential K- and L-shell REC cross sections for 295 MeV/u $U^{92+}$ on N atoms as a function of the laboratory angle. Approximate RHF wave functions have been used. For K-REC, the contributions without spin-flip and with spin-flip are shown separately. Adopted from [IcS94].

sured total capture cross section (which is mainly REC, see Fig. 8.9) of $(72 \pm 3)$ b.

In Fig. 9.9, we show the calculated photon spectrum at the fixed observation angle of $132^0$ in the laboratory system for the collision of $U^{92+}$ + N at 295 MeV/u. The maxima occur at the photon resonance $\hbar\omega_0'$, see Eq. (9.8), corresponding to zero longitudinal momentum $q_z = 0$ of the electron in the target atom. The widths of the peaks for K-REC and L-REC are caused by the electronic momentum spread in the target. The dashed lines assume a hydrogenic momentum distribution (9.52), while the solid lines are calculated from approximate Roothaan-Hartree-Fock wave functions according to Eq. (9.60). A comparison of hydrogenic and RHF results reveals that, as expected, screening effects reduce the momentum spread, in particular of outer electrons, so that photon spectra get more sharply peaked. However, the area under the peak is not much affected, leaving angular distributions (Figs 9.8 and 9.10) less sensitive to electron screening.

Figure 9.10 shows the differential REC cross section for capture into the K- and L-shells of the projectile assuming a RHF momentum distribution in the target. One notices strong deviations from a simple $\sin^2\theta$ distribution predicted by the nonrelativistic theory and observed in $Xe^{54+}$ + Be collisions, see Fig. 9.8. This difference must be attributed mainly to the higher projectile charge of $U^{92+}$ as compared to $Xe^{54+}$. Relativistic effects become more evident if one splits the K-REC cross section in Fig. 9.10 into a non-spin-flip part and a spin-flip contribution. The non-spin-flip part still follows approximately a $\sin^2\theta$ distribution, in accordance with Fig. 9.2, while the spin- flip distribution is peaked at forward angles. As discussed in Sec. 9.3, the spin degree of freedom is required to conserve angular momentum at forward (or backward angles). More details can be found in a systematic discussion of angular distributions [EiI95].

In Fig. 9.11, the subshell contributions are displayed separately for 90 MeV/u $U^{90+}$ collisions with carbon foils. At this energy, it is experimentally possible to resolve the contribution from the $L_3$-shell ($p_{3/2}$ states, lower curve) from the combined contributions of the $L_1$- and $L_2$-shells ($2s_{1/2}$ and $2p_{1/2}$ states, respectively, upper curves). The theoretical angular distributions deviate markedly from symmetry about $90^0$, with the shift of the maximum depending on the subshell. As one can see from Fig. 9.11(a), the forward peaking for REC into the into the $j = 1/2$ sublevels can be explained by transitions to the $2s_{1/2}$ state (dashed line), which can be partly attributed to spin-flip transitions in analogy to Fig. 9.10. At backward angles, the behavior is mainly determined by REC into the $2p_{1/2}$ state (dotted line). The theoretical angular distribution for REC into the $2p_{3/2}$ state, see

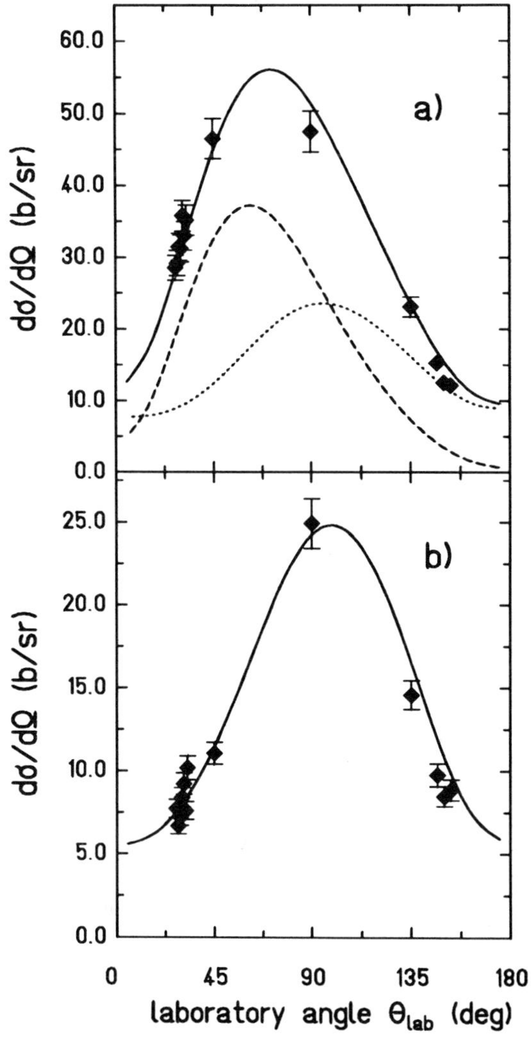


**Figure 9.11.** Differential L-shell REC cross sections for 89 MeV/u $U^{90+}$ on C atoms as a function of the laboratory photon angle. (a) REC into the $j = 1/2$ levels. Experimental data: diamonds; theoretical cross sections [IcS94]: dashed line: $2s_{1/2}$, dotted line: $2p_{1/2}$, solid line: sum of both. (b) REC into the $2p_{3/2}$ level. All data points are normalized by the common factor 0.65. From [StG94].

Fig. 9.11(b), is enhanced at backward angles similarly as the $2p_{1/2}$ curve, which is in accord with the experimental data. All data points are normalized by a common factor of 0.65, which is still within the total absolute uncertainty of the measurements [StG94].

The experimental results are in excellent agreement with theoretical predictions using an exact relativistic description. One may conclude that for high projectile charges, subshell-resolved photon angular distributions probe details of the bound-state projectile wave functions and, at forward (or backward) angles, the contribution of the electron spin caused by magnetic interactions. Accurate relativistic calculations are needed for reliable predictions.

and $\tau$ pairs), let alone exotic particles, like magnetic monopoles, W-pairs, b-quarks, and the Higgs particle, see, e.g., [GrV93]. Already for muons, the length scale connected with their mass is strongly contracted compared to the electronic length scale. This will render it difficult to clearly separate electromagnetic pair production from hadronic processes, which are beyond the scope of this book.

Following conventional usage, we adopt relativistic ("natural") units in this chapter with $\hbar = c = m_{\rm e} = 1$, (see Appendix). However, where clarity demands it, the electron mass $m_{\rm e}$, the velocity of light $c$, or the Compton wavelength $\lambda\!\!\!^{-}_{\rm c}$ are displayed explicitly. Similarly, we usually keep $\beta = v/c$ instead of replacing it by $v$. Electron and positron quantities are labelled by "e" and "p", e.g., as in $E_{\rm e}$ and $E_{\rm p}$. Only if for free electrons or positrons the momenta serve as quantum numbers, we refer to positive and negative energy states (corresponding to electrons and positrons, respectively) by the labels $p_+$ and $p_-$, e.g., for the energies, as $E_{p_+}$ and $E_{p_-}$.

## 10.2 Production of free electron-positron pairs

Figure 10.1 illustrates two competing processes for single-pair production in heavy-ion collisions. Conceptually, these processes are very similar; however, there are two distinctions which suggest a separate treatment. (1) In contrast to a free electron, an electron created in a bound state has to accommodate its momentum within the available momentum distribution of the target (if it is bound in the target) or of the projectile (if it is bound in the projectile). This entails the typical charge number dependence $Z^5_{\rm P,T}$ known from electron capture, see Chap. 8, and also an energy dependence different from that for the production of free electrons. (2) Since an electron bound in a specific state has no further degrees of freedom, bound-electron–free-positron pair production is more accessible to calculations.

The creation of free electron-positron pairs is by far the dominant pair production process for extreme relativistic energies. It had been observed already in the 1930s, while bound-free pair production has been detected only recently [BeG93, BeG94]. Both processes occur with a relatively small probability so that it is a reasonable first step to employ perturbation theory. In this section, we confine ourselves to free pair production, deferring bound-free pair production to Sec. 10.3. We start with perturbative calculations, first in lowest-order quantum electrodynamics (QED), in which electrons and positrons are represented by plane waves, and then in a first-order distorted-wave approximation. In Sec. 10.2.3, we discuss the application of the equivalent-photon or Weizsäcker-Williams method for extremely high projectile energies.

### 10.2.1 Lowest-order QED calculations

The theoretically most clear-cut way to describe the production of free electron-positron pairs is a formulation in terms of lowest-order QED with external electromagnetic fields. In the center-of-mass system, the target and projectile nuclei merely serve as sources of time-dependent electromagnetic fields. Since energy-momentum conservation rules out pair creation by a single freely moving charge, the lowest-order QED process involves two moving nuclei, each of them interacting with the electron-positron field via the exchange of a single photon. A representation in terms of Feynman diagrams is given in Fig. 10.2 to be discussed below.

The lowest-order nonvanishing term in the expansion (5.66) of the $S$-operator is therefore of second order and reads

$$\widehat{S}^{(2)} = (-i)^2 \int_{-\infty}^{\infty} dt_1 \int_{-\infty}^{t_1} dt_2 \, \widehat{H}_I(t_1) \, \widehat{H}_I(t_2), \tag{10.3}$$

where for an external electromagnetic field (5.68)

$$\widehat{H}(t) = \int \widehat{\psi}^\dagger(\mathbf{r}, t) V(t) \widehat{\psi}(\mathbf{r}, t) \, d^3r \tag{10.4}$$

with

$$V(t) = -e\gamma^0\gamma^\mu A_\mu = -e(\Phi - \boldsymbol{\alpha} \cdot \mathbf{A}). \tag{10.5}$$

Here, for brevity, we display in the potential $V(t) = V(\mathbf{r}, t)$ only the time dependence explicitly. Within QED, the space-time-dependent field operators $\widehat{\psi}$ are expanded in terms of plane waves according to Eq. (5.8).

Alternatively, we could choose target or projectile eigenstates (5.53) or (5.54) instead of plane waves. In the language of QED, such states are produced by exchanging infinitely many photons between the electron (or the positron) and the target or projectile nucleus. Since these states do not possess a definite momentum, energy-momentum conservation does not impose a constraint, so that a first-order distorted wave process as described in Eq. (5.72) and implied by Fig. 10.1 is possible. We defer a discussion to Sec. 10.2.2.

In terms of the creation and annihilation operators introduced in Sec. 4.2.2, the $S$ matrix element describing the creation from the vacuum of a single free pair with electron momentum and energy $\mathbf{p}_+$, $E_{p+}$ and positron

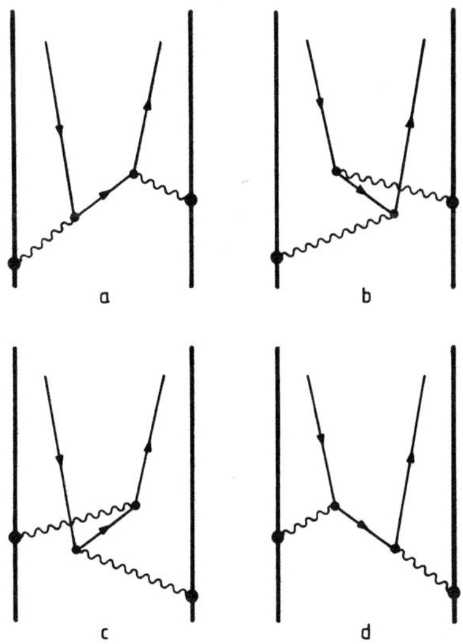


**Figure 10.2.** Lowest-order Feynman diagrams for the production of a free electron-positron pair in a collision between two heavy ions. The time arrow points upwards. The heavy vertical lines denote the nuclei 1 and 2, while photons are depicted as wavy lines and electrons and positrons are described by thin solid lines with arrows directed upwards and downwards, respectively.

momentum and energy $\mathbf{p}_-, E_{p_-}$, respectively, is obtained from Eq. (10.3) in the form

$$\begin{aligned} S_{p_+p_-} &= \langle 0|\widehat{b}_{p_+}\widehat{d}_{p_-}\widehat{S}^{(2)}|0\rangle \\ &= (-i)^2 \int_{-\infty}^{\infty} dt_1 \int_{-\infty}^{t_1} dt_2 \\ &\quad \times \langle 0|\widehat{b}_{p_+}\widehat{d}_{p_-} \int \widehat{\psi}^\dagger V(t_1)\widehat{\psi}\, d^3r_1 \int \widehat{\psi}^\dagger V(t_2)\widehat{\psi}\, d^3r_2\, |0\rangle. \end{aligned} \tag{10.6}$$

Here, as a first step in the process, a pair is created at the earlier time $t_2$ by the interaction $V(t_2)$, mediated by a photon emitted either from the target or from the projectile. Both alternatives contribute to the matrix element. In order to decompose the vacuum expectation value in Eq. (10.6) into a first and a second reaction step, we insert a complete set

$$\sum_{k_+k_-} \widehat{d}^\dagger_{k_-}\widehat{b}^\dagger_{k_+}|0\rangle\,\langle 0|\widehat{b}_{k_+}\widehat{d}_{k_-} = 1 \tag{10.7}$$

of plane-wave electron-positron states (with momenta and energies $\mathbf{k}_+, E_{k_+}$

and $\mathbf{k}_-, E_{k_-}$) between the two space integrals. The matrix element (10.6) is now reformulated by introducing the expansion (5.8) of $\widehat{\psi}$ in terms of a complete set of single-particle states. From $\widehat{\psi}^\dagger\widehat{\psi}$, we obtain four types of operator products, namely $\widehat{b}^\dagger\widehat{b}$, $\widehat{d}^\dagger\widehat{d}$, $\widehat{b}^\dagger\widehat{d}^\dagger$, and $\widehat{b}\,\widehat{d}$, corresponding to electron scattering, positron scattering, pair production, and pair annihilation, respectively. In the matrix element at the earlier time $t_2$, only the pair production term contributes, while in the matrix element at the later time $t_1$, only the electron or positron scattering terms yield a nonzero result. In a first step, the matrix element at $t_2$ can be written explicitly if, according to Eq. (4.39), we write the wave function in the usual form $\psi_{k_\pm}(\mathbf{r},t) = \varphi_{k_\pm}(\mathbf{r})\exp(-iE_{k_\pm}t)$. Then

$$\begin{aligned} S_{p_+p_-} &= (-i)^2 \sum_{k_+k_-} \int_{-\infty}^{\infty} dt_1 \int_{-\infty}^{t_1} dt_2 \\ &\quad \times \langle 0|\widehat{b}_{p_+}\widehat{d}_{p_-} \int \widehat{\psi}^\dagger V(t_1)\widehat{\psi}\, d^3r_1\; \widehat{d}^\dagger_{k_-}\widehat{b}^\dagger_{k_+}|0\rangle \\ &\quad \times \langle \varphi_{k_+}|V(t_2)|\varphi_{k_-}\rangle e^{-i(E_{k_-}-E_{k_+})t_2}, \end{aligned} \tag{10.8}$$

where the last bracket no longer involves field operators and simply denotes a space integral. In the single-particle matrix element at time $t_1$, only one state can change so that we have two nonvanishing contributions: the first one with $k_- = p_-$ describing the scattering of the electron $k_+ \to p_+$ and the second one with $k_+ = p_+$ describing the scattering of the positron $k_- \to p_-$. These possibilities are illustrated in the Feynman diagrams of Fig. 10.2. Diagrams (a) and (b) show the creation of a pair by nucleus 1 and the subsequent scattering of the electron and positron, respectively, by nucleus 2. In the "crossed" diagrams (c) and (d), the role of the two colliding nuclei is interchanged. All four diagrams contribute to the transition amplitude.

In order to distinguish the four diagrams, let us introduce the following notation. The label $s = \pm$ denotes positive- or negative-energy intermediate scattering states, while the label $n = 1$ or $n = 2$ denotes the nuclear potential that creates the electron-positron pair. Then the diagrams (a), (b), (c), and (d) are assigned to the combinations (+,1), (–,1), (+,2), and (–,2). The $S$ matrix element corresponding to the diagram (a), in which the pair is created by nucleus 1 and the intermediate scattering state is an electron state, is written as

$$\begin{aligned} S^{(+,1)}_{p_+p_-} &= (-i)^2 \sum_{k_+} \int_{-\infty}^{\infty} dt_1 \int_{-\infty}^{t_1} dt_2 \langle \varphi_{p_+}|V_2(t_1)|\varphi_{k_+}\rangle e^{-i(E_{k_+}-E_{p_+})t_1} \\ &\quad \times \langle \varphi_{k_+}|V_1(t_2)|\varphi_{p_-}\rangle e^{-i(E_{p_-}-E_{k_+})t_2}. \end{aligned} \tag{10.9}$$

lytical form and, furthermore, are characterized by well-defined asymptotic momenta. According to Bethe and Maximon [BeM54], who first applied them to the production of electron-positron pairs by real photons in the field of a nucleus, these wave functions are very good approximations to the exact continuum states for electron energies large compared to the rest energy $m_e c^2$.

The sixfold differential probability for emitting an electron-positron pair with momenta $\mathbf{p}_+, \mathbf{p}_-$ and (undetected) spin projections $s_+, s_-$, respectively, is given by [Dec91, IoE93]

$$\frac{d^6 P(\mathbf{b})}{dE_{p_+} dE_{p_-} d\Omega_{p_+} d\Omega_{p_-}} = \frac{p_+ p_- E_{p_+} E_{p_-}}{(2\pi)^6} \sum_{s_+, s_-} |A_{p_+ p_-}(\mathbf{b})|^2, \tag{10.18}$$

where, in analogy to Eqs. (6.4) and (6.6) or directly from Eq. (10.15),

$$A_{p_+ p_-}(\mathbf{b}) = i \frac{Z_P e^2}{\pi v} \int \frac{d^2 q_b}{q_b^2 + (q_0/\gamma)^2} \, e^{-i\mathbf{q}_b \cdot \mathbf{b}} M_{p_+ p_-}(\mathbf{q}_b, q_0) \tag{10.19}$$

with

$$M_{p_+ p_-}(\mathbf{q}_b, q_0) = \langle \varphi_{p_+} | (1 - \beta \alpha_z) e^{i\mathbf{q}\cdot\mathbf{r}} | \varphi_{p_-} \rangle. \tag{10.20}$$

The longitudinal momentum transfer $q_0$ resulting from the time integration is again given by Eq. (10.16).

Before evaluating the expression (10.20) with Sommerfeld-Maue wave functions, it is necessary to adopt the reformulation of the matrix element according to Eq. (5.109) in order to avoid spurious contributions. index-Matrix elements!of the alpha@of the $\alpha$-operator

The calculations [IoE93] show that even for high collision energies, the term proportional to $1/\gamma^2$ in Eq. (5.109) cannot be neglected compared to the other terms. However, when the electron wave function (4.132) with incoming spherical waves and the positron wave function (4.133) with outgoing spherical waves is inserted into Eq. (10.20), we ignore the cross term, in which the derivatives of the ${}_1F_1$ function is taken both for the electron and for the positron. This term is of higher order in $Z\alpha$, that is, of the same order as terms already omitted in the Sommerfeld-Maue wave function.

By integrating the probability (10.18) over the impact parameter plane, one obtains the sixfold differential cross section

$$\begin{aligned} \frac{d^6 \sigma(\mathbf{p}_+, \mathbf{p}_-)}{dE_{p_+} \, dE_{p_-} \, d\Omega_{p_+} \, d\Omega_{p_-}} &= 4 \frac{p_+ p_- E_{p_+} E_{p_-}}{(2\pi)^6} \left( \frac{Z_P e^2}{v} \right)^2 \\ &\times \int \frac{\sum_{s_+, s_-} |M_{p_+ p_-}(\mathbf{q}_b, q_0)|^2}{[q_b^2 + (q_0/\gamma)^2]^2} \, d^2 q_b. \end{aligned} \tag{10.21}$$

Calculations for the production of free electron-positron pairs in collisions with relativistic heavy ions using Sommerfeld-Maue wave functions have been carried out by introducing simplifying approximations [NiP82, BeB88], as well as by using fully numerical methods [Dec91]. Some inaccuracies in the latter work are corrected and the analytical reduction is pushed farther in [IoE93] by use of the trace technique for performing the calculations in spinor space. All integrals can be evaluated numerically using adaptive Monte-Carlo techniques [Lep87]. For the total cross section, seven-dimensional integrations are required.

A combination of existing perturbative results with the equivalent-photon method (Sec.10.2.3) is used by Eby [Eby89, Eby91], who also points out a connection between the two methods and analyzes it in detailed calculations.

*Results with Sommerfeld-Maue wave functions*

In the following, we present some results obtained from first-order distorted-wave calculations with Sommerfeld-Maue wave functions [IoE93]. In some cases, a comparison with other theoretical or experimental data is possible.

In a first step, the calculation of the angular correlation of pair production as a function of the relative azimuthal angle between the electron and the positron direction [DeG90, IoE93] shows that the electron and positron are preferably emitted with opposite transverse momenta yielding a distribution that is symmetric about 180°.

Next, by averaging over the azimuthal correlation and fixing the electron and positron energy, one may derive angular distributions in the laboratory system as a function of the electron and positron polar angles $\theta_\mathrm{e}$ and $\theta_\mathrm{p}$. In Figs. 10.4 and 10.5, the polar angular distributions for the collision system $\mathrm{U}^{92+} + \mathrm{Au}^{79+}$ at 960 MeV/u and for $\mathrm{S}^{16+} + \mathrm{Au}^{79+}$ at 200 GeV/u are presented. In the latter case, the increase of the Lorentz factor $\gamma$ causes a striking localization of the emitted pairs in the forward direction. This behavior is also observed in the process of pair creation by real photons [BeM54, DaB54].

Having calculated the angular distribution, one is in a position to discuss energy distributions of electrons and positrons. In Figs. 10.6 and 10.7, we consider the single-differential cross sections $d\sigma/dE_\mathrm{p}$ and $d\sigma/dE_\mathrm{e}$ (with $E_\mathrm{p} = E_{p_-}, E_\mathrm{e} = E_{p_+}$) for the collision system $\mathrm{p} + \mathrm{U}^{92+}$ for three different Lorentz factors $\gamma$. The cross sections for other projectiles may be easily obtained in perturbation theory by scaling these results with the square $Z_\mathrm{P}^2$ of the projectile charge. With increasing values of $\gamma$, the curves have a decreasing slope, showing that more and more electrons and positrons with high energies are created. While for positrons, the cross sections vanish at

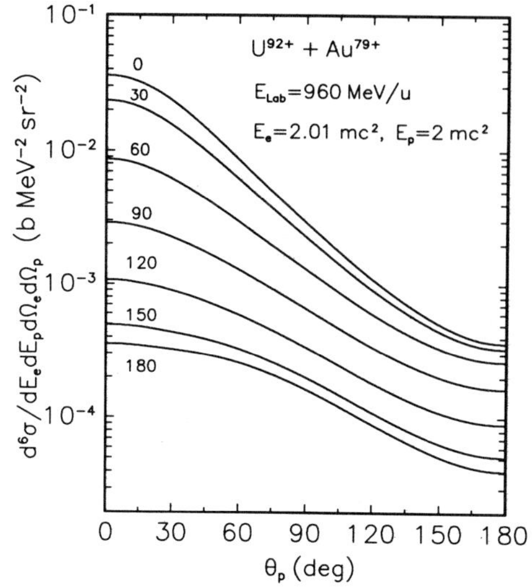


**Figure 10.4.** Angular distribution in the laboratory system of $e^+e^-$ pairs (averaged over the azimuthal distribution) as a function of the positron polar angle $\theta_p$ for different electron polar angles $\theta_e$. The collision system is $U^{92+} + Au^{79+}$ at 960 MeV/u. From [IoE93].

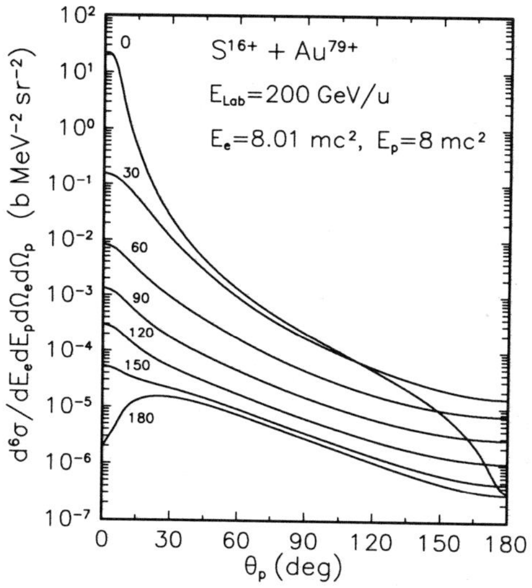


**Figure 10.5.** Same as Fig. 10.4 but for the collision system $S^{16+} + Au^{79+}$ at 200 GeV/u. From [IoE93].

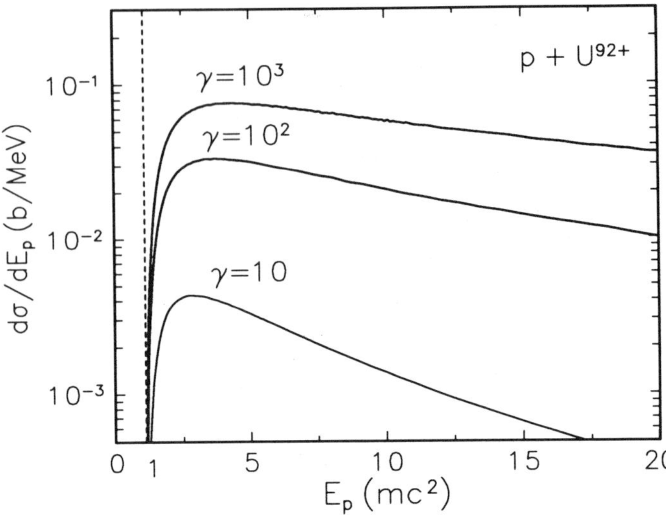


**Figure 10.6.** Single-differential cross section in p + $U^{92+}$ collisions as a function of the positron total energy $E_\mathrm{p}$ for several values of the Lorentz factor $\gamma$. For other projectiles, the cross section scales with $Z_\mathrm{P}^2$. From [IoE93].

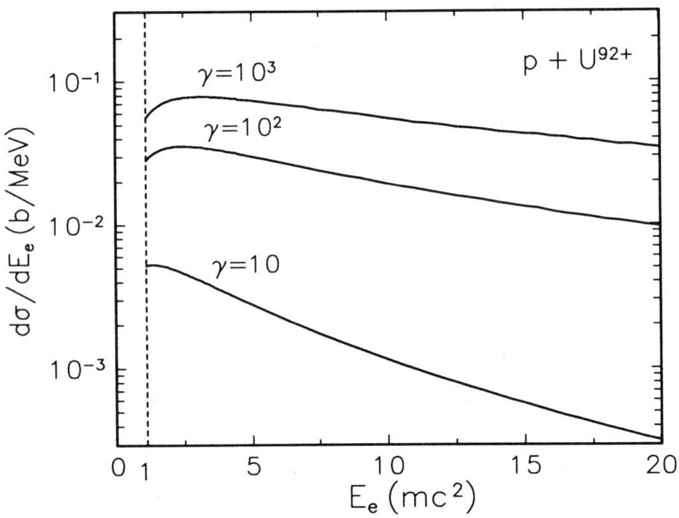


**Figure 10.7.** Single-differential cross section in p + $U^{92+}$ collisions as a function of the electron total energy $E_\mathrm{e}$ for several values of the Lorentz factor $\gamma$. For other projectiles, the cross section scales with $Z_\mathrm{P}^2$. From [IoE93].

parameter plane is illustrated in Fig. 10.11 for colliding nuclei with radii $R_1$ and $R_2$. The angle between $\mathbf{b}_1$ and $\mathbf{b}_2$ is denoted by $\varphi$, and $\sigma^{\parallel}_{\gamma\gamma}$ and $\sigma^{\perp}_{\gamma\gamma}$ are the photon-photon cross sections for parallel and perpendicular relative polarization of the photons.

The cross section for pair production in the equivalent-photon approximation is obtained by integrating over the photon frequencies and over the impact-parameter plane as

$$\begin{aligned}\sigma_{\text{pair}} &= \int d^2b \int d\omega_1 \int d\omega_2 \int d^2b_1 \int d^2b_2\, \delta(\mathbf{b}+\mathbf{b}_2-\mathbf{b}_1)\, N(\omega_1,b_1) \\ &\quad \times N(\omega_2,b_2)[\,\sigma^{\parallel}_{\gamma\gamma}(s)\,\cos^2\varphi + \sigma^{\perp}_{\gamma\gamma}(s)\,\sin^2\varphi\,]. \qquad (10.32)\end{aligned}$$

If $\sigma^{\parallel}_{\gamma\gamma}(s) = \sigma^{\perp}_{\gamma\gamma}(s) = \sigma_{\gamma\gamma}(s)$, the integration over $\mathbf{b}_1$ and $\mathbf{b}_2$ can be carried out, and one arrives at Eq. (10.25). The result, valid for extreme relativistic collisions, has also been derived in a perturbative treatment in [BaB93, ViG93].

Figure 10.12 shows the one-photon distributions entering in Eq. (10.32) in their dependence on the impact parameter $b$ for various choices of the form factor and a fixed value of the Lorentz factor $\gamma = 3500$ corresponding to the maximum energy of the Large Hadron Collider [ViG93]. The figure confirms that all form factors lead to the same result outside the extended charge distribution. For small impact parameters, the results have to be regarded with some caution since the validity of the equivalent-photon method breaks down if the assumption of a constant field across the transverse linear dimension of the system is no longer satisfied, see Sec. 3.4.1.

The cross sections for the electromagnetic production of a fermion pair in lowest order of $\alpha$ are given in the covariant form [BuG75, ViG93] as

$$\sigma^{\parallel}_{\gamma\gamma}(s) = \frac{4\pi\alpha^2}{s}\left[\left(1+\frac{4m_{\text{e}}^2}{s}-\frac{12m_{\text{e}}^4}{s^2}\right)L-\left(1+\frac{6m_{\text{e}}^2}{s}\right)M\right]\Theta(s-4m_{\text{e}}^2) \qquad (10.33)$$

and

$$\sigma^{\perp}_{\gamma\gamma}(s) = \frac{4\pi\alpha^2}{s}\left[\left(1+\frac{4m_{\text{e}}^2}{s}-\frac{4m_{\text{e}}^4}{s^2}\right)L-\left(1+\frac{2m_{\text{e}}^2}{s}\right)M\right]\Theta(s-4m_{\text{e}}^2), \qquad (10.34)$$

where

$$\begin{aligned}L &= 2\ln\left(\frac{\sqrt{s}}{2m_{\text{e}}}+\sqrt{\frac{s}{4m_{\text{e}}^2}-1}\right) \\ M &= \sqrt{1-\frac{4m_{\text{e}}^2}{s}}. \qquad (10.35)\end{aligned}$$

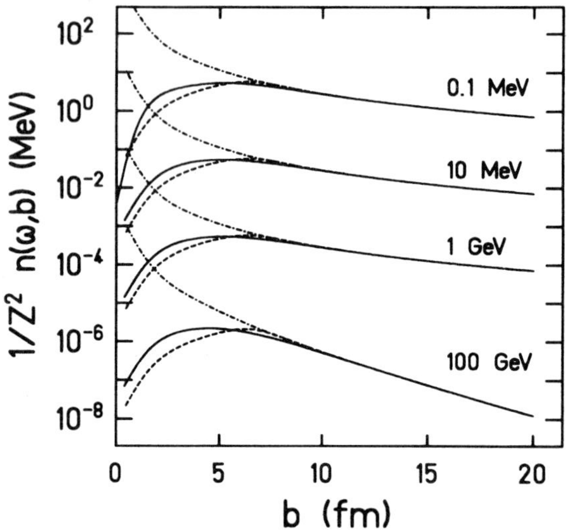


**Figure 10.12.** Photon distribution $N(\omega, b) = n(\omega, b)$ for a Pb nucleus and $\gamma = 3500$ as a function of the impact parameter $b$ for different photon energies. The dash-dotted curve corresponds to a point-like nucleus, Eq. (3.70), the other curves to photon distributions (3.69) of extended charges: the dashed line refers to the form factor (3.57) of a homogeneously charged sphere with radius $R_0 = 7.107$ fm and the solid line to the Gaussian form factor (3.58) with $Q_0 = 60$ MeV. From [ViG93].

Here, $s = 4\,\omega_1\omega_2 \geq 4m_e^2$ is ensured by the Heaviside step function $\Theta$. Equations (10.33) and (10.34) are equally valid for the electromagnetic production of other fermion pairs if the electron mass $m_e$ is replaced by the corresponding fermion mass. The cross sections refer to the production by real (i.e., on-shell) photons. Off-shell corrections can also be applied [BuG75, BaB93] and turn out to be important for $e^+e^-$ productions with invariant masses $W = \sqrt{s} \leq 1$ GeV.

As is customary, we have used here the invariant mass $W = \sqrt{s} = \sqrt{4\,\omega_1\omega_2}$ since the final dilepton state depends on a number of parameters in the collider (laboratory) system. Another convenient parameter is the rapidity $\chi$ defined in Eq. (2.15). Transverse momenta being negligible in the collider frame, one can define the total energy as $\omega_1 + \omega_2$ and the total momentum in the beam direction as $p_z = \omega_1 - \omega_2$. From Eq. (2.18) it follows that

$$\beta\gamma = \frac{p_z}{m_e c} = \frac{\omega_1 - \omega_2}{\sqrt{4\,\omega_1\omega_2}}$$

$$= \frac{1}{2}\left(\sqrt{\frac{\omega_1}{\omega_2}} - \sqrt{\frac{\omega_2}{\omega_1}}\right) = \sinh\left(\frac{1}{2}\ln\frac{\omega_1}{\omega_2}\right), \tag{10.36}$$

so that from Eq. (2.15) the rapidity is

$$\chi = \frac{1}{2}\ln\frac{\omega_1}{\omega_2}, \tag{10.37}$$

or $\omega_1 = \frac{W}{2}e^{\chi}$ and $\omega_2 = \frac{W}{2}e^{-\chi}$.

Rewriting Eq. (10.31) in terms of the variables $W$ and $\chi$ and adopting an off-shell corrected photon cross section $\overline{\sigma}_{\gamma\gamma}$ averaged over the polarizations, one obtains [BaB93]

$$\frac{d^2\sigma}{dW\,d\chi} = 2\pi\int_0^\infty b_1\,db_1\int_0^\infty b_2\,db_2\int_0^{2\pi} d\varphi\,\Theta(b-2R_0)\frac{2\overline{\sigma}_{\gamma\gamma}}{W}$$
$$\times N(\tfrac{W}{2}e^{\chi}, b_1)\,N(\tfrac{W}{2}e^{-\chi}, b_2). \tag{10.38}$$

Here, it is explicitly assumed that the nuclei with radii $R_0$ do not penetrate each other so that no nuclear debris and other hadronic background arises. Figure 10.13 shows the differential cross section for $e^+e^-$ pair production as a function of the invariant mass $M \equiv W$ for Lorentz factors $\gamma = 10$ corresponding to the SPS, $\gamma = 100$ corresponding to RHIC, and $\gamma = 3400$ corresponding to LHC. The rapidity is taken as $\chi = 0$, that is, $\omega_1 = \omega_2$. It is seen that with increasing $\gamma$, more collision energy becomes available for producing the invariant mass $W$. The cross sections for $b \geq 2R_0$ exceed the cross sections for electromagnetic pair production at $b < 2R_0$ by almost one order of magnitude [BaB93]. If the sharp cut-off function $\Theta(b-2R_0)$ is replaced with a soft cut-off Fermi function $\{1+\exp[(b-2R_0)/a]\}^{-1}$, where $a$ is a thickness parameter, the cross sections are not affected significantly.

*Cautioning remarks*

In concluding this section, it is appropriate to recall the limitations of the equivalent-photon method. When replacing the transient electromagnetic field of a fast-moving charge, two decisive approximations are made: (a) the virtual (off-shell) photons produced by the moving charge are replaced with real (on-shell) photons and (b) the impact-parameter-dependent electromagnetic fields produced by the moving charge are replaced with plane waves whose amplitudes do not vary in the transverse direction. That is, even if the photon spectrum is taken to be $b$ dependent, there is no complete equivalence between the original field and its substitute for perturbed systems that occupy a transverse space region comparable or larger than the range $\Delta b$, over which the original field changes appreciably.

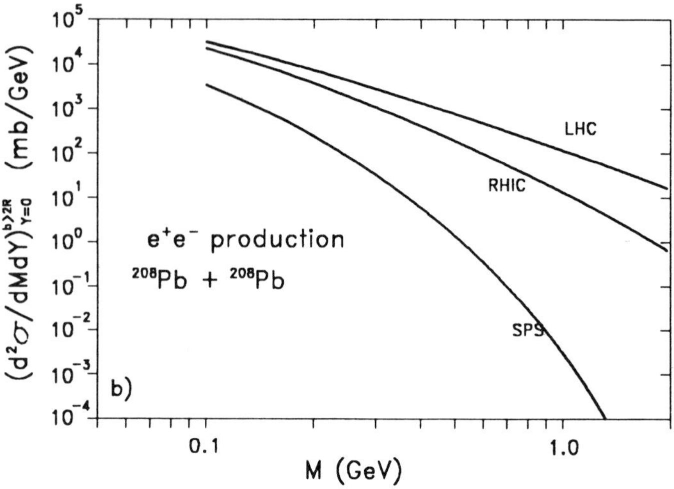


**Figure 10.13.** Double differential cross section for $e^+e^-$ production as a function of the invariant mass $M \equiv W$ for $Y \equiv \chi = 0$ in Pb + Pb collisions. The impact parameter $b > 2R_0$ is chosen so that the nuclei do not penetrate each other during the collision. The Lorentz factors are $\gamma = 10$ for the CERN Super Proton Synchrotron (SPS), $\gamma = 100$ for the Relativistic Heavy-Ion Collider (RHIC), and $\gamma = 3400$ for the Large Hadron Collider (LHC). From [BaB93].

This limits the applicability of the method to impact parameters $b \gg \Delta b$. For pair production, one may take $\Delta b \simeq \lambda\!\!\!{}^-_c$, since the localization of a Dirac electron to better than the Compton wavelength $\lambda\!\!\!{}^-_c$ in itself requires the use of negative-energy states [NeW49]. In the absence of rigorous tests of the viability of the equivalent-photon method for impact parameters $b \ll \lambda\!\!\!{}^-_c$, results for $b$ comparable to the nuclear radii should be viewed with some caution.

In a quantitative way, the applicability of the equivalent-photon method for electron-positron pair production is examined by Hencken et al. [HeT94] in a comparison with exactly calculated second-order cross sections for impact parameters $b = 0$. In Eq. (10.32), then $\phi = 0$ and only $\sigma^{||}_{\gamma\gamma}(s)$ contributes to the cross section. It turns out that the differential probabilities calculated with the equivalent-photon method are by more than an order of magnitude too large for small emission angles ($< 0.1^0$) of the positron and only begin to agree with the exactly evaluated second-order probabilities at angles of more than a few degrees. As the total probability is dominated by forward emission, the equivalent-photon method is not suitable for a description of electron-positron creation at small impact parameters.

## 10.3 Bound-electron – positron pair production

When an electron-positron pair is created in a relativistic heavy-ion collision, the electron may find itself in a bound state of the target or of the projectile, provided its momentum can be accommodated in the bound-state momentum distribution. Since K-shell wave functions offer the broadest momentum spread, production in the K-shell is dominant.

In this subsection, we consider electron-positron pair formation by the transient electromagnetic field of a moving charge for the case where the electron ends up in a bound state. This production mechanism is not a capture or transfer process which is characterized by a transition between two inertial frames moving with respect to each other. However, in a full two-center description, in principle it could happen that the electron is asymptotically bound to one of the nuclei while the positron is in a continuum state of the other nucleus. We may denote this transfer-like reaction, which has not been theoretically investigated so far, as "charge transfer from the negative-energy continuum". Presumably, this mechanism would not have the charge dependence $Z^2$ characteristic for a perturbing nuclear charge $Ze$ and would be less important at high energies than the ionization-like bound-free pair production discussed here.

Bound-electron – free-positron (or bound-free) pair production changes the charge state of one of the colliding ions, therefore, this process is of great interest for the design and operation of relativistic heavy-ion colliders. The reason is that heavy ions whose charge state is decreased by one unit get lost from the beam in the storage ring, thus limiting its luminosity and lifetime. The potential harm is even greater because, in contrast to all electron capture processes, the cross section for bound-free pair production increases rather than decreases with the collision energy.

### 10.3.1 First-order perturbation theory

Let us consider the process in which the electron is created in a K-shell orbit of the target. Since creation in a bound projectile state may be analogously described in the projectile system, this choice can be made without any loss of generality.

In first-order time-dependent perturbation theory, the differential probability for the production of an electron in a bound state of the target and a positron with an energy $E_{\mathrm{p}}$ can be written as

$$\frac{dP(b)}{dE_{\mathrm{p}}} = \sum_{\kappa_{\mathrm{p}},\mu_{\mathrm{p}},\mu_{\mathrm{e}}} |A_{\mathrm{ep}}(b)|^2 \tag{10.39}$$

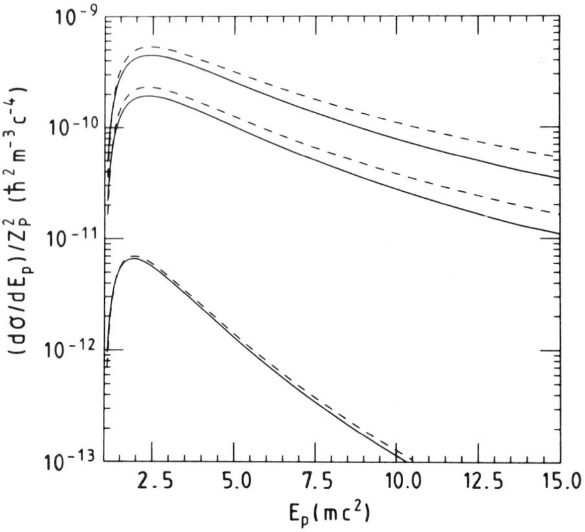


**Figure 10.14.** Differential cross section for bound-free pair production into the K-shell in $Z_{\mathrm{P}} + \mathrm{S}^{16+}$ collisions as a function of the positron energy $E_{\mathrm{p}}$ for the laboratory projectile energies $E = 1$ (bottom), 15 (middle), and 100 (top) GeV/u. Solid curves: partial-wave analysis; dashed curves: Darwin and Sommerfeld-Maue wave functions. All quantities are given in relativistic units such that $\hbar^2/m_{\mathrm{e}}^3c^4 = 2.9$ barn/keV and the kinetic energy of the positron is $(E_{\mathrm{p}} - 1)m_{\mathrm{e}}c^2$. From [Bec87].

with $A_{\mathrm{ep}}(b)$ being defined by Eq. (10.15), subject to the only change that $\varphi_{p_+}$ now is a bound-state Coulomb-Dirac wave function $\varphi_{\mathrm{e}}$, see Eqs. (4.100) and (4.101), while the exact positron continuum wave function is given by the partial wave (4.113) and (4.115). The evaluation of Eq. (10.39) proceeds along the same lines as for ionization, see Sec. 6.3.

As an alternative to exact Dirac wave functions [BeG87a, Bec87], one may also use approximate electron and positron states [Bec87] and thus avoid the partial-wave analysis. As an approximation to the K-shell wave function, one may choose the Darwin form (4.109) and for the positron state the Sommerfeld-Maue wave function (4.133). These "quasirelativistic" wave functions are valid if $\alpha Z \ll 1$ and, consequently, only the leading terms of $\alpha Z_{\mathrm{T}}$ are retained in the derivation. Similarly, as in Sec. 10.2.2, care has to be taken at which stage the approximation is introduced in evaluating Eq. (10.15). Rewriting Eq. (5.109) for bound-electron – positron pair production, we obtain from Eq. (10.15)

$$\langle\varphi_{\rm e}|(1-\beta\alpha_z)e^{i\mathbf{q}\cdot\mathbf{r}}|\varphi_{\rm p}\rangle = \frac{1}{\gamma^2}\langle\varphi_{\rm e}|e^{i\mathbf{q}\cdot\mathbf{r}}|\varphi_{\rm p}\rangle + \frac{\beta^2}{E_{\rm p}+E_{\rm e}}\langle\varphi_{\rm e}|(q_x\alpha_x+q_y\alpha_y)\,e^{i\mathbf{q}\cdot\mathbf{r}}|\varphi_{\rm p}\rangle. \quad (10.40)$$

The first term on the right-hand side of Eq. (10.40) proportional to $1/\gamma^2 = 1-\beta^2$ comes about as the difference between two matrix elements, which almost cancel for large $\gamma$ only if they are evaluated between *exact* eigenstates. With the reformulation (10.40), the correct asymptotic behavior of the cross section as $(\ln\gamma)$ for $\gamma\to\infty$ is ensured even when approximate wave functions are used in evaluating the right-hand side.

Using Darwin and Sommerfeld-Maue wave functions (Secs. 4.3.3 and 4.4.3) for an electron and a positron in the target potential and keeping the lowest-order terms in $\alpha Z_{\rm T}$, the angle-differential cross section $d^2\sigma/dE_{\rm p}\,d\Omega_{\rm p}$ and the angle-integrated cross section $d\sigma/dE_{\rm p}$ can be expressed as one-dimensional integrals over the momentum $q$. For the resulting – somewhat lengthy formulas – the reader is referred to the original paper by Becker [Bec87].

Figure 10.14 shows the angle-integrated cross section for bound-free pair production in $Z_{\rm P}$ + $S^{16+}$ collisions at 1, 15, and 100 GeV/u as a function of the positron total energy. The solid curves are calculated with approximate wave functions while the dashed curves represent results obtained from a partial-wave analysis. The results scale with $Z_{\rm P}^2$. For high projectile energies, the curves differ because the partial-wave expansion is cut off for $|\kappa_{\rm p}| > 10$. The positron spectra are seen to follow a similar behavior as in Fig. 10.3 for a fixed impact parameter $b$ and in Fig. 10.6 for the $b$-integrated cross section. With higher target charges, the discrepancy between the methods of calculation increases, so that for $Z_{\rm T} = 92$ the treatment using an $\alpha Z$ expansion underestimates the cross section by about a factor of 4 compared to the truncated partial-wave treatment.

Combining the projectile and the target charge dependence, the cross sections scale approximately as

$$\frac{d\sigma}{dE_{\rm p}} \propto (\alpha Z_{\rm P})^2(\alpha Z_{\rm T})^5 \quad (10.41)$$

for $\alpha Z_{\rm P,T} \ll 1$. Here, the $Z_{\rm T}^5$ dependence reflects the availability of high-momentum components in a bound-state wave function and also occurs for charge transfer. In order to exhibit this dependence in Fig. 10.15, the total cross sections obtained by numerically integrating over the positron spectra are plotted in such a way that Eq. (10.41) gives a horizontal line

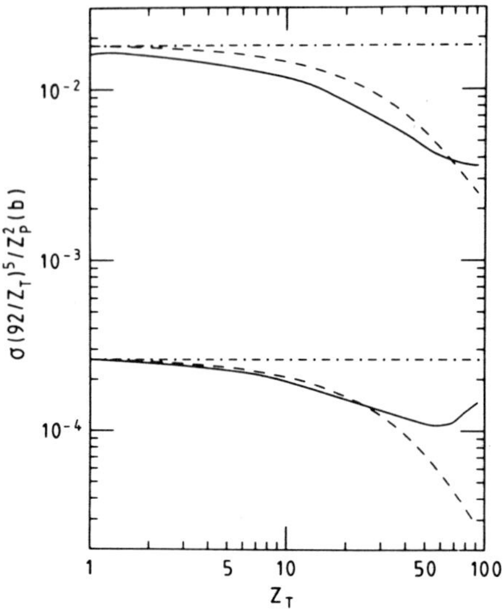


**Figure 10.15.** Cross section for bound-electron (K-orbit) – positron pair production as a function of the target charge for laboratory projectile energies $E = 1$ (bottom) and $E = 15$ (top) GeV/u. Solid curves: partial-wave expansion; dashed curves: Darwin and Sommerfeld-Maue approximations; and dash-dot curves: $Z_T^5$ scaling. From [Bec87].

[Bec87]. The results derived from the $\alpha Z$ expansion (solid line) and from the partial wave expansion (dashed line) are in rather good agreement with one another for not too high target charges. The simple scaling behavior of Eq. (10.41) does not hold except for the lowest charges $Z_P$. A charge dependence different from Eq. (10.41) has been obtained by a convolution prescription [RhB89]; however, it has been criticized in [Eic90, Bau91].

Perturbative estimates of the cross sections for bound-free pair production have been obtained in a closed analytical form by Bertulani and Baur [BeB88] for $\alpha Z \ll 1$ (in order to be able to use quasirelativistic wave functions). For the dominant K-shell capture they get

$$\sigma_{\mathrm{ep}}^{\mathrm{bound}} \approx \frac{33\pi}{10} (\alpha Z_P)^2 (\alpha Z_T)^6 \, \lambda_c^2 \, \frac{1}{e^{2\pi\alpha Z_T} - 1} \left[ \ln\left(\frac{d\gamma}{2}\right) - \frac{5}{3} \right], \qquad (10.42)$$

where $d = 0.681$ is a number related to Euler's constant and $\lambda_c = \hbar/m_e c$ is the Compton wavelength. Equation (10.42) is valid [BeB88] for $\gamma > 50$ and has the same dependence on target and projectile charges for $\alpha Z_T \ll 1$ as expressed by Eq. (10.41). The ratio of the total cross section for bound-free

proximately as $Z_{\rm T}^{2.8\pm0.25}$ as compared to the $Z_{\rm T}^2$ dependence expected from perturbation theory.

Experiments [BeG94] including $\rm La^{57+}$ projectiles yield a target charge dependence as $Z_{\rm T}^{2.95\pm0.40}$ for 0.956 GeV/u and $Z_{\rm T}^{2.65\pm0.35}$ at 1.3 GeV/u, while the projectile charge dependence is $Z_{\rm P}^{6.54\pm0.65}$. This is roughly in line with Eq. (10.41) if $Z_{\rm T}$ and $Z_{\rm P}$ are interchanged, because the bound electron is created in the projectile, not in the target.

# 10.4 Multiple pair production

In the discussion of bound-free pair production, Sec.10.3, indications emerge that for collisions between high-$Z$ ions at small impact parameters $b$, perturbation theory breaks down and higher-order effects have to be taken into account. These calculations, while still burdened with the problem of gauge dependence, give a hint that similar effects must also be expected for the production of free electron-positron pairs.

## 10.4.1 Limits of perturbative probabilities for single-pair production

Limitations become already evident in perturbative calculations following the early work of Bethe, Maximon, and Davies [BeM54, DaB54]. Disregarding Coulomb corrections proportional to the logarithm $\ln\gamma$ of the Lorentz factor and assuming $E_{\rm e}, E_{\rm p} \gg m_{\rm e}$, one obtains an approximate analytical expression [Bau90b] for the first-order impact-parameter-dependent pair production probability in the form

$$P^{(1)}(b) = \frac{14}{9\pi^2}\,(\alpha Z_{\rm T})^2(\alpha Z_{\rm P})^2\,\frac{\lambda\!\!\!^{-}_{\rm c}{}^2}{b^2}\,\ln^2\left(\frac{\gamma d \lambda\!\!\!^{-}_{\rm c}}{2b}\right), \tag{10.49}$$

where $d = 0.681$ [compare Eq. (10.42) for bound-free pair production]. This expression is valid for $\lambda\!\!\!^{-}_{\rm c} \le b \le \gamma d \lambda\!\!\!^{-}_{\rm c}$. For Pb + Pb collisions at $\gamma_{\rm FT} = 3\times10^7$, see Eq. (10.2), and $b = \lambda\!\!\!^{-}_{\rm c}$, one obtains a probability $P^{(1)}(\lambda\!\!\!^{-}_{\rm c}) = 5.3$, which clearly shows the need for a nonperturbative treatment.

Another violation of unitarity is observed in the total cross section at asymptotically large values of $\gamma$. Froissard has derived an upper bound on the total cross section of two strongly interacting particles colliding with the invariant energy $W = \sqrt{s}$, see Eq. (2.20), by using the Mandelstam representation [Fro61] (for a review see [Val91]) as

$$\sigma_{\rm tot} \le {\rm const.} \times \ln^2\left(\frac{s}{s_0}\right) \qquad \text{as } s \to \infty, \tag{10.50}$$

where $s_0$ is a constant. It is usually assumed that this relation is also valid for electromagnetic processes. If this is so, and since $\ln s \propto \ln \gamma$ for large values of $\gamma$, we have $\sigma_{\text{tot}} \leq \text{const.} \times \ln^2(\gamma/\gamma_0)$, in contradiction to the asymptotic $\gamma$ dependence of the partial cross section (10.1), proportional to $\ln^3(\gamma/\gamma_0)$, for single-pair production. Another brief discussion of the asymptotic behavior of pair creation can be found in [Bau90c], where also further references are given. While such discussions of unitarity may be only of academic interest as long as it is not specified what asymptotically large values of $\gamma$ mean, they point in the same direction as Eq. (10.49), namely that higher-order effects become important.

### 10.4.2 Multiple production of independent electron-positron pairs

Violation of unitarity in perturbative results for extreme relativistic collisions between high-$Z$ ions indicates that some of the probability flux is diverted into other reaction channels, in particular into multiple pair production.

We give here a very simple discussion of multiple production of free electron positron pairs,[2] which, in fact, is the common root of various more sophisticated treatments that have been published so far. The basic assumptions, without which a theory would become exceedingly complicated, are the following: (a) The energy of the colliding high-$Z$ ions is so large that even the production of several pairs does not change their energies or trajectories. (b) The pairs are created completely independently of each other. This means that electrons and positrons do not interact with one another (this is a $1/Z$ correction) and are not subject to the Pauli principle. Both assumptions should be well justified for extreme relativistic collisions of, say, Au on Au or Pb on Pb.

Let us now suppose that we have calculated an impact-parameter-dependent transition amplitude $A_{p_+p_-}(b) = A_{ep}(b)$, for example from Eqs. (10.15) and (10.19), where $e = \{\mathbf{p}_+, s_+\}$ and $p = \{\mathbf{p}_-, s_-\}$ denote a set of momentum and spin quantum numbers of electron and positron, respectively. Then the first-order probability of finding an electron in a state $e$ after the collision is

$$\mathcal{P}_e(b) = \sum_p \mathcal{P}_{ep}(b) = \sum_p |A_{ep}|^2. \tag{10.51}$$

As long as $\mathcal{P}_e(b) \ll 1$, there is no violation of unitarity and we may take $[1 - \mathcal{P}_e(b)]$ as the probability of *not* having created an electron in the state

[2]Multiple bound-free pair production is not expected to compete with free pair creation because of the restricted phase space.

$e$. In general, however, it is necessary to obtain the single-electron probabilities from more exact methods like coupled-channel calculations, which rigorously satisfy the unitarity requirement. The probability that only *one* electronic state $e$ and *no other* positive-energy state is occupied is then

$$P_e(1,b) = \mathcal{P}_e(b) \prod_{e' \neq e} [1 - \mathcal{P}_{e'}(b)], \tag{10.52}$$

where the number $N = 1$ in $P_e(1,b)$ indicates that only one electron has been created. In writing down the product (10.52), the assumption (b), listed above, of independent probabilities has been used. From here we get the probability that exactly one electron (i.e., because of summation over positron states, one pair) is created in any state as

$$P(1,b) = \sum_e \mathcal{P}_e(b) \prod_{e' \neq e} [1 - \mathcal{P}_{e'}(b)]. \tag{10.53}$$

Rewriting the product term in Eq. (10.53), we find

$$\begin{aligned} P(1,b) &= \sum_e \mathcal{P}_e(b) \exp\left( \sum_{e' \neq e} \ln[1 - \mathcal{P}_{e'}(b)] \right) \\ &\approx \sum_e \mathcal{P}_e(b) \exp\left( - \sum_{e'} \mathcal{P}_{e'}(b) \right), \end{aligned} \tag{10.54}$$

where we have used the first term in the expansion $\ln(1+x) = x - \frac{1}{2}x^2 + \cdots$ of the logarithm and no longer exclude the term $e' = e$ from the sum, owing to the smallness of $\mathcal{P}_{e'}(b)$ for any given state $e'$.

Proceeding to two electrons $e_1, e_2$, irrespective of their permutation, we have

$$P_{e_1 e_2}(2,b) = \frac{1}{2!} \mathcal{P}_{e_1}(b) \mathcal{P}_{e_2}(b) \prod_{e' \neq e_1,\, e' \neq e_2} [1 - \mathcal{P}_{e'}(b)], \tag{10.55}$$

and, discarding higher-order terms in the small quantities $\mathcal{P}_{e'}(b)$,

$$P(2,b) \approx \frac{1}{2!} \left( \sum_e \mathcal{P}_e(b) \right)^2 \exp\left( - \sum_{e'} \mathcal{P}_{e'}(b) \right). \tag{10.56}$$

In general, for $N$ pairs, with

$$\mathcal{P}(b) = \sum_e \mathcal{P}_e(b) = \sum_{e,p} |A_{ep}|^2, \tag{10.57}$$

we obtain the Poisson distribution

$$P(N, b) = \frac{[\mathcal{P}(b)]^N}{N!} \, e^{-\mathcal{P}(b)}, \tag{10.58}$$

which one would expect for independent events. One verifies that the total probability for creating $0, 1, 2, \cdots$ pairs in a single collision is in agreement with the unitarity requirement. The average number of pairs

$$\sum_{N=1}^{\infty} N \, P(N, b) = \mathcal{P}(b) \tag{10.59}$$

provides an interesting interpretation of the perturbative (or nonperturbative) single-pair probability $\mathcal{P}(b)$. Since the derivation of Eqs. (10.58) and (10.59) is based on the assumption that each individual $\mathcal{P}_e(b) \ll 1$ but does not require $\mathcal{P}(b) = \sum_e \mathcal{P}_e(b) \ll 1$, Eq. (10.59) offers a possibility to give meaning to unitarity-violating perturbative total pair creation probabilities. Under all the assumptions outlined above, creation probabilities larger than one express, in an approximate fashion, that – on the average – more than one electron-positron pair is produced. In this sense, perturbation theory can still be used when it yields probabilities exceeding unity.

The Poisson distribution (10.58) has been derived in several ways. Baur [Bau91] used a " sudden approximation" to the $S$-matrix and a quasi-Boson approximation for describing electron-positron pairs, thus discarding the Pauli principle. Rhoades-Brown and Weneser [RhW91] perform a summation of a perturbational subseries of selected Feynman diagrams including the interaction between the colliding ions and the electron-positron pairs to all orders but dropping the interaction between the leptons. They also discuss coupled-channel calculations. Best et al. [BeG92] adopt a similar nonperturbative approach neglecting quantum mechanical interference terms. While these works differ in the theoretical methods used, they are based essentially on the same assumptions, namely the complete independence of the individual pair production processes. In each case, it is the goal to derive the structure of Eq. (10.58), not the input probability $\mathcal{P}(b)$.

The single-pair probabilities for extreme relativistic collisions may be estimated in several ways, as discussed in Sec. 10.2. One may use lowest-order QED calculations (Sec. 10.2.1), first-order perturbation theory with distorted waves (Sec. 10.2.2), or the equivalent-photon method (Sec. 10.2.3). Coupled-channel calculations for free electron-positron pairs have not been performed so far, owing to computational difficulties.

As a simple application, we show in Figs. 10.17 and 10.18 the $N$-pair probability distributions [RhW91] calculated from Eq. (10.58) and

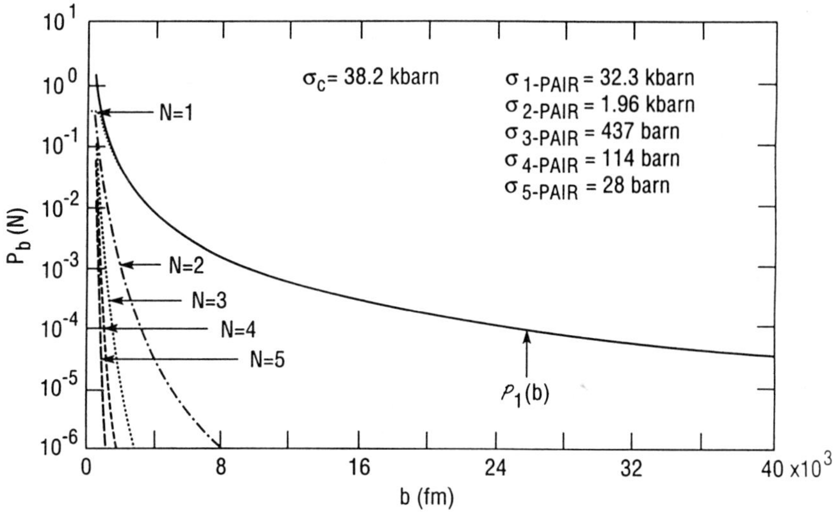


**Figure 10.17.** Probability distribution for $N$-pair production in Au + Au collisions at the highest RHIC energy with $\gamma_{\text{FT}} = 2 \times 10^4$ as a function of the impact parameter $b$ (in units of $10^3$). Here, $P_1(b)$ is the perturbative result corresponding to Eq. (10.49). It deviates from $P(1,b)$ of Eq. (10.58) only at the smallest impact parameters shown. From [RhW91].

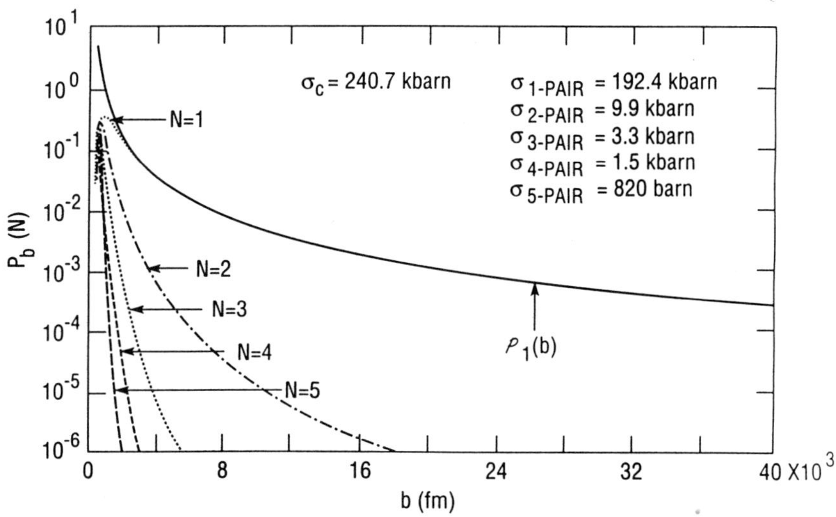


**Figure 10.18.** Same as Figure 10.17, but for Pb + Pb collisions at the highest LHC energy with $\gamma_{\text{FT}} = 3 \times 10^7$. From [RhW91].

from Eq. (10.49) for the approximate single-pair production probability $\mathcal{P}(b) = P^{(1)}(b)$. Also shown in these figures are the original lowest-order perturbative results (10.49). The curves are plotted for the highest envisioned beam energies and charge numbers of the respective facilities. It can be seen that single-pair creation dominates the picture at RHIC as well as at LHC. The $N$-pair production probabilities with $N \geq 2$ are seen to decrease very rapidly for the depicted impact-parameter range $b \geq \lambda\!\!\!^{-}_c$.

The method outlined in this section will certainly have limitations when the impact parameters become very small, that is when they get close to the sum of the nuclear radii. Then the probability for single-pair production has to be calculated nonperturbatively and may become very large. It is not certain whether the basic assumptions for deriving the Poisson distributions (10.58) are justified any longer. However, for the total pair multiplicity, the regime of very small impact parameters carries little weight.

# Part III

# Experimental Methods

measured in number of atoms/cm$^2$, if all $\sigma$ are given in cm$^2$:

$$x = \frac{t_\mathrm{T}\,\rho_\mathrm{T}\,N_\mathrm{A}}{A_\mathrm{T}}\,, \tag{11.2}$$

where $t_\mathrm{T}$ is the penetration distance in cm, $\rho_\mathrm{T}$ is the stripper density in g/cm$^2$, $N_\mathrm{A}$ is Avogadro's number, and $A_\mathrm{T}$ is the atomic weight of the stripper. For (ideal) gases, the equivalent expression is

$$x = \frac{t_\mathrm{T}\,N_\mathrm{A}\,P}{R\,T}\,, \tag{11.3}$$

where $P$ is the pressure of the gas, $T$ its absolute temperature, and $R$ the gas constant.

It is difficult to solve Eqs. (11.1) in the general case. The most ambitious calculation to date has been made by Betz and Höppler [BeH88]. Their model includes ions with up to seven electrons, distributed over 20 states among which 157 different transitions can occur. Less ambitious excited-state models are worked out in [NiC90]. Fortunately, to provide guidance in the selection of stripper foils, it is sufficient to use the ground-state model, mentioned above. In this model, each ionic state is characterized only by the number of electrons on the ion. A modification of this model, called the quasi ground-state model, takes into account some solid-state effects through the use of effective cross sections [AnM86].

It has been suggested that multiply excited ions might, upon leaving a solid, undergo Auger decay [BeG70]. In that case, the charge distribution outside of a foil would differ from that just inside the foil. For relativistic projectiles, such effects, if present, appear to be very small.

### 11.2.2 Ground-state model

Although it is common to label ionic ground states by their charge $q$, it is more convenient for the present purpose to label them by the number of electrons $n = Z_\mathrm{P} - q$ on the ion. The cross section for changing the number of electrons on the ion from $n$ to $n'$ in a single collision is denoted by $\sigma(n, n')$. These cross sections can be related to the cross sections for ionization, discussed in Chaps. 6 and 7, and for nonradiative and radiative capture, in Chaps. 8 and 9. Restricting consideration to ions filled at most up to the M shell, i.e., with 28 or fewer electrons, the relations are

$$\sigma_\mathrm{s1}(n) \equiv \sigma(n, n-1) = n_\mathrm{K}\sigma_\mathrm{K}^\mathrm{l} + n_\mathrm{L}\sigma_\mathrm{L}^\mathrm{l} + n_\mathrm{M}\sigma_\mathrm{M}^\mathrm{l}\,, \tag{11.4}$$

$$\sigma_\mathrm{a1}(n) \equiv \sigma(n, n+1) = \frac{2-n_\mathrm{K}}{2}\sigma_\mathrm{K}^\mathrm{c} + \frac{8-n_\mathrm{L}}{8}\sigma_\mathrm{L}^\mathrm{c} + \frac{18-n_\mathrm{M}}{18}\sigma_\mathrm{M}^\mathrm{c} + \sigma_\mathrm{H}^\mathrm{a}\,. \tag{11.5}$$

Here, $\sigma_{\text{s1}}$ is called the one-electron stripping and $\sigma_{\text{a1}}$ the one-electron attachment cross section; $\sigma_{\text{S}}^{\text{l}}$ is the loss cross section per electron and $n_{\text{S}}$ is the initial number of electrons in shell S (K, L, or M); $\sigma_{\text{S}}^{\text{c}}$ is the nonradiative plus radiative capture cross section from all filled target states into a completely vacant projectile shell S and $\sigma_{\text{H}}^{\text{a}}$ is the attachment cross section into all shells higher than M. In principle, each $n_{\text{S}}$ can take values from zero up to the maximum number of electrons in the shell.

Equation (11.4) assumes that the one-electron loss cross section for any shell is proportional to the number of electrons in that shell. A more accurate expression would decompose the cross section for each shell into partial cross sections for the subshells and also would take into account the fact that each partial cross section is a function of $n$ because of screening effects. The same considerations apply to Eq. (11.5) which assumes that the capture cross section into each shell is proportional to the fractional number of vacancies in the shell.

Similar to Eqs. (11.4) and (11.5), one can write expressions for $\sigma(n, n-m)$ in terms of the $m$-fold multiple-electron stripping cross sections and for $\sigma(n, n+m')$ in terms of the $m'$-fold multiple-attachment cross sections. Unless simplifying assumptions are made, these expressions are quite complicated because multiple processes (in a single collision) can involve more than one shell [Eq. (7.36)]. Fortunately, for relativistic ions, cross sections decrease rapidly with the multiplicity, as shown in Figs. 7.8 and 8.14, typically by an order of magnitude for each unit increase in multiplicity, so that multiple processes can be neglected for the present purpose.

Once the cross sections $\sigma(n, n')$ are known, the charge state yields of ions emerging from stripper foils can be evaluated by a set of equations similar to Eq. (11.1). Inside the foil, the fraction $Y_n$ of ions with $n$ electrons ($\sum_n Y_n = 1$) increases through attachment and stripping processes from other states and decreases if $n$ is altered by any charge changing collision:

$$\frac{dY_n(x)}{dx} = \sum_{n' \neq n} \sigma(n', n)\, Y_{n'}(x) - \sigma_{\text{tot}}(n)\, Y_n(x)\,, \tag{11.6}$$

where the total charge changing cross section for ions with $n$ electrons is

$$\sigma_{\text{tot}}(n) = \sum_{n' \neq n} \sigma(n, n')\,. \tag{11.7}$$

Even this simplified set of equations is not easily solved in general, because for projectiles of atomic number $Z_{\text{P}}$ it involves $(Z_{\text{P}}+2)(Z_{\text{P}}+1)$ cross sections, assuming the negative ion with $(Z_{\text{P}}+1)$ electrons is stable. Therefore, it is appropriate to examine approximate solutions.

*Pure stripping*

For heavy ions of several hundred MeV/u energy and for light ions of less energy, and for light stripper foils approximately up to Al, attachment is negligible compared to loss from the M-, L-, and possibly K-shells of the ions. This can be seen from the cross sections discussed in Chaps. 6 to 9, see also Fig. 1.1. If one also neglects multiple stripping compared to one- electron stripping, a simple statistical argument provides an analytical solution to Eq. (11.6). From this solution, the choice of stripper thickness for maximum yield of the desired charge state, as well as the width of the charge state distribution, can be obtained. If stripping takes place only within one shell (M, L, or K), the expressions are particularly simple.

Using as basis Eq. (11.4), we start with a schematic one-shell model in which a $n$ electron ion has a one-electron stripping cross section

$$\sigma_{s1}(n) = n\,\sigma^{l}, \tag{11.8}$$

where $\sigma^{l}$ is the average loss cross section per electron. On traversing a target foil of thickness $x$ (in atoms/cm$^2$), each electron has the same probability

$$\tau = e^{-\sigma^{l}x} \tag{11.9}$$

of surviving on the ion and has the probability $(1-\tau)$ to be stripped. Hence, if the ion enters with $N$ electrons, the probability that just $n$ electrons survive on the ion is given by the binomial expression

$$Y_n(x) = \frac{N!}{n!\,(N-n)!}\,\tau^{n}(1-\tau)^{N-n}. \tag{11.10}$$

By substitution, one finds that Eq. (11.10) is a solution of Eq. (11.6) which satisfies $\sum_{n=0}^{N} Y_n = 1$.

From Eq. (11.10) one concludes that the mean number of electrons $\overline{n}$ on the ion is

$$\overline{n} = N\tau\,. \tag{11.11}$$

By choosing the stripper foil thickness $x = x_0$ so that $\overline{n}$ corresponds to a desired charge $q_0 = Z_P - \overline{n}$, one profits from the fact that, for a given value of $\tau$, $Y_n$ passes through a maximum at $n = \overline{n}$. The corresponding peak yield is

$$Y_{\overline{n}} = \frac{N!}{\overline{n}!\,(N-\overline{n})!}\,\frac{\overline{n}^{\,\overline{n}}(N-\overline{n})^{N-\overline{n}}}{N^{N}} \tag{11.12}$$

$$\simeq \left[\frac{N}{2\pi\overline{n}\,(N-\overline{n})}\right]^{1/2}. \tag{11.13}$$

**Table 11.1.** Optimum stripper foil ($Z_T \leq 13$) thicknesses $x_0$ for producing ions listed in the first column with maximum yield $Y$. It is assumed that the K and L, or K, shells are initially filled, but otherwise the values of $x_0$ are appropriate for any ion species or energy above a few hundred MeV/u. The calculations are based on $g_L = 0.3, g_K = 0$. Ten percent variations in stripper thickness will not alter the yield significantly.

| Desired ion | $x_0$(optimum) (atoms/cm$^2$) | $Y$(maximum) |
|---|---|---|
| C-like | $0.7/\sigma_L^1$ | 0.27 |
| B-like | $0.9/\sigma_L^1$ | 0.27 |
| Be-like | $1.3/\sigma_L^1$ | 0.30 |
| Li-like | $1.9/\sigma_L^1$ | 0.36 |
| He-like | $3.3/\sigma_L^1$ | 0.59 |
| H-like | $0.7/\sigma_K^1$ | 0.50 |

In the last expression, Stirling's approximation has been used for each factorial, which is accurate to at least 2% for $N, \overline{n}, N - \overline{n} \geq 4$.

For the more realistic case of an ion with electrons in several shells S, it is convenient to write the cross section for one-electron stripping, Eq. (11.4), for the M- or L-shell in the form

$$\sigma_{s1}(n) = (n_S + g_S)\,\sigma_S^1\,, \tag{11.14}$$

where, for the M-shell,

$$g_M = (8\sigma_L^1 + 2\sigma_K^1)/\sigma_M^1\,, \tag{11.15}$$

and, for the L-shell,

$$g_L = 2\sigma_K^1/\sigma_L^1\,, \tag{11.16}$$

and $g_K = 0$. Equation (11.10) still gives the correct result, if one substitutes

$$N \to N_S + g_S \quad \text{and} \quad n \to n_S + g_S\,. \tag{11.17}$$

Here, $N_S$ is the initial number of electrons in shell S. For typical relativistic ions, one finds $g_M \simeq 2 - 3$ and $g_L \simeq 0.2 - 0.3$. Substitution into Eq.

(11.12) then gives peak values $Y_{\overline{n}} \simeq 0.2$ or 0.3 for stripping to the middle of the M- or L-shells, respectively. For stripping to $\overline{n}_M = 0$ or $\overline{n}_L = 0$, peak values $Y_{\overline{n}} \simeq 0.6$ can be obtained. These maximum yields are independent of stripper atomic number ($Z_{\mathrm{T}} \leq 13$), ion atomic number $Z_{\mathrm{P}}$, or ion velocity $v$, except through a slight $Z_{\mathrm{P}}$ and $v$ dependence of $g_{\mathrm{S}}$. Of course, for each value of $\overline{n}$, $\tau$ must be chosen to be approximately equal to

$$\tau = \frac{\overline{n}}{N} = \frac{\overline{n}_{\mathrm{S}} + g_{\mathrm{S}}}{N_{\mathrm{S}} + g_{\mathrm{S}}} \,. \tag{11.18}$$

Table 11.1 gives values of the stripper thicknesses $x_0$ calculated with Eqs. (11.9) and (11.18), with the assumptions $N_{\mathrm{L}} = 8$, $g_{\mathrm{L}} = 0.3$ or $N_{\mathrm{K}} = 2$, $g_{\mathrm{K}} = 0$. The expected charge state distributions can be computed from Eq. (11.10). Figure 11.2 shows an example for 300-MeV/u $\mathrm{Xe}^{45+}$ ($N_{\mathrm{L}} = 7$) stripped by 2.5 mg/cm$^2$ Be. Generally, one finds for the L-shell a full width at half maximum of the charge distribution $\Delta n_{\mathrm{L}} \simeq 3$, practically independent of $n_{\mathrm{L}}$. For the M-shell, one finds $\Delta n_{\mathrm{M}} \simeq 5$.

*General case: equilibrium. Few electron ion yields.*

If one wishes to produce bare, relativistic, heavy ions, generally one has to use strippers of atomic number so high that attachment can no longer be ignored. One then has to solve Eq. (11.6) in a less restrictive form than above. Analytical solutions exist for two-electron ions [All58, NiC90], but already these are not too transparent because six charge changing cross sections are involved, or at least four, if double stripping and attachment are omitted. For more than two electrons, it is easiest to solve the equations by numerical integration, for which the Runge-Kutta method is appropriate. One finds that, as in Eq. (11.10), each charge fraction $Y_n(x)$, except $Y_N$ and $Y_0$, passes through a maximum as a function of $x$. At large enough values of $x$, $dY_n/dx = 0$ and $Y_n(x)$ reaches an equilibrium value called $F_n$ below. If attachment for the charge state with $n$ electrons is small compared to stripping, the value of the peak yield is close to that given by Eq. (11.12) or in Table 11.1. But, if the opposite is true, the maximum in $Y_n(x)$ is reached only at equilibrium thicknesses. For the production of bare nuclei, the latter situation always occurs. Hence, we examine the equilibrium case only, but for purpose of illustration we show in Fig. 11.3 calculated, and some measured, charge fractions for 962-MeV/u $\mathrm{U}^{68+}$ ions (24 electrons) penetrating various Cu foils. The curves have been computed with a model which partially takes into account solid-state effects. For the largest thicknesses, energy loss is appreciable and has to be taken into account in the data analysis.

We proceed to derive the bare, one-electron, and two-electron ion yields

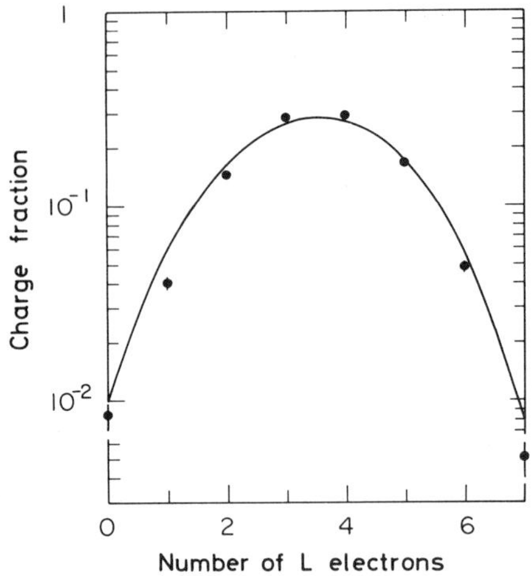


**Figure 11.2.** Charge state distribution for 300-MeV/u $Xe^{45+}$ ions emerging from a 2.5-mg/cm$^2$ Be foil. The number $n_L$ of L electrons on the ion is plotted on the abscissa. A fit to Eq. (11.10) has been made, using $N_L = 7, g_L = 0.3$. From [MeA87b].

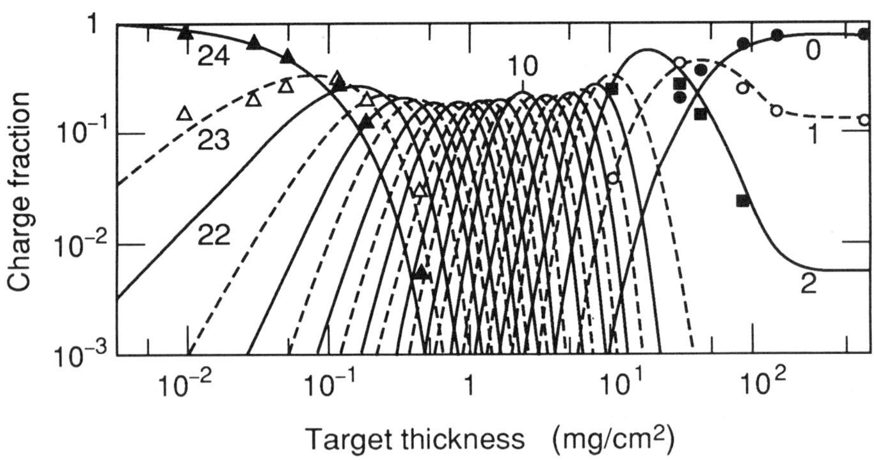


**Figure 11.3.** Charge state fractions for 962-MeV $U^{68+}$ ions (24 electrons) emerging from Cu foils as a function of the foil thickness. The number on certain curves indicates the number of electrons on the ions. Curves in solid lines are for even and, in dashed lines, for odd numbers of electrons. For clarity, data points are shown only for ions with 0, 1, 2, 23, and 24 electrons. From [GoG84] and [MeA87a].

at equilibrium. The general equilibrium solution of Eq. (11.6), including all processes, is discussed by Betz [Bet72]. As a first simplification, for relativistic ions, multiple-electron processes can usually be neglected with respect to single-electron processes, as noted above. One then finds that the equilibrium solutions $F_n$ of Eq. (11.6) obey the relation

$$F_{n+1}\,\sigma_{\mathrm{s1}}(n+1) = F_n\,\sigma_{\mathrm{a1}}(n)\,, \tag{11.19}$$

where the definitions of Eqs. (11.4) and (11.5) for the one-electron stripping and attachment cross sections have been used. The relation (11.19) is most easily derived by starting with $n = 0$ or $n = Z_{\mathrm{P}}$ and iterating Eq. (11.6) step by step, with the condition $dY_n(x)/dx = 0$.

To solve Eq. (11.19), it is convenient to write it in the form

$$F_n = r_n\,F_{n+1}\,, \tag{11.20}$$

where

$$r_n = \frac{\sigma_{\mathrm{s1}}(n+1)}{\sigma_{\mathrm{a1}}(n)}\,. \tag{11.21}$$

Iterating Eq. (11.20) to $F_{Z_{\mathrm{P}}}$ (which represents the equilibrium fraction of neutral projectile atoms), one finds

$$F_n = r_n\,r_{n+1}\ldots r_{Z_{\mathrm{P}}-1}\,F_{Z_{\mathrm{P}}}\,, \tag{11.22}$$

$$F_{Z_{\mathrm{P}}} = \left(1 + \sum_{n'=0}^{Z_{\mathrm{P}}-1} r_{n'}\,r_{n'+1}\ldots r_{Z_{\mathrm{P}}-1}\right)^{-1}. \tag{11.23}$$

Two further simplifications can now be made. First, for any ion with $Z_{\mathrm{P}} \geq 10$, one expects the equilibrium fraction of neutral atoms, $F_{Z_{\mathrm{P}}}$, to be extremely small after passage through any reasonable stripper foil. Hence, the sum in the denominator of expression (11.23) must be much larger than unity. Then

$$(F_{Z_{\mathrm{P}}})^{-1} \simeq r_{Z_{\mathrm{P}}-1}\,(1 + r_{Z_{\mathrm{P}}-2}\,(1 + r_{Z_{\mathrm{P}}-3}\,(\ldots$$

$$(1 + r_2\,(1 + r_1\,(1 + r_0\,)))\ldots)))\,. \tag{11.24}$$

Second, for relativistic ions with many electrons, electron stripping by far exceeds attachment, down to stripping into the L- or K-shells. This means that $r_n \gg 1$, until some low value of $n$ is reached. In many cases, it is

**Table 11.2.** Calculated zero-, one-, and two-electron equilibrium yields for uranium projectiles passing through various strippers. For each projectile energy $E_\mathrm{P}$ in MeV/u, the equilibrium stripper thickness $t_\mathrm{eq}\rho$ in mg/cm$^2$ is given in parentheses above the equilibrium fractions $F_0 : F_1 : F_2$. The most intense fraction is given in bold face. Fractions of less than 1% are denoted by 0.00. If the fraction sum is less than unity, ions with more than two electrons are produced.

| $E_\mathrm{P}$ | Stripper | | | |
|---|---|---|---|---|
| | Mylar | Al | Cu | Au |
| 100 | (40)<br>0.00:0.05:**0.70** | (25)<br>0.00:0.06:**0.57** | (10)<br>0.00:0.05:**0.20** | (6)<br>0.00:0.00:**0.10** |
| 200 | (150)<br>0.03:0.30:**0.60** | (100)<br>0.07:0.34:**0.54** | (30)<br>0.02:0.20:**0.63** | (7)<br>0.01:0.13:**0.53** |
| 430 | (400)<br>0.18:**0.48**:0.32 | (200)<br>0.34:**0.49**:0.17 | (100)<br>0.43:**0.44**:0.13 | (50)<br>0.20:**0.48**:0.30 |
| 960 | (700)<br>**0.53**:0.40:0.07 | (400)<br>**0.73**:0.25:0.02 | (200)<br>**0.84**:0.15:0.01 | (60)<br>**0.79**:0.20:0.01 |

sufficient to set this lowest value of $n$ equal to two, although the results given below are easily generalized to higher values. Then one can write

$$(F_{Z_\mathrm{P}})^{-1} \simeq r_{Z_\mathrm{P}-1}\, r_{Z_\mathrm{P}-2}\, r_{Z_\mathrm{P}-3} \cdots r_2\,(1 + r_1\,(1 + r_0\,))\,. \tag{11.25}$$

Using Eq. (11.22), one now obtains for the two-electron, one-electron, and bare-ion yields,

$$F_2 = 1/\,D\,, \quad F_1 = r_1/\,D\,, \quad F_0 = r_1\,r_0/\,D\,, \tag{11.26}$$

where

$$D = 1 + r_1 + r_1\,r_0\,. \tag{11.27}$$

These expressions can be evaluated using Eqs. (11.4), (11.5), and (11.21) :

$$\begin{aligned} r_1 &= \frac{2\sigma_\mathrm{K}^\mathrm{I}}{\sigma_\mathrm{a1}(0) - \sigma_\mathrm{K}^\mathrm{c}/2}\,, \\ r_0 &= \frac{\sigma_\mathrm{K}^\mathrm{I}}{\sigma_\mathrm{a1}(0)}\,. \end{aligned} \tag{11.28}$$

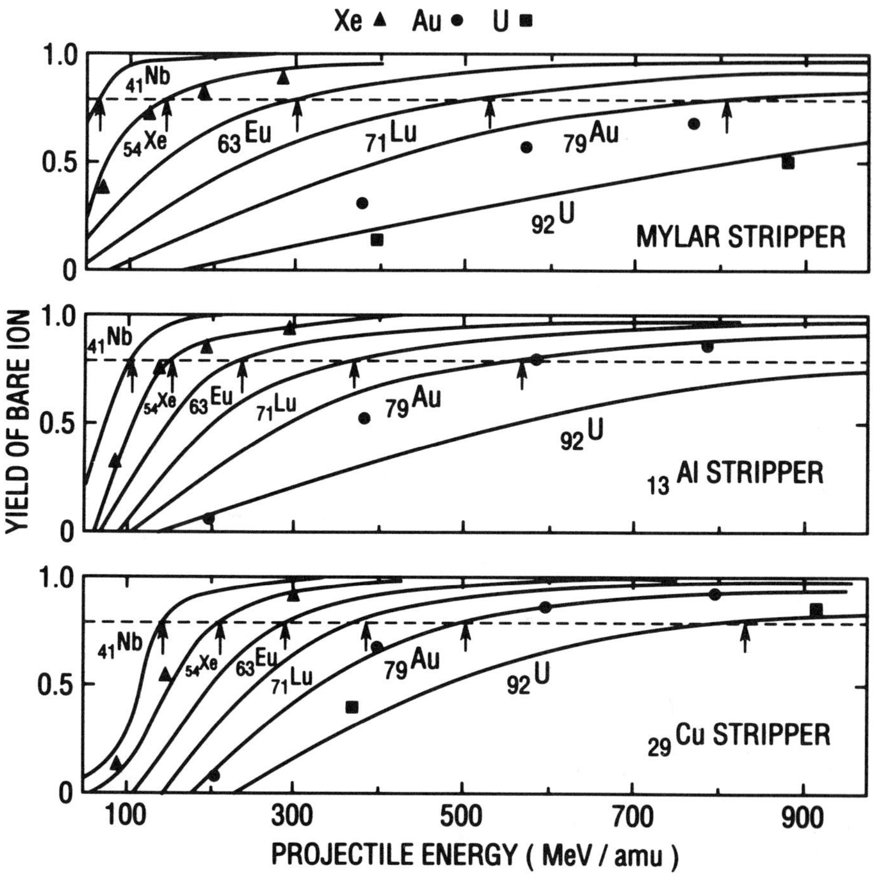


**Figure 11.4.** Equilibrium fractions for various bare ions as a function of projectile energy in Mylar, Al, and Cu stripper foils. The Xe-, Au-, and U-projectile data are represented by solid triangles, circles, and squares, respectively. Arrows indicate the minimum projectile energy for 80% bare-ion yield (Table 11.3). From [ThW85]; see also [StG91].

**Table 11.3.** Projectile energies for 80% bare-ion yield and equilibrium thicknesses for various strippers. For each projectile energy $E_P$ in MeV/u, the equilibrium stripper thickness $t_{eq}\rho$ in mg/cm$^2$ is given. One-half of this thickness will still provide a 60 – 70% bare-ion yield. From [ThW85].

| $Z_P$ | Stripper | | | | | |
|---|---|---|---|---|---|---|
| | Mylar | | Al | | Cu | |
| | $E_P$ | $t_{eq}\rho$ | $E_P$ | $t_{eq}\rho$ | $E_P$ | $t_{eq}\rho$ |
| $_{26}$Fe | <50 | <5 | <50 | ~3 | 60 | 1.6 |
| $_{41}$Nb | 70 | 25 | 110 | 15 | 140 | 8 |
| $_{54}$Xe | 160 | 80 | 150 | 45 | 210 | 25 |
| $_{63}$Eu | 310 | 170 | 240 | 85 | 300 | 45 |
| $_{71}$Lu | 530 | 270 | 370 | 140 | 380 | 70 |
| $_{79}$Au | 760 | 400 | 570 | 210 | 500 | 100 |
| $_{92}$U | >1000 | >600 | ~1100 | ~360 | 820 | 180 |

Here, $\sigma_K^l$ is the one-electron loss cross section in the projectile K-shell, $\sigma_{a1}(0)$ denotes the one-electron attachment cross section into the bare projectile from all filled target shells, and $\sigma_K^c$ is the capture cross section into the empty K-shell. These cross sections can be obtained from information given in Chaps. 6 to 9 or in [AnB87].

Equations (11.26) have to be modified if solid-state effects are important. Then, the stripping cross section $\sigma_{s1}(1)$ is increased, because formation of the metastable 2s state allows ionization with a higher cross section than $\sigma_K^l$. Similarly, $\sigma_a(1)$ is reduced because capture can lead to the metastable 1s2sS$_{0,1}$ and 1s2pP$_{0,2}$ states which can be ionized before decay. Also, $\sigma_{a1}(0)$ is reduced because of possible capture to the 2s state. Quantitatively, the modifications are rather complex [AnM86]. They affect mainly the $n = 2$ charge fraction and are not worthwhile considering if one wishes only to choose an appropriate stripper foil.

For general guidance, Table 11.2 gives equilibrium fractions for uranium ions, evaluated with a program which takes into account some solid-state effects and which is generally in agreement with measured equilibrium fractions. Computed and measured bare equilibrium fractions are also given in Fig. 11.4.

Finally, we note that the minimum foil thickness necessary to produce charge state equilibrium, i.e., maximum bare-ion yield, is given to a good approximation by the expression [ThW85]

$$x_{\mathrm{eq}} \simeq \frac{4.6}{\sigma_{\mathrm{K}}^{1} + \sigma_{\mathrm{a}}(0)/2}, \tag{11.29}$$

although the relation $x_{\mathrm{eq}} \simeq 1/\,\sigma_{\mathrm{K}}^{1}$ has also been proposed [Bet83]. Table 11.3 shows suggested equilibrium thicknesses at various energies to produce $\sim$ 80% bare-ion fractions. In general, to produce bare, high-$Z_{\mathrm{P}}$ ions, it is not necessary or desirable to use very high-$Z_{\mathrm{T}}$ strippers because the then dominant nonradiative capture increases approximately as $Z_{\mathrm{T}}^{5}$, whereas loss increases as $Z_{\mathrm{T}}^{2}$. On the other hand, if $Z_{\mathrm{T}}$ is too low, the stripper thickness will cause appreciable energy loss. A Cu stripper represents a good compromise for producing bare high-$Z_{\mathrm{P}}$ ions.

# Chapter 12

# Target Arrangements

## 12.1 General considerations

In relativistic atomic-collision experiments, once the incident ion has been chosen, one typically determines cross sections which depend on one or more of the following quantities: charge and angle of the outgoing projectile; charge, angle, and energy of the target recoil ion; and spectral and angular distributions of emitted photons, electrons, and positrons (the latter two are called leptons here and in Chap. 13). The present discussion is limited to suitable target arrangements for the simplest of these measurements.

### 12.1.1 Target thickness

Cross sections for relativistic atomic collisions can be determined by "thin-target" or "thick-target" methods, as for nonrelativistic ions. In the thin-target method, one tries to assure single-collision conditions for the process under investigation by making the targets thin enough. In the thick-target method, one typically uses a range of thicknesses up to that at which the relevant projectile charge states reach their equilibrium values. Both methods have advantages and disadvantages which are discussed in Sec. 13.3.1.

*Thin targets*

In the thin-target method, one makes measurements with a range of target thicknesses small enough to obtain a linear dependence of the yield $Y_{\mathrm{D}}$ of detected particles or photons on the target thickness $x$ [Eqs. (11.2, 11.3)]. The cross section then is proportional to the slope $\Delta Y_{\mathrm{D}}/\Delta x$. The presence of a nonlinear component in the yield, usually a quadratic $x$ dependence, indicates the admixture of secondary processes such as double collisions, or in the case of photon measurements, electronic pileup or photon production by electrons emitted in the primary collisions. The general target thickness dependence of the thin-target yield can be written

$$Y_{\mathrm{D}} = Y_{\mathrm{B}} + ax + bx^2 , \tag{12.1}$$

where $Y_{\mathrm{B}}$ is the background yield due to slit scattering or to restgas in the beam line. Depending on the detector configuration and units chosen for $Y$ and $x$, the cross section is proportional or equal to the coefficient $a$.

Difficulty in the interpretation of Eq. (12.1) arises if the target induces background. Unwanted electrons or photons produced in the target could strike the detector directly or release radiations from the target chamber walls, which are then detected. From energy and momentum conservation one can show that the kinetic energy $T_{\mathrm{e}}$ in MeV of forward "knock-on" target electrons emitted by heavy projectiles is given by

$$T_{\mathrm{e}} = 1.02\,(\gamma_{\mathrm{P}}^2 - 1) \quad \mathrm{MeV} , \tag{12.2}$$

where $\gamma_{\mathrm{P}}$ is the Lorentz factor of the projectile. This energy effectively is independent of the heavy-projectile mass.[1] For 1-GeV/u projectiles, the energy is 1.7 MeV, but for 10-GeV/u projectiles, electrons up to 140 MeV will be emitted which cannot easily be kept away from a detector. If target-induced background is present, $Y_{\mathrm{B}}$ is a function of $x$, and any linear dependence can falsify the coefficient $a$ in Eq. (12.1). There is no simple way to deal with this kind of background. Application of a magnetic field close to the target, change of target geometry, and judicious use of absorbers can be helpful [AnS86].

As a general rule, a target can be considered as "thin," if

$$x \leq 0.05/\sigma^* , \tag{12.3}$$

where $\sigma^*$ is the largest cross section which the projectile can induce in the target, in principle, even if it is not measured. In Eq. (12.3), $x$ is expressed in atoms/cm$^2$ and $\sigma^*$ in cm$^2$ . For relativistic ions, thin-target thicknesses lie in the $\mu$g/cm$^2$ to mg/cm$^2$ range.

*Thick targets*

The thick-target method of determining cross sections depends on making yield measurements at various target thicknesses up to equilibrium or

[1] The exact value of the Lorentz factor $\gamma$ of forward knock-on electrons is $\gamma = (\Gamma + 1)/(\Gamma - 1)$, where $\Gamma = (\gamma_{\mathrm{P}} - \mu)^2/(\gamma_{\mathrm{P}}^2 - 1)$ and $\mu$ is equal to the ratio of the electron to projectile rest masses. If $\mu \ll \gamma_{\mathrm{P}}$, Eq. (12.2) is obtained.

only at equilibrium. In the former case, one uses the ground- or excited-state model equations for the charge fractions (Sec. 11.2) and attempts to extract as many cross sections as possible by iterative numerical integration of the equations and least-squares fitting to the data [BeH88]. It is necessary to have considerable redundancy: the number of target thicknesses used, multiplied by the number of charge fractions determined at each thickness (typically 3 – 10), must exceed the number of cross sections to be extracted. The approximate dependence of the charge fractions on the parameter $\tau$ [Eq. (11.9)] suggests that the target thicknesses should be spaced approximately logarithmically, i.e., in ratios such as 1:2:4:10:20:40. If only equilibrium charge fractions are determined, one cannot obtain more than cross section ratios, such as Eq. (11.21).

With the thick-target method, targets with thickness up to those given by Eq. (11.29) must be used. These lie in the range from mg/cm$^2$ to g/cm$^2$ (Tables 11.2 and 11.3). For the latter thicknesses, energy loss of the projectile may have to be taken into account in order to extract accurate cross sections (Sec. 11.1).

### 12.1.2 Target area

Although beams from storage rings are collimated to diameters of a few mm or less, direct beams from relativistic accelerators may be surrounded by a halo of particles and photons which can strike the target frame. It may not even be possible to focus the main beam sufficiently to avoid hitting the target frame. In such cases, one must collimate the projectiles by electronic means. Figure 12.1 shows one such arrangement, useful for solid targets. After passing through the target, the beam traverses two thin transmission-mounted silicon surface barrier particle detectors. The size and spacing of the detectors are chosen such that any particle passing though both detectors must have passed through the target foil. A triple coincidence between signals from these detectors and from a detector for electrons or photons assures that the latter were produced by projectiles passing through the target. One still has to guard against a target-induced background though (Sec. 12.1.1). Typical photon spectra from a solid target are shown in Fig. 13.10.

The arrangement of Fig. 12.1 is suitable only if one is not interested in the projectile charge state emerging from the target. Otherwise, the electronic collimation must be provided by detectors which have holes aligned in such a way that any particle passing through the holes must have passed through the target. An anticoincidence between either detector and the detector for radiations emitted from the target eliminates unwanted events. This type of arrangement has also been used to electronically collimate pro-

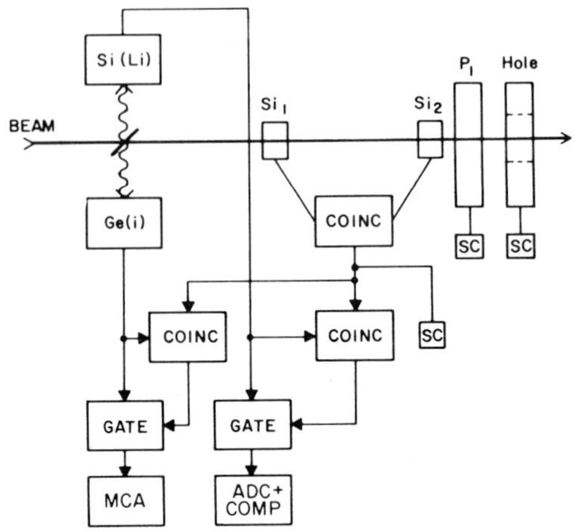


**Figure 12.1.** Simplified schematic diagram for photon cross section determination with electronically collimated relativistic projectiles. Passage of an ion through the detectors $S_1$ and $S_2$ assures prior passage through the target. Triple coincidences between a Si(Li) or Ge(Li) photon detector and $S_1$ and $S_2$ signal a valid event. The scintillation detectors $P_1$ and 'Hole' are set up only to check the counting method. Other abbreviations are: COINC, coincidence circuit; MCA, multichannel analyzer; ADC, analog-to-digital converter; COMP, computer; and SC, scaler. From [AnM84b].

jectiles through a gas cell by making the entrance and exit holes of the cell out of scintillating material [GrB90]. In the experimental arrangement of Fig. 12.1, the hole detector and the scintillator $P_1$ serve only to check the triple-coincidence counting method.

Figure 12.2 sketches a modern relativistic heavy-ion storage ring. Here, the beam diameter is of the order of 1 mm, so that electronic collimation of the beam is not necessary. The experimental facilities available at his particular ring are as follows. A dense electron target is provided by the electron beam cooler (2), allowing for electron recombination studies using the photon detectors (3). A gas target (4) (details are shown in Fig. 12.4) can be used with subdivided photon detectors for accurate angular measurements, needed to correct for Doppler shifts. A ample spectrum is shown in Fig. 13.11. There is a detecting system (5) for charge changed projectiles, permitting coincidence measurements with the photon detectors. Radiative collision processes in solid targets are conveniently studied with a magnetic nuclear-fragment separator system, sketched in Fig. 13.2.

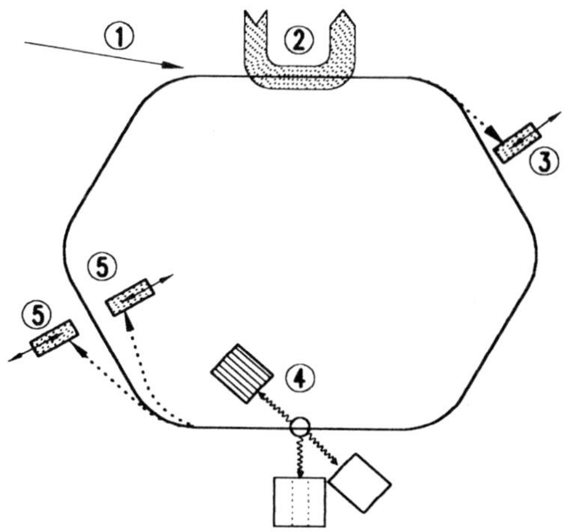


**Figure 12.2.** Sketch of a relativistic storage ring (ESR, at GSI, Darmstadt). The beam bending magnets are not shown. (1) Injection beam line from accelerator. (2) Electron beam cooler, usable as an electron target. (3) Electron recombination photon detector. (4) Gas target (see Fig. 12.4), surrounded by angle-subdivided x-ray detectors. (5) Detectors for charge changed projectiles. The diameter of the ring is approximately 30 m. In the upper and lower straight portions of the ring, laser-induced processes can be studied (arrangement not shown). From [Mok94].

## 12.2 Solid targets

The choice of target, whether solid or gaseous, depends on the type of experiment to be performed. With relativistic ions, one needs to be less concerned about solid-state effects (Sec. 11.2.1) than with nonrelativistic ions because typically one deals with few-electron ions in which excited-state lifetimes are short compared to the time between collisions. Therefore, unless experiments have to be performed with hydrogen, helium, or other gaseous elements, one can choose solid targets, which are much less cumbersome to use than gaseous ones (Sec. 12.3).

The fabrication of solid targets is a science and an art to which the meetings of the International Nuclear Target Development Society give ample testimony. Their proceedings [Fol89] describe some techniques useful in the target preparation for relativistic projectiles. With extreme care, a target uniformity of 1% can be reached, in which case the target thickness can be determined to the same accuracy by weighing. Five to 10% uncertainties in target thickness are more typical, which are then reflected in the accuracy of cross section determinations. In principle, one can also calibrate cross sections by comparison with Rutherford scattering. As the projectile energy increases, this method becomes more difficult because nuclear processes set in at progressively smaller scattering angles.

In general, it is desirable to use self-supporting targets, but below $\sim 0.1$ mg/cm$^2$ most solid foils become extremely fragile. Light elements have been used as backings; they have to be sufficiently thick to withstand the target deposition process. Commercial foils of Be down to $\sim 2$ mg/cm$^2$, C foils down to $\sim 5$ $\mu$g/cm$^2$, and Al foils down to $\sim 1$ mg/cm$^2$ are available, the former two only with appreciable nonuniformity. Evaporated Al foils can be made to thicknesses of $\sim 10$ $\mu$g/cm$^2$. The presence of backings may require careful consideration in the derivation of cross sections from measured yields (Chap. 13).

Although relativistic ion beams are hardly ever intense enough to deteriorate solid targets, or indeed may not be too intense if electronic collimation is used, one may have to be concerned with lifetimes of foils under heavy-ion bombardment. Estimates for C and Ni foils are given in [Bar79] and [MaY68], respectively, which may be extrapolated to relativistic projectiles by assuming that the lifetime is inversely proportional to the nuclear stopping power. Target wheels and wobblers have been used to extend foil lifetimes by an order of magnitude [MaN79, BoF89].

## 12.3 Gas targets

Gas targets in the form of cells and supersonic jets have been used in relativistic heavy-ion experiments. The advantages of a gaseous target are that the target thickness is easily adjusted down to restgas pressures, that the thickness can be accurately determined if gas cells are used, and that the targets are essentially indestructible. Figure 12.3 sketches a cell which has been used with hydrogen, helium, and nitrogen in various charge changing experiments with beams of Au and U ions between $\sim 100$ and 400 MeV/u. Since the expected charge changing cross sections are small, the cell is 2.4 m long. For shorter, continuous-flow cells, one has to compute the pressure drop across the entrance and exit collimators to obtain the effective cell length [Dus62].

Supersonic jets are useful for internal targets where space is limited, but a good vacuum outside of the target has to be maintained [ScB88]. Figure 12.4 sketches the gas jet production region of one such target, from which the complexity of such a device can be appreciated. In this jet, the target region has a thickness up to $\sim 10^{13}$ molecules/cm$^2$. The jet is used as an internal target with heavy-ion beams up to 560 MeV/u and circulating

which allows for a crude identification of the triggering particle. The latter has the advantage that fast timing can be obtained, which is useful for coincidence measurements.

Semiconductor detectors, charge-coupled devices [LuH88], and microchannel plates [Wiz79] can also serve as two-dimensional detectors with position resolution down to tens of microns. Such resolutions are too fine for beams from relativistic accelerators, but may be useful in relativistic storage rings as beams become more compact.

Solid scintillation detectors are most useful as veto counters (Sec. 12.1.2). If it is desired to measure the total energy of a particle, solid or liquid organic scintillators are suitable detectors, or sodium iodide or bismuth germanate, if compactness is essential.

### 13.2.3 Recoil-ion detection

As noted above, a measurement of the differential cross section as a function of projectile scattering-angle provides theoretically interesting details of ionization and capture processes. Figure 13.4 shows the scattering angles relevant for Ne+Xe to U+He collisions (right and left sloped solid lines, respectively), below which approximately 95% of the total ionization cross section is comprised. Hence, one should explore the differential ionization cross sections as a function of scattering angle at angles smaller than those indicated, but direct measurements of angles below $10^{-5}$ rad (0.5 mm at 10 m) are extremely difficult. A method has been developed to investigate collisions with scattering angles as small as $10^{-8}$ rad, which is applicable to relativistic projectiles.

The method relies on the fact that the momentum component of a recoiling target ion perpendicular to the beam direction is a direct measure of the projectile laboratory-scattering angle, which is practically independent of the inelasticity (Q-value) of the collision.[1] Hence, by measuring the velocity component of the recoiling ion perpendicular to the beam direction by time of flight, the projectile-scattering angle can be inferred for close collisions. At large impact parameters, with small momentum transfer, the summed momenta of the emitted electrons may influence the target recoil momentum appreciably [DoU91]. If the flight path normal to the beam direction has a length $l$ and the flight time is $t$, the corresponding projectile

[1]For an elastic, relativistic collision at large impact parameters, the target recoil momentum, equal to the momentum change of the projectile, lies nearly perpendicular to the beam direction. For an inelastic collision, the target recoil momentum is no longer perpendicular to the beam direction, but its perpendicular component is practically independent of the Q-value [UlS88].

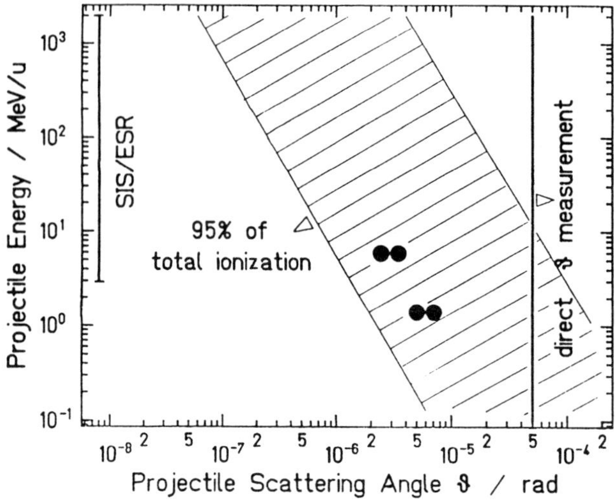


**Figure 13.4.** Projectile laboratory-scattering angle versus projectile kinetic energy for Ne+Xe (right sloped line) to U+He collisions (left sloped line). Most other collision systems lie in the shaded region between these lines. This region corresponds to a target recoil energy of 0.1 eV, below which approximately 95% of the ionization cross section in these collisions is comprised. The vertical line on the left indicates the range of kinetic energies available at the GSI heavy-ion synchrotron (SIS). The vertical line on the right indicates the lower limit of scattering angles typically accessible by direct measurement. The solid dots denote systems which had been studied by the recoil method at the time of publication. From [UlD91].

scattering angle in the laboratory system is given by

$$\theta = \frac{M_{\mathrm{T}}\, l/t}{\gamma_{\mathrm{P}} M_{\mathrm{P}}\, v}\,, \tag{13.1}$$

where $M_{\mathrm{T}}$ and $M_{\mathrm{P}}$ are the target and projectile masses, respectively, and $\gamma_{\mathrm{P}}$ and $v$ are the projectile Lorentz factor and velocity. By using gases cooled to 20°K or less, recoil energies can be determined to ±5 meV accuracy, but careful geometrical restrictions must be imposed on the recoil-ion beam.

An efficient recoil-ion spectrometer design is sketched in Fig. 13.4, which allows the full vector momentum of the recoiling target atoms to be determined. The intrinsic recoil velocity $\mathbf{v}_{\mathrm{coll}}$ is superposed on the velocity $\mathbf{v}_{\mathrm{jet}}$ of a cryo-pump cooled (14 °K), supersonic jet [ScB88]. Applying a small, very homogeneous, electric field of approximately 1 V/cm, a large portion

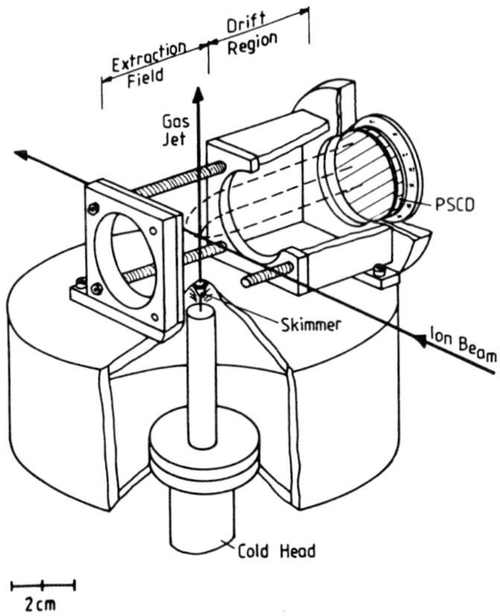


**Figure 13.5.** Recoil-ion spectrometer with a large collection efficiency. A supersonic jet with a velocity $\mathbf{v}_{\text{jet}}$ serves as the target. After collision with the projectile, the velocity of the recoiling target ion is $\mathbf{v}_{\text{jet}} + \mathbf{v}_{\text{coll}}$. An external electric field extracts the ions which then drift toward a position-sensitive detector PSCD. From time of flight and position measurements, the original momentum components of the recoil ions can be calculated. From [MeD94].

of the recoiling target ions is projected toward a two-dimensional position-sensitive channel-plate detector. From the impact coordinates on the detector, two velocity components can be inferred. The third component is obtained from a time of flight measurement, with respect to an electron or photon released in the collision.

With such a spectrometer, the inelasticity of an atomic collision can be determined, as manifested in the recoil momentum component parallel to the incident beam. A momentum resolution of $\sim 0.1$ a.u. is expected even for relativistic beams. From the transverse component, which has a resolution of $\sim 0.02$ a.u., the projectile-scattering angle can be inferred for close collisions, as mentioned above.

### 13.2.4 Electron, positron, and pair detectors

In relativistic atomic collision experiments, one may wish to determine the energy and angular distribution of electrons, positrons, or pairs. Gaseous

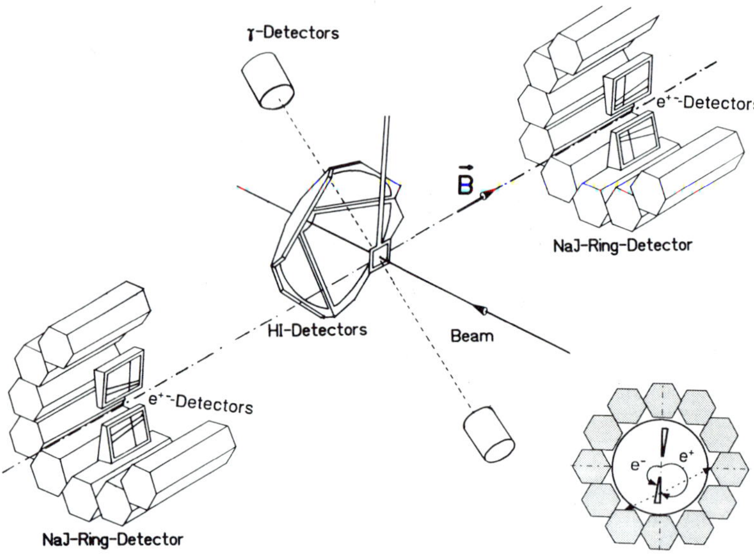


**Figure 13.6.** Solenoidal spectrometer for the detection of electron-positron pairs with the target near the center of the spectrometer. Electrons and positrons are detected at both ends of the spectrometer in semiconductor detectors to increase the efficiency. Separation is assured by the opposite spiraling of the leptons. Scattered ions are detected in parallel-plate avalanche counters marked "HI-Detectors." From [BoF91].

detectors and semiconductor detectors which are unable to stop these leptons are not particularly useful, except as focal-plane detectors for spectrographs. Scintillation detectors are ideal for electron detection, especially in conjunction with electrostatic or magnetic spectrometers. By assuring that the energy signal in the detectors matches the energy expected on the basis of the spectral selection, background signals due to photons or scattered electrons can be eliminated. Various spectrometers are described in [Sie68, Bro79, Tso83]. For the detection of electrons with energies exceeding $\sim$ 100 MeV, electron-photon shower counters are useful [Fer86, Kle86].

Efficient solenoidal spectrometers for the determination of the energy and angular distributions of electron-positron pairs have been constructed. The source of the pairs, i.e., the target, is placed either at the center of a magnetic solenoidal spectrometer or toward one end [BoF91,CoC92,BeG93]. Figures 13.6 and 13.7 provide sketches of typical designs. In the design of Fig. 13.6, segmented semiconductor detectors are placed at each end of

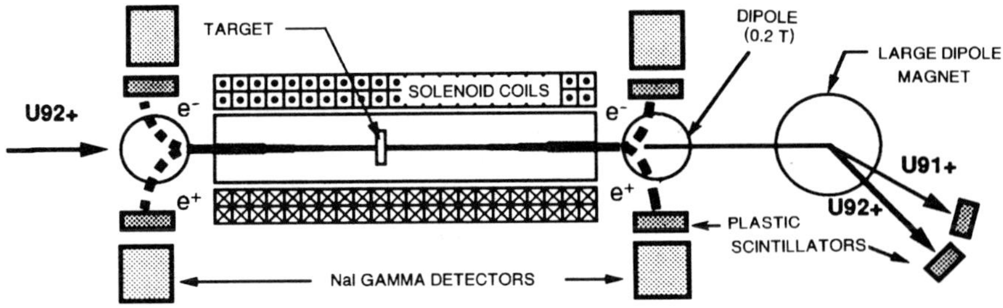


**Figure 13.7.** Advanced positron spectrometer. Electrons and positrons are efficiently guided to dipole magnets outside of the spectrometer by soft-iron endplates (not shown) provided with holes. There, they are separated and detected in plastic scintillators backed by sodium iodide detectors for the detection of annihilation radiation. Events are detected in coincidence with the ion charge state, determined by a large dipole magnet and by separate detectors for each charge state. From [BeG93].

the spectrometer for detection of electrons and positrons. To distinguish these particles, the respective semiconductor detectors face the oppositely directed spiral trajectories on either side of the solenoid axis. Also, they are surrounded by scintillation detectors in which signals from 0.51-MeV annihilation photons are put in coincidence with the positron-detector signals. For a given lepton energy, the flight time between the source and the semiconductor detector can be used to determine the length of the spiral trajectory, and hence the angle of inclination of the orbit with respect to the solenoid axis. By ray tracing, the emission angle from the source can then be obtained. In this device, the scattered ions are detected in position-sensitive avalanche detectors.

In the spectrometer of Fig. 13.7, the magnetic field is arranged so that at each end of the spectrometer, electrons and positrons are efficiently guided through a hole in a soft-iron endplate (not shown) into a dipole magnet. There, the leptons are separated and detected by scintillation detectors with annihilation radiation identification, if needed. Time of flight is used to determine the lepton emission angles from the target with respect to the beam axis. The events are detected in coincidence with the charge state of the ion, determined by magnetic separation and appropriate detection.

### 13.2.5 Photon detectors

In relativistic atomic collisions, one may wish to detect target or projectile x-rays, photons from radiative electron capture or primary bremsstrahlung, or secondary bremsstrahlung. The laboratory energy of radiative capture photons emitted at an angle $\theta$ with respect to the incident beam is given by Eq. (9.9) in terms of the projectile Lorentz factor $\gamma$. If $\gamma \gg 1$, the photon energy becomes

$$\hbar\omega \simeq \frac{m_e c^2}{1 - \beta \cos\theta}. \tag{13.2}$$

Typically, energies of these or other photons of interest will not exceed ~10 MeV. Hence, photon detectors normally used in nuclear physics experiments are appropriate, such as Si(Li) detectors for x-rays, with typical resolutions of 0.1 – 0.2 keV, and Ge(Li), NaI, or bismuth germanate detectors for MeV photons, with typical resolutions of one, tens, or hundreds of keV, respectively. These detectors are described in the references given at the beginning of Sec. 13.2.

## 13.3 Cross section measurements

This section discusses common features of atomic-collision cross section measurements. Subsequent sections refer to particular processes.

Atomic collision cross sections typically are obtained by detecting photons or leptons (line or continuum spectra) from the target, the projectile, or the collision system, by measuring the charge states of the projectile or target recoil ions, or by a combination of these methods. Absolute cross sections can be obtained by normalizing the detected photon or lepton yield to the integrated beam charge and target thickness, by normalizing to Rutherford scattered projectiles or recoil ions, or by counting individual projectiles. If cross sections are obtained from charge-changing methods, they are normalized to the total number of incident projectiles (which is very closely equal to the total number of emerging projectiles) and the target thickness. Thin and thick targets can be used for these measurements (Sec. 12.1.1).

### 13.3.1 Thick- versus thin-target method

In subsequent sections, only the thin-target method for cross section determination is discussed, in which targets are chosen thin enough so that the yield of detected photons, leptons, or charge-changed projectiles varies essentially linearly with the target thickness [Eqs. (12.1) and (12.3)]. Nevertheless, a few remarks about the thick-target method are in order.

In this method, yields are determined as a function of target thickness up to equilibrium thicknesses [Eq. (11.29)]. The yield equations, such as Eqs. (11.1) or (11.6) if charge change is measured, or the equivalent equations for the emitted radiations [AnM86], are integrated after an initial set of cross sections is substituted. Individual cross sections or groups of cross sections are multiplied by fitting factors which are varied until the best fit with the experimental data is obtained. The variation is guided by a least-squares fitting procedure such as that described by Cohen et al. [CoC57] or the program CURFIT of Bevington [Bev69]. It is necessary to start with the best available theoretical cross sections; otherwise, the fitting procedure may end up in an incorrect minimum in the multiparameter fitting space.

Application of this method to charge-state yields from relativistic heavy ions penetrating various targets between C and Au typically yields cross sections consistent to within factors 1.3 and 2. At nonrelativistic energies, cross sections with accuracies of 5 – 15% have been reported [BeH88]. In this work, charge fractions, characteristic K- and L-x-ray yields, and radiative electron capture yields were fitted simultaneously. The difference in accuracy between the relativistic and nonrelativistic cross section determinations can be traced simply to the beam time availability at the corresponding accelerators. At the former, typically 10 or fewer target thicknesses could be examined, at the latter 20 to 30 target thicknesses were used. Under these circumstances, the thin-target method is preferred. Even though for a given incident projectile charge only a few cross sections can be determined, accuracies of the order of 10% can be reached with five target thicknesses.

### 13.3.2 Measurement of photon cross sections

We describe here the experimental techniques which have been used to obtain cross sections by photon detection, and, next, by projectile charge change, with self-supporting and backed targets.

Figures 12.1, 12.2, and 13.2 sketch various arrangements suitable to cross section measurement by photon (or lepton) detection. We assume first that the photon spectrum is a continuum and that a differential cross section is to be obtained by normalization through beam charge integration (Fig. 13.8a). With an energy-sensitive detector, one measures the yield per projectile per unit photon energy centered around the photon energy $E_x$ and per unit photon solid angle centered around the photon emission angle

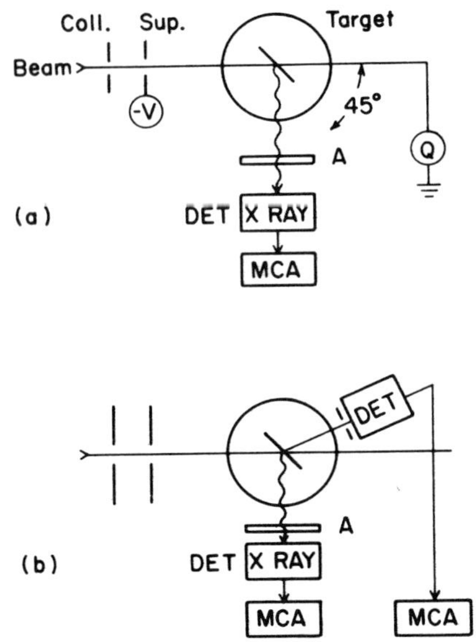


**Figure 13.8.** Schematic detector arrangements for photon (or electron) detection. (a) Normalization by beam current integration. (b) Normalization by Rutherford scattering. A third normalization method by projectile counting is sketched in Fig. 12.1. Abbreviations: Coll., beam collimator; Sup., electron suppressor (-$V$=-200 to -1000 V); DET X-RAY, photon detector; A, photon attenuator; DET, particle detector; MCA, multichannel analyzer or computer. From [Anh85b].

$\theta_{\rm x}$ (all quantities refer to the laboratory system):

$$\frac{d^2 Y_{\rm x}(E_{\rm x}, \theta_{\rm x})}{dE_{\rm x}\, d\Omega_{\rm x}} = \frac{\Delta C_{\rm x}(E_{\rm x}, \theta_{\rm x})}{\Delta E_{\rm x}\; N_{\rm P}\; \epsilon_{\rm x}(E_{\rm x})\; \Delta\Omega_{\rm x}} \,. \tag{13.3}$$

Here, $\Delta C_{\rm x}(E_{\rm x}, \theta_{\rm x})$ is the number of counts in the energy interval $\Delta E_{\rm x}$ around $E_{\rm x}$, corrected for background, deadtime effects, self-absorption in the target, and external attenuation, $N_{\rm P}$ is the total number of projectiles, and $\epsilon_{\rm x}(E_{\rm x})$ and $\Delta\Omega_{\rm x}$ are the efficiency and solid angle of the photon detector. Equation (13.3) assumes azimuthal symmetry about the beam direction. The cross section in the laboratory system then is given by

$$\frac{d^2 \sigma_{\rm x}(E_{\rm x}, \theta_{\rm x})}{dE_{\rm x}\, d\Omega_{\rm x}} = \frac{1}{x} \frac{d^2 Y_{\rm x}(E_{\rm x}, \theta_{\rm x})}{dE_{\rm x}\, d\Omega_{\rm x}} \,, \tag{13.4}$$

where $x$ is the target thickness in atoms per unit area [Eqs. (11.2) or (11.3)].

This equation is valid only if the target thickness is small enough so that the yield varies linearly with the thickness.

If line spectra such as characteristic x-rays are measured, the yield expressions are replaced by $dY_{\mathrm{x}}(\theta_{\mathrm{x}})/d\Omega_{\mathrm{x}}$ and the ratio $\Delta C_{\mathrm{x}}/\Delta E_{\mathrm{x}}$ is replaced by the number of counts in a line, $C_{\mathrm{x}}(\theta_{\mathrm{x}})$, corrected for background, deadtime, and absorption. Correspondingly, one obtains the differential cross section per unit solid angle $d\sigma_{\mathrm{x}}(\theta_{\mathrm{x}})/d\Omega_{\mathrm{x}}$. To obtain total cross sections, one has to take into account possible intrinsic anisotropic emission of the radiation, as well as any anisotropy induced by the Doppler effect (Secs. 2.3.2, 2.4.2). To correct for the latter, one has to know whether the radiation is emitted by the target, the projectile, or some other moving system, such as the center-of-charge system.

*Normalization by Rutherford scattering*

If it is possible to set a particle detector at sufficiently small forward angles $\theta_{\mathrm{P}}$ to assure pure Rutherford scattering of the projectiles (Fig. 13.8b), one can normalize the photon or lepton cross section to the Rutherford cross section. This avoids the need to determine the beam charge or target thickness, and even takes care of possible target thickness nonuniformity, as long as the thickest target spots still lie in the thin-target regime. For a continuum photon spectrum, the differential cross section is given by

$$\frac{d^2\sigma_{\mathrm{x}}(E_{\mathrm{x}},\theta_{\mathrm{x}})}{dE_{\mathrm{x}}\,d\Omega_{\mathrm{x}}} = \frac{d\sigma_{\mathrm{R}}(\theta_{\mathrm{P}})}{d\Omega_{\mathrm{P}}}\,\frac{\Delta C_{\mathrm{x}}(E_{\mathrm{x}},\theta_{\mathrm{x}})}{\Delta E_{\mathrm{x}}\;C_{\mathrm{P}}(\theta_{\mathrm{P}})}\,\frac{\Delta\Omega_{\mathrm{P}}}{\epsilon_{\mathrm{x}}(E_{\mathrm{x}})\;\Delta\Omega_{\mathrm{x}}}\,. \tag{13.5}$$

Here, $d\sigma_{\mathrm{R}}(\theta_{\mathrm{P}})/d\Omega_{\mathrm{P}}$ is the Rutherford cross section at the laboratory angle $\theta_{\mathrm{P}}$, $C_{\mathrm{P}}(\theta_{\mathrm{P}})$ is the deadtime-corrected number of counts in the elastic scattering peak of the particle spectrum, and $\Delta\Omega_{\mathrm{P}}$ is the solid angle of the particle detector (unit detection efficiency is assumed). For an accurate measurement of the Rutherford cross section, one needs to make measurements on both sides of the beam and possibly correct for energy loss in the target. For total cross section determinations, the same remarks apply as for the beam integration method.

*Projectile counting*

For relativistic heavy ions, cross sections are sometimes large enough so that the beam intensity can be reduced to a point at which projectile counting becomes practical. Referring to Fig. 12.1, if only those photons or leptons are counted which are in coincidence with the projectile detector signals, or in Figs. 12.2 and 13.2, if only photons in coincidence with charge-changed projectiles are counted, Eq. (13.4) applies. The ratio $\Delta C_{\mathrm{x}}/\Delta E_{\mathrm{x}}$ has to be interpreted as the background, deadtime, and absorption-corrected photon counts per unit photon energy interval, and $N_P$ as the number of projectiles counted by the particle detectors. The quantity $(\Delta C_x/\Delta E_x)/N_P$ is independent of the particle detection efficiency. The method has been applied successfully to determine projectile and target photon emission cross sections for 80 to 670-MeV/u Ne to U projectiles bombarding a range of targets between Be and U [AnS86].

In photon cross section determinations, particularly in solid targets, a careful evaluation of background effects is needed. Not only are there possible external effects, mentioned in Sec. 12.1.1, but secondary electron production in the target foil can generate electron-induced characteristic x-rays, as well as bremsstrahlung. The latter can be intense enough to produce additional target inner-shell vacancies by the photoelectric effect.

### 13.3.3 Measurement of charge-changing cross sections

In the cross section determination by projectile charge change with thin targets one integrates Eq. (11.6). With thin targets, excited-state effects are negligible and it is justified to use the ground-state model described in Sec. 11.2.2.

To illustrate the method, assume that a beam contains initially fractions $f_0, f_1, \text{and} f_2$ of zero-, one-, and two-electron ions. In the notation of Eq. (11.6),

$$f_0 = Y_0(0), \quad f_1 = Y_1(0), \quad f_2 = Y_2(0) . \tag{13.6}$$

In a normal situation, one of these fractions is unity, and the others are zero, but to discuss backed targets a more general treatment is useful. In order to avoid the cumbersome notation $\sigma(n, n')$ for the cross section for changing from $n$ to $n'$ projectile electrons in a single collision (Sec. 11.2.1), we introduce the notation $s_{\mathrm{a}}^n, d_{\mathrm{a}}^n$, and $t_{\mathrm{a}}^n$ for single-, double-, and triple-electron attachment and $s_{\mathrm{l}}^n$ and $d_{\mathrm{l}}^n$ for single- and double-electron stripping from a $n$-electron ion. We neglect processes of higher multiplicity. Then, the only cross sections occurring in the present illustration are as follows:

$$s_{\mathrm{a}}^0 \equiv \sigma(0,1), \quad d_{\mathrm{a}}^0 \equiv \sigma(0,2), \quad t_{\mathrm{a}}^0 \equiv \sigma(0,3) , \tag{13.7}$$

and similarly for $n = 1, 2$, and

$$\begin{aligned} s_{\mathrm{l}}^1 &\equiv \sigma(1,0) , \\ s_{\mathrm{l}}^2 &\equiv \sigma(2,1), \quad d_{\mathrm{l}}^2 \equiv \sigma(2,0) . \end{aligned} \tag{13.8}$$

Upon integrating Eq. (11.6) to first order in $x$, one finds [MeA85]

$$Y_0 = f_0 + [-\sigma_{\mathrm{tot}}(0)\, f_0 + s_{\mathrm{l}}^1\, f_1 + d_{\mathrm{l}}^2\, f_2]x + O(x^2) ,$$

$$
\begin{aligned}
Y_1 &= f_1 + [s_{\rm a}^0\, f_0 - \sigma_{\rm tot}(1)\, f_1 + s_{\rm l}^2\, f_2]x + O(x^2)\ , \\
Y_2 &= f_2 + [d_{\rm a}^0\, f_0 + s_{\rm a}^1\, f_1 - \sigma_{\rm tot}(2)\, f_2]x + O(x^2)\ , \\
Y_3 &= (t_{\rm a}^0\, f_0 + d_{\rm a}^1\, f_1 + s_{\rm a}^2\, f_2)x + O(x^2)\ , \\
Y_4 &= (t_{\rm a}^1\, f_1 + d_{\rm a}^2\, f_2)x + O(x^2)\ , \\
Y_5 &= t_{\rm a}^2\, f_2\, x + O(x^2)\ . \qquad (13.9)
\end{aligned}
$$

Here, $x$ is the target thickness in atoms per unit area, Eqs. (11.2) or (11.3). In the first three expressions, we have used the total charge-changing cross sections, Eq. (11.7),

$$
\begin{aligned}
\sigma_{\rm tot}(0) &= s_{\rm a}^0 + d_{\rm a}^0 + t_{\rm a}^0\ , \\
\sigma_{\rm tot}(1) &= s_{\rm a}^1 + d_{\rm a}^1 + t_{\rm a}^1 + s_{\rm l}^1\ , \\
\sigma_{\rm tot}(2) &= s_{\rm a}^2 + d_{\rm a}^2 + t_{\rm a}^2 + s_{\rm l}^2 + d_{\rm l}^2\ . \qquad (13.10)
\end{aligned}
$$

*Self-supporting targets*

With self-supporting targets and an entering beam in a single charge state, one can extract cross sections from Eq. (13.9) by plotting the experimental values of $Y_n/x$ against $x$, and extrapolating to $x = 0$. For example, if $f_1 = 1$ and $f_0 = f_2 = 0$, the intercept $(Y_0/x)_{x=0}$ yields $s_{\rm l}^1$ and the intercept $(Y_2/x)_{x=0}$ yields $s_{\rm a}^1$. Figure 13.9 illustrates this case for 200-MeV/u $\mathrm{Xe}^{53+}$+ Be collisions.

Unfortunately, self-supporting targets often are thick enough so that terms even cubic in $x$ contribute noticeably to the yield. In that case, one has to use the same integration method as for thick targets (Sec. 13.3.1). If the number of available target thicknesses is limited, one is forced to make assumptions about cross sections $\sigma(n, n')$ removed from the values $n = n_0, n' = n_0'$ of interest. Writing $n_0' = n_0 + m$, one can reduce the number of unknown cross sections by assumptions such as [Bet72]

$$
\sigma(n_0, n_0 + m) = g(n_0)\, h(m)\ , \qquad (13.11)
$$

where $g$ and $h$ are functions of $n_0$ and $m$ alone, respectively. Such relations may be valid for limited regions of $m$.

*Backed targets*

If self-supporting targets are too thick for a particular measurement, one can use backed targets. Carbon and aluminum foils form suitable backings (Sec. 12.2). It is advantageous to let the backing face the beam because then one can measure separately the charge fractions emerging from a pure

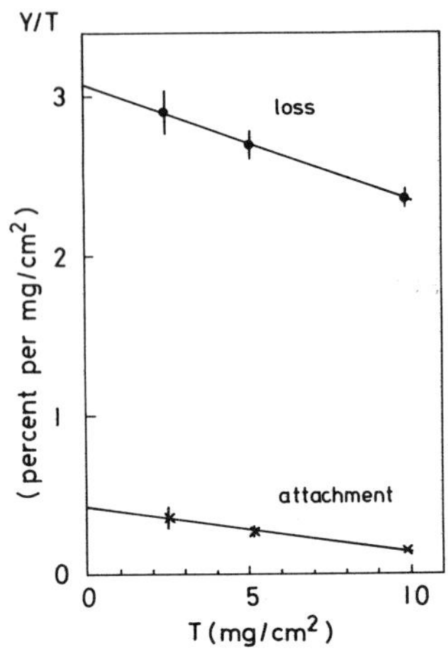


**Figure 13.9.** Dependence of the yield to target thickness ratio $Y_n/t_T$ for 200-MeV/u $Xe^{53+}$ (one electron) + Be collisions for $n = 0$ (one-electron loss) and $n = 2$ (one-electron attachment). Cross sections are obtained from the intercepts at $t_T = 0$. From [MeA85].

backing foil. These fractions can be used as initial values for the charge fractions entering the target material.

In the illustration of Eq. (13.9), assume that $f_0, f_1,$ and$f_2$ are the charge fractions emerging from a pure backing foil. One sees that even with these fractions known, 12 unknown cross sections occur in the yield equations, but only 6 intercepts $(Y_n/x)_{x=0}$, $(n = 0$ to $5)$ can be determined experimentally. Hence, one is forced once more to make assumptions similar to Eq. (13.11) for the cross sections of little interest, or one has to use theoretical estimates for these cross sections.

## 13.3.4 Cross sections for specific processes

In the following sections, the preceding general techniques of photon, lepton, and charge-changing cross section measurements are applied to specific processes, such as ionization, excitation, capture, and electron-positron pair emission.

*Ionization cross sections*

Inner-shell ionization cross sections at relativistic bombarding energies can be determined from target x-ray yields and from projectile charge change measurements. From a theoretical point of view, each method has advantages and disadvantages.

If target x-ray yields are measured, the ionizing agent can be a bare projectile, for which the basic theory of ionization is formulated (Chaps. 6 and 7). The method has disadvantages though. First, since the target typically is a multi-electron atom, screening effects on electron energies and wave functions must be considered, which can be treated only approximately (Sec. 5.6). Second, capture from inner target shells to the projectile can occur. This generates inner-shell target vacancies and subsequent x-rays which cannot be distinguished from the x-rays produced by ionization unless the x-ray measurement is made in coincidence with a particular projectile charge state.

The advantages of using charge change for (projectile) ionization measurements are that one can choose the number of electrons on the ion to be ionized and can even investigate hydrogenic ions which are ideal for theoretical studies. Furthermore, there is no disturbing influence of capture since all the target states are filled. The disadvantage of the method is that the ionizing agent is a neutral atom whose electrons also contribute to ionization. The resulting theoretical cross section is somewhat complex and is not fully derivable from first principles (Sec. 7.4).

Theoretically, it is of interest to investigate ionization cross sections for separate atomic shells. This can be done directly by measuring the corresponding target x-rays (K, L, or M) or indirectly by subtracting from a given projectile electron loss cross section the contribution from the other shells. Using as an example one-electron stripping [Eq. (11.4)], one can identify the cross section $\sigma(1,0)$ as the loss cross section $\sigma^1_{\mathrm{K}}$ per electron in the K-shell of the projectile. The cross section difference

$$\sigma(3,2) - \sigma(2,1) = \sigma^1_{\mathrm{L}_1} \tag{13.12}$$

can be identified with the one-electron loss cross section in the projectile $\mathrm{L}_1$−subshell. Similar relations can be established for higher shells and for multiple ionization, but must be used with care because screening effects on the subtracted electron shells may differ with the number of outer electrons.

*Capture cross sections*

Relativistic electron capture cross sections are most easily obtained from charge change measurements with bare or few-electron projectiles [MeA85]. With such measurements, though, nonradiative and radiative capture cannot be separated. The analysis of the data must rely on theoretical calculations, such as [Anh85a], see Figs. 8.9 and 8.10.

An important feature of the capture process is charge transfer to excited projectile states. In fact, it would be of theoretical interest to determine cross sections for transitions between given initial and final states. In principle, this could be done by measuring target x-rays from the vacancy generated in the initial shell in coincidence with the x-ray produced by capture into an excited projectile state, and to assure by a further coincidence that these x-rays are correlated with a projectile charge change. Such experiments have been performed with relativistic projectiles [StM93]. A cruder selection of final shells, but not initial shells, is possible if one uses the difference between electron attachment cross sections [Eq. (11.5)] obtained with different number of electrons on the projectile. For example, the cross section for capture into an empty K-shell is given by the difference

$$\sigma(0,1) - \sigma(2,3) = \sigma_{\mathrm{K}}^{\mathrm{c}} , \qquad (13.13)$$

where the notation of Eq. (11.5) has been used. An application of this equation is shown in Fig. 8.12. As for ionization, this type of relationship can be extended to higher projectile shells and to multiple electron capture, but again care must be taken to consider screening effects properly.

Radiative electron capture cross sections can be obtained directly from photon measurements by integrating the area under the REC peak. Equation (9.9) gives the centroid energy of the peak. The exact determination of the peak area is difficult because of the presence of primary bremsstrahlung. This radiation, which is the radiative electron capture to projectile continuum states, has a continuous spectrum with an endpoint given by Eq. (9.9) with $\epsilon_{\mathrm{f}} = 0$. Spreading of the REC peak due to the initial electron momentum distribution (Sec. 9.1) causes some overlap with the primary bremsstrahlung spectrum. With solid targets, there is an additional background of secondary bremsstrahlung. This is clearly brought out in Figs. 13.10 a and b which show the photon spectra from 82-MeV/u Xe+Be and 422-MeV/u U+U collisions. The short-dashed and dash-dotted curves are the calculated secondary and primary bremsstrahlung spectra, respectively, and the long-dashed curves are the K- and L-REC peaks. Primary and secondary bremsstrahlung produce continuum spectra under the K- and L-REC peaks, and Doppler-shifted projectile K x-rays can be seen (Sec. 2.4.1). For low-$Z$ targets, the secondary bremsstrahlung is so small that the spectrum should not differ from one taken with a gas target. For high-$Z$ targets, secondary bremsstrahlung dominates the spectrum.

With a gas target, very clean spectra can be obtained because the secondary processes are largely suppressed. Figure 13.11 shows a photon spectrum from 295-MeV/u $U^{92+}$ + Ar at a laboratory angle of 132°, obtained in coincidence with $U^{91+}$ ions. The REC peaks and the Lyman (K) x-ray peaks from capture into excited states of $U^{91+}$ are nearly background free.

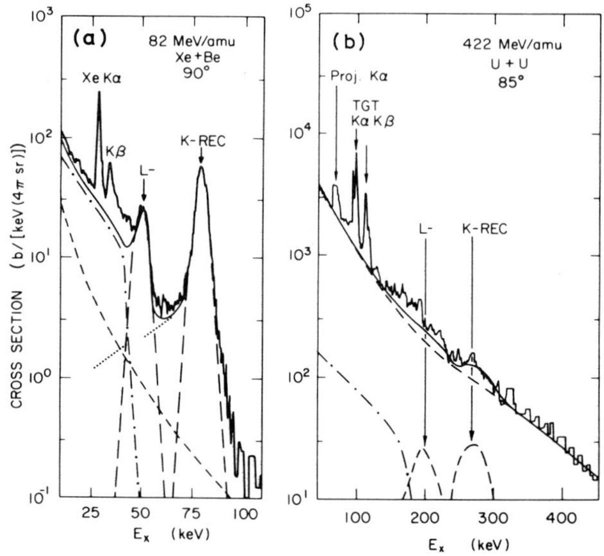


**Figure 13.10.** Photon spectra from a solid target. (a) Spectrum from 82-MeV/u U+Be collisions, at 90° to the beam. The photons are detected in an intrinsic Ge detector covered by a 3-g/cm$^2$ Be absorber to block out secondary electrons. The solid curve is the sum of all the computed curves, but fails to completely reproduce the experimental spectrum. (b) Same for 422-MeV/u U+U collisions at 85° to the beam. From [AnS86].

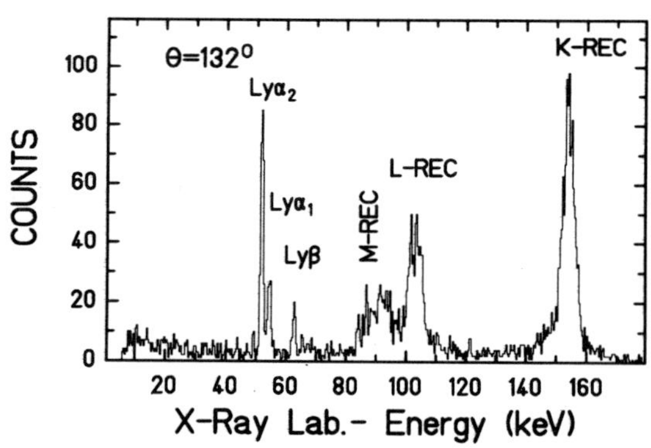


**Figure 13.11.** Photon spectrum from a gas target: 295-MeV/u $U^{92+}$+Ar collisions, at 132° to the beam. The photons are detected in an intrinsic Ge detector. The REC peaks and the Lyman x-ray peaks from capture into excited states of $U^{91+}$ can be seen. From [StM93].

*Excitation cross sections*

In principle, it is possible to study excitation cross sections by measuring the x-rays emitted by a projectile which has *not* undergone a charge change. For example, one could measure the K x-rays emitted in coincidence with a one-electron projectile which has kept its charge state upon emerging from the target. In the analysis, one would have to be aware that the emitted x-rays can be due not only to 1s→2p excitation, but also due to 1s→2s excitation followed by a collisional 2s→2p transition, for which the cross section can be very large [Anh85a]. The projectile x-ray lines in Figs. 13.10 and 13.11 are partly due to excitation.

*Electron, positron, and pair cross sections*

The study of electron continuum spectra is of great interest because it would test directly a differential cross section expression like (6.4). Such measurements are best undertaken with gas targets in order to avoid secondary electron emission from the target. For highly stripped ions, the Auger effect, which forms a background in nonrelativistic collisions, will be small. In principle, one could use any efficient electron spectrometer mentioned in Sec. 13.2.4. For projectile ionization, the differential cross section measurements in the laboratory system will have to be properly transformed into the projectile frame (Sec. 2.3).

The measurement of positrons and of electron-positron pairs emitted from relativistic collisions is and will be one of the most interesting research areas in the field (Chap. 10). Measurements of pair production cross sections from highly relativistic projectiles have been undertaken [VaD92]. The detection of positrons in coincidence with projectiles which have gained an electron indicates the occurrence of bound-free pair production (Sec. 10.3). This process has been detected and cross sections have been determined [BeG93]. As projectile energies approaching the TeV/u range become available, multiple-pair production will be copious [ViG93], and detection techniques of elementary particle physics will have to be adapted to atomic-physics experiments.

# Part IV

# Appendix

# Fundamental Constants and Units

## Physical constants

| Physical quantity | Symbol | Numerical value |
|---|---|---|
| speed of light in vacuum | $c$ | 299 792 458 m $s^{-1}$ |
| Planck's constant | $\hbar$ | 1.054 572 66 (63) $10^{-34}$ J s |
| | | 6.582 122 0 (20) $10^{-16}$ eV s |
| elementary charge | $e$ | 1.602 177 33 (49) $10^{-19}$ C |
| electron mass | $m_e$ | 9.109 389 7 (54) $10^{-31}$ kg |
| proton mass | $M_p$ | 1.672 623 1 (10) $10^{-27}$ kg |
| atomic mass unit | u | 1.660 540 2 (10) $10^{-27}$ kg |
| electron Compton wavelength | $\lambda\!\!\!^{-}_c$ | 3.861 593 23 (35) $10^{-13}$ m |
| proton Compton wavelength | $\lambda\!\!\!^{-}_p$ | 2.103 089 37 (19) $10^{-16}$ m |
| classical electron radius | $r_0$ | 2.817 940 92 (38) $10^{-15}$ m |
| Bohr radius | $a_0$ | 5.291 772 49 (24) $10^{-11}$ m |
| atomic mass unit in $m_e$ | $u/m_e$ | 1822. 888 5 |
| atomic mass unit in MeV | u/MeV | 931.494 32 (28) |
| inverse fine-structure constant | $1/\alpha$ | 137.035 989 5 (61) |
| Hartree energy | $E_h$ | 27.211 396 1 (81) eV |

From [CoT87]. The number in parentheses denotes the uncertainty in the last digits.

## Atomic units: $\hbar = 1$, $m_e = 1$, $e = 1$

| Physical quantity | Symbol | Numerical value |
|---|---|---|
| unit of length | $a_0 = \hbar^2/m_e e^2$ | 5.291 772 49 $10^4$ fm |
| unit of time | $\hbar^3/m_e e^4$ | 2.418 884 3 $10^{-17}$ s |
| unit of velocity | $e^2/\hbar = \alpha c$ | 2.187 691 4 $10^6$ m $s^{-1}$ |
| unit of energy | $e^2/a_0 = m_e e^4/\hbar^2$ | 27.211 396 1 eV |
| unit of electric charge | $e$ | 1.602 177 33 $10^{-19}$ C |
| unit of electric potential | $e/a_0$ | 27.211 396 1 V |
| unit of electric field | $e/a_0^2$ | 5.142 208 24 $10^{-4}$ V/fm |

## Natural (relativistic) units: $\hbar = 1$, $m_e = 1$, $c = 1$

| Physical quantity | Symbol | Numerical value |
|---|---|---|
| unit of length | $\lambda\!\!\!^{-}_c = \hbar/m_e c$ | 386.159 323 fm |
| unit of time | $\lambda\!\!\!^{-}_c/c = \hbar/m_e c^2$ | 1.288 088 52 $10^{-21}$ s |
| unit of velocity | $c$ | 2.997 924 58 $10^8$ m $s^{-1}$ |
| unit of energy | $m_e c^2$ | 0.510 999 06 MeV |
| unit of electric charge | $e$ | 1.602 177 33 $10^{-19}$ C |
| unit of electric potential | $e/\lambda\!\!\!^{-}_c = \alpha\, m_e c^2/e$ | 3.728 940 6 $10^3$ V |
| unit of electric field | $e/\lambda\!\!\!^{-2}_c$ | 9.656 482 0 V/fm |

# Graphs and Tables

## Graphs

# Tables

# Reference Books

M. Abramowitz and I.A. Stegun, *Handbook of Mathematical Functions* (Dover, New York, 1965).

K. Alder and A. Winther, *Electromagnetic Excitation. Theory of Coulomb Excitation with Heavy Ions* (North Holland, Amsterdam, 1975).

A. Barone, *Superconductive Particle Detectors* (World Scientific, Singapore, 1988).

H.A. Bethe and R. Jakiw, *Intermediate Quantum Mechanics* (W. A. Benjamin, New York, 1968).

H.A. Bethe and E.E. Salpeter, *Quantum Mechanics of One- and Two-Electron Atoms* (Academic, New York, 1957).

P. R. Bevington, *Data Reduction and Error Analysis for the Physical Sciences* (McGraw-Hill, New York, 1969).

J.D. Bjorken and S.D. Drell, *Relativistic Quantum Mechanics* (McGraw-Hill, New York, 1964).

B.H. Bransden and M.R.C. McDowell, *Charge Exchange and the Theory of Ion-Atom Collisions* (Clarendon Press, Oxford 1992).

D. A. Bromley, ed., *Detectors in Nuclear Science*, Nucl. Inst. Meth. **162**, 1 (1979).

E.R. Cohen, K.M. Crowe, and J.W.M. DuMond, *Fundamental Constants of Physics: Linear Least-Squares Fitting*, Chap. 7 (Interscience, New York, 1957).

E.U. Condon and G.H. Shortley, *The Theory of Atomic Spectra* (University Press, Cambridge, 1951).

S. Dushman, *Scientific Foundations of Vacuum Technique* (Wiley, New York, 1962).

A.R. Edmonds, *Angular Momentum in Quantum Mechanics* (Princeton University Press, Princeton, 1957).

A. Erdélyi, *Tables of Integral Transforms*, Vol. I (McGraw-Hill, New York, 1954).

R. C. Fernow, *Introduction to Experimental Particle Physics* (Cambridge University Press, Cambridge, 1986).

H. Goldstein, *Classical Mechanics*, 2nd ed. (Addison Wesley, Reading, 1981).

M.L. Goldberger and K.M. Watson, *Collision Theory* (Wiley, New York, 1964).

I.S. Gradshteyn and I.M. Ryzhik, *Table of Integrals, Series, and Products* (Academic, New York, 1980).

W. Heitler, *The Quantum Theory of Radiation* (Oxford, 1954).

J.D. Jackson, *Classical Electrodynamics*, 2nd ed. (Wiley, New York, 1975).

K. Kleinknecht, *Detection for Particle Radiation* (Cambridge University Press, Cambridge, 1986).

G. F. Knoll, *Radiation Detection and Measurement* (Wiley, New York, 1989).

W.R. Leo, *Techniques for Nuclear and Particle Physics Experiments* (Springer, Berlin, 1987).

J.B. Marion and F.C. Young, *Nuclear Reaction Analysis, Graphs and Tables* (North Holland, Amsterdam, 1968).

M.R.C. McDowell and J.P. Coleman, *Introduction to the Theory of Ion-Atom Collisions* (North-Holland, Amsterdam, 1970).

A. Messiah, *Quantum Mechanics* (Wiley, New York, 1962).

R.G. Newton, *Scattering Theory of Waves and Particles* (McGraw-Hill, New York, 1966).

M.E. Rose, *Elementary Theory of Angular Momentum* (Wiley, New York, 1957).

M.E. Rose, *Relativistic Electron Theory* (Wiley, New York, 1961).

J.J. Sakurai, *Advanced Quantum Mechanics* (Addison-Wesley, Reading, 1967).

S.S. Schweber, *An Introduction to Relativistic Quantum Field Theory* (Row, Peterson and Co., Evanston, 1961).

L.I. Schiff, *Quantum Mechanics* (McGraw-Hill, New York, 1968).

K. Siegbahn, *Alpha-, Beta-, and Gamma-Ray Spectroscopy*, Vol. 1 (North-Holland, Amsterdam, 1968).

J.C. Slater, *Quantum Theory of Atomic Structure*, Vol. 1 (McGraw-Hill, New York, 1960).

A. Sommerfeld, *Atombau und Spektrallinien*, Bd. I, II (Friedr. Vieweg & Sohn, Braunschweig, 1931, 1939).

N. Tsoulfanidis, *Measurement and Detection of Radiation* (McGraw-Hill, New York, 1983).

G. Waysand and C. Chardin, eds., *Superconducting and Low-Temperature Particle Detectors* (North-Holland, Amsterdam, 1989).

# List of Abbreviations

BEVALAC Bevatron-Linear Accelerator (at Lawrence Berkeley Laboratory, inoperative)

B1B Boundary-corrected Born approximation

CDW Continuum distorted wave approximation

CTMC Classical trajectory Monte-Carlo method

DWBA Distorted-wave Born approximation

ESR Experimental storage ring (at GSI, Darmstadt)

GTO Gaussian-type orbitals

LHC Large Hadron Collider (at CERN, Geneva)

MCDW Matrix continuum distorted-wave approximation

OBK Oppenheimer-Brinkman-Kramers approximation

OBK1 First-order Oppenheimer-Brinkman-Kramers approximation

OBK2 Second-order Oppenheimer-Brinkman-Kramers approximation

PWBA Plane-wave Born approximation

QED Quantum electrodynamics

REC Radiative electron capture

RHF Roothaan-Hartree-Fock wave function

RHIC Relativistic Heavy-Ion Collider (at Brookhaven National Laboratory)

RR Radiative recombination

SCA Semiclassical approximation

SE Symmetric eikonal approximation

SIS Heavy-ion storage ring (Schwerionen-Speichering at GSI, Darmstadt)

SPS Super Proton Synchrotron (at CERN, Geneva)

STO Slater-type orbital

TCDW Target continuum distorted wave approximation

# Bibliography


[AbS65] M. Abramowitz and I.A. Stegun, *Handbook of Mathematical Functions* (Dover, New York, 1965).

[Ahl80] S.P. Ahlen, Rev. Mod. Phys. **52**, 121 (1980).

[AlJ65] W.R. Alling and W.R. Johnson, Phys. Rev. **139**, A1050 (1965).

[All58] S.K. Allison, Rev. Mod. Phys. **30**, 1137 (1958).

[Als88] S. Alston, Phys. Rev. A **38**, 6092 (1988).

[AlW75] K. Alder and A. Winther, *Electromagnetic Excitation. Theory of Coulomb Excitation with Heavy Ions* (North-Holland, Amsterdam, 1975)

[AmA81] P.A. Amundsen and K. Aashamar, J. Phys. B **14**, 4047 (1981).

[Amu78] P.A. Amundsen, J. Phys. B. **11**, 3197 (1978).

[AnA84] R. Anholt, S.A. Adriamonje, E. Morenzoni, Ch. Stoller, J.D. Molitoris, W.E. Meyerhof, H. Bowman, J.-S. Xu, Z.-Z. Xu, J.O. Rasmussen, and D.H.H. Hoffmann, Phys. Rev. Lett. **53**, 234 (1984).

[AnB87] R. Anholt and U. Becker, Phys. Rev. A **36**, 4628 (1987).

[AnE85] R. Anholt and J. Eichler, Phys. Rev. A **31**, 3505 (1985).

[AnG86] R. Anholt and H. Gould, in *Advances in Atomic and Molecular Physics*, Vol. **22**, p. 315 (Academic, New York, 1986).

[Anh79] R. Anholt, Phys. Rev. A **19**, 1004 (1979).

[Anh85a] R. Anholt, Phys. Rev. A **31**, 3579 (1985).

[Anh85b] R. Anholt, Rev. Mod. Phys. **57**, 995 (1985).

[Anh86] R. Anholt, Phys. Lett. **114A**, 126 (1986).

[AnM84a] R. Anholt, W.E. Meyerhof, Ch. Stoller, E. Morenzoni, S.A. Adriamonje, J.D. Molitoris, D.H.H. Hoffmann, H. Bowman, J.-S. Xu, Z.-Z. Xu, and J.O. Rasmussen, Phys. Rev. Lett. **53**, 234 (1984).

[AnM84b] R. Anholt, W.E. Meyerhof, Ch. Stoller, E. Morenzoni, S.A. Adriamonje, J.D. Molitoris, O.K. Baker, D.H.H. Hoffmann, H. Bowman, J.-S. Xu, Z.-Z. Xu, K. Frankel, D. Murphy, K. Crowe, and J.O. Rasmussen, Phys. Rev. A **30**, 2234 (1984).

[AnM86] R. Anholt and W.E. Meyerhof, Phys. Rev. A **33**, 1556 (1986).

[AnM87] R. Anholt, W.E. Meyerhof, X.-Y. Xu, H. Gould, B. Feinberg, R.J. McDonald, H.E. Wegner, and P. Thieberger, Phys. Rev. A **36**, 1586 (1987).

[AnS86] R. Anholt, Ch. Stoller, J.D. Molitoris, D.W. Spooner, E. Morenzoni, S.A. Adriamonje, W.E. Meyerhof, H. Bowman, J.-S. Xu, Z.-Z. Xu, J.O. Rasmussen, and D.H.H. Hoffmann, Phys. Rev A **33**, 2270 (1986).

[BaB92] N. Baron and G. Baur, Phys. Rev. D **46**, R3695 (1992).

[BaB93] G. Baur and N. Baron, Nucl. Phys. **A561**, 628 (1993).

[BaC72] W. Bambynek, B. Crasemann, R.W. Fink, H.-U. Freund, H. Mark, C.D. Swift, R. E. Price, and P. Venugopala, Rev. Mod. Phys. **44**, 716 (1972), err. **46**, 853 (1974).

[BaG53] D.R. Bates and G. Griffing, Proc. Phys. Soc. (London) A **66**, 96 (1953).

[BaG54] D.R. Bates and G. Griffing, Proc. Phys. Soc. (London) A **67**, 663 (1954).

[BaG55] D.R. Bates and G. Griffing, Proc. Phys. Soc. (London) A **68**, 90 (1955).

[BaH59] J. Bang and J.M. Hansteen, Mat. Fys. Medd. Vid. Selsk. **31**, No. 13 (1959).

[Bar79] E. Baron, IEEE Trans. Nucl. Sci. **NS-26**, 2411 (1979).

[Bar88] A. Barone, *Superconductive Particle Detectors* (World Scientific, Singapore, 1988).

[BaR91] A.J. Baltz, M.J. Rhoades-Brown, and J. Weneser, Phys. Rev. A **44**, 5569 (1991).

[BaR93a] A.J. Baltz, M.J. Rhoades-Brown, and J. Weneser, Phys. Rev. A **48**, 2002 (1993).

[BaR93b] A.J. Baltz, M.J. Rhoades-Brown, and J. Weneser, Phys. Rev. A **47**, 3444 (1993).

[BaR94] A.J. Baltz, M.J. Rhoades-Brown, and J. Weneser, Phys. Rev. A **50**, 4842 (1994).

[BaS85] H.J. Bär and G. Soff, Physica **128C**, 225 (1985).

[Bau90a] G. Baur, Physica Scripta **T32**, 76 (1990).

[Bau90b] G. Baur, Phys. Rev. A **42**, 5736 (1990).

[Bau90c] G. Baur, Phys. Rev. D **41**, 3535 (1990).

[Bau91] G. Baur, Phys. Rev. A **44**, 4767 (1991).

[Bau92a] G. Baur, Z. Phys. C **54**, 419 (1992).

[BeB88] C.A. Bertulani and G. Baur, Phys. Rep. **163**, 299 (1988).

[Bec86] U. Becker, Ph.D. Thesis, Giessen (1986), see in *Physics of Strong Fields*, ed. W. Greiner, NATO Advanced Study Institute Series B: Physics, Vol. **153**, p. 609 (Plenum, New York, 1987).

[Bec87] U. Becker, J. Phys. B **20**, 6563 (1987).

[BeF84] R.L. Becker, A. L. Ford, and J.F. Reading, Phys. Rev. A **29**, 3111 (1984).

[BeG70] H.D. Betz and L. Grodzins, Phys. Rev. Lett. **25**, 903 (1970).

[BeG83] U. Becker, N. Grün, and W. Scheid, J. Phys. B **16**, 1967 (1983).

[BeG85] U. Becker, N. Grün, and W. Scheid, J. Phys. B **18**, 4589 (1985).

[BeG86a] U. Becker, N. Grün, and W. Scheid, J. Phys. B **19**, 1347 (1986).

[BeG86b] U. Becker, N. Grün, W. Scheid, and G. Soff, Phys. Rev. Lett. **56**, 2016 (1986).

[BeG87a] U. Becker, N. Grün, and W. Scheid, J. Phys. B **20**, 2075 (1987).

[BeG87b] U. Becker, N. Grün, K. Momberger, and W. Scheid, in *Physics of Strong Fields*, ed. W. Greiner, NATO Advanced Study Institute Series B: Physics, Vol. **153**, p. 609 (Plenum, New York, 1987).

[BeG92] Ch. Best, W. Greiner, and G. Soff, Phys. Rev. A **46**, 261 (1992).

[BeG93] A. Belkacem, H. Gould, B. Feinberg, R. Bossingham, and W.E. Meyerhof, Phys. Rev. Lett. **71**, 1514 (1993).

[BeG94] A. Belkacem, H. Gould, B. Feinberg, R. Bossigham, and W.E. Meyerhof, Phys. Rev. Lett. **73**, 2432 (1994).

[BeH88] H.D. Betz, R. Höppler, R. Schramm, and W. Oswald, Nucl. Inst. Meth. Res. B **33**, 185 (1988).

[BeJ68] H.A. Bethe and R. Jakiw, *Intermediate Quantum Mechanics*, Chap. 11 (W. A. Benjamin, New York, 1968).

[BeK78] O. Benka and A. Kropf, Ato. Data Nucl. Data Tables **22**, 219 (1978).

[BeL82] V.B. Berestetskii, E.M. Lifshitz, and L.P. Pitaevskii, *Quantum Electrodynamics* (Pergamon, Oxford, 1982).

[BeM54] H.A. Bethe and L. Maximon, Phys. Rev. **93**, 768 (1954).

[BeS57] H.A. Bethe and E.E. Salpeter, *Quantum Mechanics of One- and Two-Electron Atoms* (Academic, New York, 1957).

[Bet30] H.A. Bethe, Ann. Phys. (Leipzig) **5**, 325 (1930).

[Bet72] H.D. Betz, Rev. Mod. Phys. **44**, 465 (1972).

[Bet83] H.D. Betz, in *Applied Atomic Collision Physics*, Vol. 4, ed. S. Datz, p. 2 (Academic, New York, 1983).

[Bev69] P. R. Bevington, *Data Reduction and Error Analysis for the Physical Sciences* (McGraw-Hill, New York, 1969).

[Bha35] H.J. Bhabha, Proc. R. Soc. London Ser. A **152**, 559 (1935).

[Bie68] J.P. Biersack, Z. Phys. **211**, 495 (1968).

[BiM75] F. Biggs, L.B. Mendelsohn, and J.B. Mann, At. Data Nucl. Data Tables **16**, 201 (1975).

[BjD64] J.D. Bjorken and S.D. Drell, *Relativistic Quantum Mechanics* (McGraw-Hill, New York, 1964).

[BoF89] H. Bokemeyer and H. Folger, Nucl. Inst. Meth. Res. A **282**, 40 (1989).

[BoF91] H. Bokemeyer, H. Folger, K. Bethge, H. Kaper, D. Kraft, W. Schön, K. Sakaguchi, K. E. Stiebing, P. Thee, H. Backe, M. Samek, D. Schwalm, A. Balanda, and P. Salabura, *GSI Annual Report 1990*, Gesellschaft für Schwerionenforschung Report GSI-91-1, p. 162 (1991).

[BoG82] D.J. Botto and M. Gavrila, Phys. Rev. A **26**, 237 (1982).

[Boh48] N. Bohr, Kgl. Dan. Vid. Selsk. Mat.-Fys. Medd. **18**, No. 8 (1948).

[BoK75] C. J. Borkowski and N. M. Kepp, Rev. Sci. Inst. **46**, 951 (1975).

[BoL54] N. Bohr and J. Lindhard, Kgl. Danske Videnskab. Selskab, Mat.-Fys. Medd. **28**, No. 7 (1954).

[BoS87] C. Bottcher and M.R. Strayer, Ann. Phys. (N.Y.) **175**, 64 (1987).

[BoS89] C. Bottcher and M.R. Strayer, Phys. Rev. D **39**, 1330 (1989).

[BoS92] C. Bottcher and M.R. Strayer, Nucl. Phys. A **544**, 255c (1992).

[Bot84] C. Bottcher, in *Physics of Electronic and Atomic Collisions*, eds. J. Eichler et al., p.187 (North-Holland, Amsterdam, 1984).

[BrD74] J.S. Briggs and K. Dettmann, Phys. Rev. Lett. **33**, 1123 (1974).

[BrK30] H.C. Brinkman and H.A. Kramers, Proc. Acad. Sci. Amsterdam **33**, 973 (1930).

[BrM92] B.H. Bransden and M.R.C. McDowell, *Charge Exchange and the Theory of Ion-Atom Collisions* (Clarendon Press, Oxford 1992).

[Bro79] D. A. Bromley, ed., *Detectors in Nuclear Science*, Nucl. Inst. Meth. **162**, 1 (1979).

[BrP86] U. G. Briel and E. Pfefferman, Nucl. Inst. Meth. Phys. Res. A **242**, 376 (1986).

[BrR51] G.E. Brown and D.G. Ravenhall, Proc. R. Soc. London, Ser. A **208**, 552 (1951).

[BrT78] J.S. Briggs and K. Taulbjerg, in *Structure and Collisions of Ions and Atoms*, ed. I. A. Sellin, p. 105 (Springer, Berlin, 1978).

[BuG75] V.M. Budnev, I.F. Ginzburg, G.V. Meledin, and V.G. Serbo, Phys. Rep. **15**, 183 (1975).

[Che64] I.M. Cheshire, Proc. Phys. Soc. (London) **84**, 89 (1964).

[ChE79a] F.T. Chan and J. Eichler, Phys. Rev. Lett. **42**, 58 (1979).

[ChE79b] F.T. Chan and J. Eichler, Phys. Rev. A **20**, 1841 (1979).

[ChE89] V.A. Chechin and V.K. Ermilova, Z. Phys. D **13**, 33 (1989).

[ChK66] I.M. Cheshire and H.L. Kyle, Proc. Phys. Soc. (London) **88**, 785 (1966).

[ChK79] K.T. Cheng, Y.–K. Kim, and J.P. Desclaux, Ato. Data Nucl. Data Tables **24**, 111 (1979).

[ChM73] B.H. Choi, E. Merzbacher, and G.S. Khandelwal, At. Data **5**, 291 (1973).

[ClR74] E. Clementi and C. Roetti, At. Data Nucl. Data Tables **14**, 177 (1974).

[CoC57] E.R. Cohen, K.M. Crowe, and J.W.M. DuMond, *Fundamental Constants of Physics: Linear Least-Squares Fitting*, Chap. 7 (Interscience, New York, 1957).

[CoC92] T.E. Cowan, M.W. Clark, J.H. Hartley, R.H. Howell, J.L. McDonald, J.W. McDonald, R.R. Rohatgi, D.H.G. Schneider, R.L. Watson, and W.H. Trzaska, Bull. Am. Phys. Soc. **37**, 1262 (1992).

[CoS51] E.U. Condon and G.H. Shortley, *The Theory of Atomic Spectra* (University Press, Cambridge, 1951).

[CoT87] E.R. Cohen and B.N. Taylor, Rev. Mod. Phys. **59**, 1121 (1987).

[Cra79] H.J. Crawford, Ph.D. thesis, University of California, 1979, available as Lawrence Berkeley Laboratory Report No. LBL-8807(1979).

[CrF70] A. Crispin and G.N. Fowler, Rev. Mod. Phys. **42**, 290 (1970).

[CrG83] H. Crawford, H. Gould, D. Greiner, P. Lindstrom, and T. J. M. Symons, in *Proc. 6th High-Energy Heavy-Ion Study, 2nd Workshop on Anomalons*, Lawrence Berkeley Laboratory Report LBL-16281 (1983), and Lawrence Berkeley Laboratory Report LBL-16241 (1983).

[DaB54] H. Davies, H.A. Bethe, and L.C. Maximon, Phys. Rev. **93**, 788 (1954).

[DaM78] D.M. Davidovic, B.L. Moiseiwitsch, and P.H. Norrington, J. Phys. B **11**, 847 (1978).

[Dar28] C.G. Darwin, Proc. Roy. Soc. London, Ser. A **118**, 654 (1928).

[Dec90] F. Decker, Ph.D. Thesis, Free University of Berlin, 1990; see also Phys. Rev. A **41**, 6552 (1990).

[Dec91] F. Decker, Phys. Rev. A **44**, 2883 (1991).

[Ded62] K.G. Dedrick, Revs. Mod. Phys. **34**, 429 (1962).

[DeE85] D.P. Dewangan and J. Eichler, J. Phys. B **18**, L65 (1985).

[DeE86] D.P. Dewangan and J. Eichler, J. Phys. B **19**, 2939 (1986).

[DeE91a] F. Decker and J. Eichler, Phys. Rev. A **44**, 377 (1991).

[JoB79] D.E. Johnson, G. Basbas, and F. McDaniel, At. Data Nucl. Data Tables **24**, 1 (1979).

[Kaw86] S. Kawabata, Comp. Phys. Comm. **41**, 127 (1986).

[KeU86] S. Kelbch, J. Ullrich, W. Rauch, H. Schmidt-Böcking, M. Horbatsch, R.M. Dreizler, S. Hagmann, R. Anholt, A.S. Schlachter, A. Müller, P. Richard, Ch. Stoller, C.L. Cocke, R. Mann, W.E. Meyerhof, and J.D. Rasmussen, J. Phys. B **19**, L47 (1986).

[KhC69] G.S. Khandelwal, B.H. Choi, and E. Merzbacher, At. Data **1**, 103 (1969).

[KiK73] P. Kienle, M. Kleber, B. Povh, R.M. Diamond, F.S. Stephens, E. Grosse, M.R. Maier, and D. Proetel, Phys. Rev. Lett. **31**, 1099 (1973).

[Kle86] K. Kleinknecht, *Detection for Particle Radiation* (Cambridge University Press, Cambridge, 1986).

[KlJ75] M. Kleber and D.H. Jakubassa, Nucl. Phys. A **252**, 152 (1975).

[Kno89] G. F. Knoll, *Radiation Detection and Measurement* (Wiley, New York, 1989).

[Kol75] H. Kolbenstvedt, J. App. Phys. **46**, 2771 (1975).

[LaL34] L.D. Landau and E.M. Lifshitz, Phys. Zs. Sowjetunion **6**, 244 (1934).

[LeB88] A. Lennon, K.L. Bell, H.B. Gilbody, J.G. Hughes, A.E. Kingston, M.J. Murray, and J.F. Smith, J. Phys. Chem. Ref. Data **17**, 1285 (1988).

[Lee77] J.S. Lee, J. Chem. Phys. **66**, 4906 (1977).

[Leo87] W.R. Leo, *Techniques for Nuclear and Particle Physics Experiments* (Springer, Berlin, 1987).

[Lep87] P. Lepage, Comput. Phys. Commun. **27**, 192 (1987).

[Lev51] M. Lévy. Proc. Roy. Soc. Lond. Ser. A **204**, 145 (1950).

[LeZ92] D.H. Lee, T.J.M. Zuoros, J.M. Sanders, P. Richard, J.M. Anthony, J.D. Wang, and J.H. McGuire, Phys. Rev. A **46**, 1374 (1992).

[LuH88] D. H. Lumb and A. D. Holland, Nucl. Inst. Meth. Phys. Res. A **273**, 696 (1988).

[MaF90] R. Matzdorf, B. Fricke, and G. Soff, Z. Phys. D **12**, 233(1990)

[Mag54] W. Magnus, Commun. Pure Appl. Math. **7**, 649 (1954).

[MaN79] D. Marx, F. Nickel, G. Münzenberg, K. Güttner, H. Ewald, W. Faust, S. Hofmann, H.J. Schött, and W. Thalheimer, Nucl. Inst. Meth. Res. **163**, 15 (1979).

[MaR84] J.M. Maidagan and R.D. Rivarola, J. Phys. B **17**, 2477 (1984).

[MaS87] R. Matzdorf, G. Soff, and G. Mehler, Z. Physik D **6**, 5 (1987).

[MaY68] J.B. Marion and F.C. Young, *Nuclear Reaction Analysis, Graphs and Tables*, p. 71 (North-Holland, Amsterdam, 1968).

[McC70] M.R.C. McDowell and J.P. Coleman, *Introduction to the Theory of Ion-Atom Collisions* (North-Holland, Amsterdam, 1970).

[Mcc85] J.F. McCann, J.Phys. B **18**, L569 (1985).

[McG77] J. McEnnan and M. Gavrila, Phys. Rev. A **15**, 1537 (1977).

[Mcg82] J.M. McGuire, Phys. Rev. A **26**, 143 (1982).

[McS68] R. McCarroll and A. Salin, J.Phys. B **1**, 163 (1968).

[McS81] J.H. McGuire, N. Stolterfoht, and P.R. Simony, Phys. Rev. A **24**, 97 (1981).

[MeA85] W.E. Meyerhof, R. Anholt, J. Eichler, H. Gould, Ch. Munger, J. Alonso, P. Thieberger, and H.E. Wegner, Phys. Rev. A **32**, 3291 (1985).

[MeA87a] W.E. Meyerhof, R. Anholt, X.-Y. Xu, H. Gould, B. Feinberg, R.J. McDonald, H.E. Wegner, and P. Thieberger, Nucl. Inst. Meth. Res. A **262**, 10 (1987).

[MeA87b] W.E. Meyerhof, R. Anholt, and X.-Y. Xu, Phys. Rev. A **35**, 1055 (1987).

[MeD94] V. Mergel, R. Dörner, J. Ullrich, O. Jagutzki, S. Lencinas, S. Nüttgens, L. Spielberger, M. Unverzagt, C.L. Cocke, R.E. Olson, M. Schultz, U. Buck, and H. Schmidt-Böcking, to be published.

[MeH91] W.E. Meyerhof, H.-P. Hülskötter, Qing Dai, and J. H. McGuire, Phys. Rev. A **43**, 5907 (1991).

[MeL58] E. Merzbacher and H.W. Lewis, in *Handbuch der Physik*, Vol. XXXIV, ed. S. Flügge, p. 166 (Springer, Berlin, 1958).

[MeR85] G. Mehler, T. deReus, U. Müller, J. Reinhardt, B. Müller, W. Greiner, and G. Soff, Nucl. Instr. Meth. A **20**, 559 (1985).

[Mes62] A. Messiah, *Quantum Mechanics* (Wiley, New York, 1962).

[MeS89] G. Mehler, G. Soff, K. Rumrich, and W. Greiner, Z. Phys. D **13**, 193 (1989).

[Mit64] M.H. Mittleman, Proc. Phys. Soc. (London) **84**, 453 (1964).

[MoG80] C.E. Moores, L.B. Golden, and D.H. Sampson, J. Phys. B **13**, 385 (1980).

[MoG87] K. Momberger, N. Grün, W. Scheid, U. Becker, and G. Soff, J. Phys. B **20**, 281 (1987).

[MoG89] K. Momberger, N. Grün, W. Scheid, and U. Becker, J. Phys. B **22**, 3269 (1989).

[MoG90] K. Momberger, N. Grün, W. Scheid, and U. Becker, J. Phys. B **23**, 2293 (1990).

[MoG93] K. Momberger, N. Grün, and W. Scheid, J. Phys. B **26**, 1851 (1993).

[Moi85] B.L. Moiseiwitsch, Phys. Reports, **118**, 135 (1985).

[Moi86] B.L. Moiseiwitsch, J. Phys. B **19**, 3733 (1986).

[Moi87] B.L. Moiseiwitsch, J. Phys. B **20**, L171 (1987).

[Moi88] B.L. Moiseiwitsch, J. Phys. B **21**, 603 (1988).

[Moi89] B.L. Moiseiwitsch, Phys. Rev. A **39**, 5609 (1989).

[Moi92] B.L. Moiseiwitsch, J. Phys. B **25**, 3015 (1992).

[Mok94] P.H. Mokler, Phys. Bl. **50**, 145 (1994) [in German].

[Mol32] C. Møller, Ann. Phys. (Leipzig) **14**, 531 (1932).

[Mol47] G. Molière, Z. Naturforsch. **2a**, 133 (1947).

[MoM91a] E.C. Montenegro and W.E. Meyerhof, Phys. Rev. A **43**, 2289 (1991).

[MoM91b] E.C. Montenegro and W.E. Meyerhof, Phys. Rev. A **44**, 7229 (1991).

[MoM92a] E.C. Montenegro and W.E. Meyerhof, Phys. Rev. A **46**, 5506 (1992).

[MoM92b] E.C. Montenegro, W.S. Melo, W.E. Meyerhof, and A.G. de Pinho, Phys. Rev. Lett. **69**, 3033 (1992).

[MoM94] E.C. Montenegro, W.E. Meyerhof, and J.H. Mcguire, Adv. Ato. Mol. Opt. Phys. **34**, 249 (1994).

[Mos66] S.A. Moszkowski, in *Alpha, Beta and Gamma Ray Spectroscopy*, ed. K. Siegbahn, Vol. 2, p.863f (North Holland, Amsterdam, 1966).

[MoS79] B.L. Moiseiwitsch and S.G. Stockman, J. Phys. B **12**, L695 (1979).

[MoS80] B.L. Moiseiwitsch and S.G. Stockman, J. Phys. B **13**, 2975 (1980).

[MuG76] B. Müller and W. Greiner, Z. Naturforsch. **31a**, 1 (1976).

[MuR73] B. Müller, J. Rafelski, and W. Greiner, Nuov. Cim. **18A**, 551 (1973).

[Nag60] B.C.H. Nagel, Ark. Fysik **18**, 1 (1960).

[NeW49] T.D. Newton and E.P. Wigner, Rev. Mod. Phys. **21**, 400 (1949).

[New66] R.G. Newton, *Scattering Theory of Waves and Particles* (McGraw-Hill, New York, 1966).

[NiC90] P. Nicolai, M. Chabot, J.P. Rozet, M.F. Politis, A. Chetoui, C. Stéphan, A. Touati, D. Vernhet, and K. Wohrer, J. Phys. B **23**, 3609 (1990).

[NiP82] A.L. Nikishov and N.V. Pichkurov, Sov. J. Nucl. Phys. **35**, 561 (1982).

[NiT35] Y. Nishina, S. Tomonaga, and M. Kobayashi, Sci. Pap. Inst. Phys. Chem. Res. **27**, 137 (1935).

[Opp28] J.R. Oppenheimer, Phys. Rev. **31**, 349 (1928).

[PeL66] P. Pechukas and J.C. Light, J. Chem. Phys. **44**, 3897 (1966).

[PrL64a] R.H. Pratt, R.D. Levee, R.L. Pexton, and W. Aron, Phys. Rev. **134**, A898 (1964).

[PrL64b] R.H. Pratt, R.D. Levee, R.L. Pexton, and W. Aron, Phys. Rev. **134**, A916 (1964).

[PrR73] R.H. Pratt, A. Ron, and H.K. Tseng, Revs. Mod. Phys. **45**, 273 (1973).

[Rac37] G. Racah, Nuovo Cimento **14**, 93 (1937).

[RaY71] G. Raisbeck and F. Yiou, Phys. Rev. Lett. **4**, 1858 (1971).

[RhB89] M.J. Rhoades-Brown, C. Bottcher, and M.R. Strayer, Phys. Rev. A **40**, 2831 (1989).

[RhW91] M.J. Rhoades-Brown and J. Weneser, Phys. Rev. A **44**, 330 (1991).

[RiB77] R. Rice, G. Basbas, and F. McDaniel, At. Data Nucl. Data Tables **20**, 503 (1977).

[RoG94] A. Ron, I.B. Goldberg, J. Stein, S.T. Manson, R.H. Pratt, and R.Y. Yin, Phys. Rev. A **50**, 1312 (1994).

[Roh79] F. Rohrlich, Ann. Phys. **117**, 292 (1979).

[Ros57] M.E. Rose, *Elementary Theory of Angular Momentum* (Wiley, New York, 1957).

[Ros61] M.E. Rose, *Relativistic Electron Theory* (Wiley, New York, 1961).

[Rub48] A. Rubinowitz, Phys. Rev. **73**, 1330 (1948).

[RuG89] K. Rumrich, W. Greiner, G. Soff, K.H. Wietschorke, and P. Schlüter, J. Phys B **22**, 165 (1989).

[RuM91] K. Rumrich, K. Momberger, G. Soff, W. Greiner, N. Grün, and W. Scheid, Phys. Rev. Lett. **66**, 2613 (1991).

[RuS93] K. Rumrich, G. Soff, and W. Greiner, Phys. Rev. A **47**, 215 (1993).

[RyS93] P. Rymuza, Th. Stöhlker, C.L. Cocke, H. Geissel, C. Kozhuharov, P.H. Mokler, R. Moshammer, F. Nickel, C. Scheidenberger, Z. Stachura, J. Ullrich, and A. Warczak, J. Phys. B **26**, L169 (1993).

[RyW78] H. Ryufuku and T. Watanabe, Phys. Rev. A **18**, 2005 (1978).

[SaA51] R.G. Sachs and N. Austern, Phys. Rev. **81**, 705 (1951).

[Sak67] J.J. Sakurai, *Advanced Quantum Mechanics* (Addison Wesley, Reading, 1967).

[SaM87] F. Salvat, J.D. Martínez, R. Mayol, and J. Parellada, Phys. Rev. A **36**, 467 (1987).

[Sau31] F. Sauter, Ann. Physik **9**, 217 (1931); ibid. **11**, 454 (1931).

[SaW55] F. Sauter and H.O. Wüster, Z. Physik **141**, 83 (1955).

[ScB72] H.W. Schnopper, H. Betz, J.P. Devaille, K. Kalata, A.R. Sohval, K.W. Jones, and H.E. Wegner, Phys. Rev. Lett. **29**, 898 (1972).

[ScB88] G. Scoles, D. Bassi, U. Buck, and D. Lainé, eds., *Atomic and Molecular Beam Methods,* Vol. 1, p. 14 (Oxford University Press, Oxford, 1988).

[ScB91] J. Schweppe, A. Belkacem, L Blumenfeld, N. Clayton, B. Feinberg, H. Gould, V.E. Kostroun, L. Levy, S. Misawa, J.R. Mowat, and M.L. Prior, Phys. Rev. Lett. **66**, 1434 (1991).

[ScD74] H. W. Schnopper, J.P. Delvaille, K. Kalata, A.R. Sohval, M. Abdulvahab, K.W. Jones, and H.E. Wegner, Phys. Lett. **47A**, 61 (1974).

[Sch61] S.S. Schweber, *An Introduction to Relativistic Quantum Field Theory* (Row, Peterson and Co., Evanston, 1961).

[Sch68] L.I. Schiff, *Quantum Mechanics* (McGraw-Hill, New York, 1968).

[Sch91] T. Schwab, GSI-Report 91-10 (Gesellschaft für Schwerionenforschung, Darmstadt, Germany, 1991).

[ScK89] R. Schmidt, B. Kämpfer, H. Feldmeyer, and O. Knospe, Phys. Lett. B **220**, 107 (1989).

[Sco73] J.H. Scofield, University of California Report No. UCLR-51326 (1973).

[Sco78] J.H. Scofield, Phys. Rev. A **18**, 963 (1978).

[Sha79] R. Shakeshaft, Phys. Rev. A **20**, 779 (1979).

[ShS77] R. Shakeshaft and L. Spruch, Phys. Rev. Lett. **38**, 175 (1977).

[ShS79] R. Shakeshaft and L. Spruch, Rev. Mod. Phys. **51**, 369 (1979).

[Sie68] K. Siegbahn, *Alpha-, Beta-, and Gamma-Ray Spectroscopy*, Vol. 1, Chap. 3 (North-Holland, Amsterdam, 1968).

[Sla60] J.C. Slater, *Quantum Theory of Atomic Structure,* Vol. 1, Sec. 15-7 (McGraw-Hill, New York, 1960).

[SoB94] A.H. Sørensen and A. Belkacem, Phys. Rev. A **49**, 81 (1994).

[Sof80] G. Soff, in *18th Winter School*, eds. A. Balanda and Z. Stachura, p. 201 (Bielsko-Biala, Poland, 1980).

[Som31] A. Sommerfeld, *Atombau und Spektrallinien*, Bd. I (Friedr. Vieweg & Sohn, Braunschweig, 1931).

[SoM35] A. Sommerfeld and A.W. Maue, Ann. Physik **22**, 629 (1935).

[Som39] A. Sommerfeld, *Atombau und Spektrallinien*, Bd. II, 2nd ed. (Friedr. Vieweg & Sohn, Braunschweig, 1939).

[Sor87] A.H. Sorensen, Phys. Rev. A **36**, 3125 (1987).

[SoU86] A.H. Sorensen and E. Uggerhoj, Comments Ato. Mol. Phys. **17**, 285 (1986).

[SpB79] E. Spindler, H.-D. Betz, and F. Bell, Phys, Rev. Lett. **42**, 832 (1979).

[Spi79] E. Spindler, Ph.D. Thesis, München, 1979.

[SpM94] D.W. Spoonner, W.E. Meyerhof, J.J. Kuffner, E.C. Montenegro, K. Ishii, S.E. Kuhn, D.M. Kawall, D.G. Jensen, and Z.-E. Meziani, Z. Phys. D **29**, 265 (1994).

[StB84] R.M Sternheimer, M.J. Berger, and S.M. Seltzer, At. Data Nucl. Data Tables **30**, 261 (1984).

[StB90] M.R Strayer, C. Bottcher, V.E. Oberacker, and A.S. Umar, Phys. Rev. A **41**, 1399 (1990).

[StG91] Th. Stöhlker, H. Geissel, H. Folger, C. Kozhuharov, P.H. Mokler, G. Münzenberg, D. Schardt, Th. Schwab, M. Steiner, H. Stelzer, and K. Sümmerer, Nucl. Inst. Meth. Res. B **61**, 408 (1991).

[StG94] Th. Stöhlker, H. Geissel, H. Irnich, T. Kandler, C. Kozhuharov, P.H. Mokler, G. Münzenberg, F. Nickel, C. Scheidenberger, T. Suzuki, M. Kucharski, A. Warczak, P. Rymuza, Z. Stachura, A. Kriessbach, D. Dauvergne, B. Dunford, J. Eichler, A. Ichihara, and T. Shirai, Phys. Rev. Lett. **73**, 3520 (1994)..

[StK92] Th. Stöhlker, C. Kozhuharov, A.E. Livingston, P.H. Mokler, Z. Stachura, and A. Warczak, Z. Phys. D **23**, 121 (1992).

[StK95] Th. Stöhlker, C. Kozhuharov, P.H. Mokler, F. Bosch, B. Franzke, H. Geissel, R. Moshammer, C. Scheidenberger, A. Warczak, M. Kucharski, Z. Stachura, P. Rymuza, J. Eichler, A. Ichihara, and T. Shirai, Phys. Rev. A **51** (1995), in press.

[StM92] T. Stöhlker, P.H. Mokler, H. Geissel, R. Moshammer, P. Rymuza, E.M. Bernstein, C.L. Cocke, C. Kozhuharov, G. Münzenberg, F. Nickel, C. Scheidenberger, J. Ullrich, and A. Warczak, Phys. Lett. A **168**, 285 (1992).

[StM93] T. Stöhlker, P.H. Mokler, K. Beckert, F. Bosch, H. Eickhoff, B. Franzke, M. Jung, T. Kandler, O. Klepper, C. Kozhuharov, R. Moshammer, F. Nolden, H. Reich, P. Rymuza, P. Spädtke, and M. Steck, Phys. Rev. Lett. **71**, 2184 (1993).

[Sto30] M. Stobbe, Ann. Phys. (Leipzig) **7**, 661 (1930).

[TeT80] E. Teubner, G. Terlecki, N. Grün, and W. Scheid, J. Phys. B **13**, 523 (1980).

[ThB92] J. Thiel, A. Bunker, K. Momberger, N. Grün, and W. Scheid, Phys. Rev. A **46**, 2607 (1992).

[Tho27] L.H. Thomas, Proc. Roy. Soc. London Ser. A **114**, 561 (1927).

[ThW85] P. Thieberger, H.E. Wegner, J. Alonso, H. Gould, C. Munger, R. Anholt, and W.E. Meyerhof, IEEE Trans. Nucl. Sci. **NS-32**, 1767 (1985).

[ToE88a] N. Toshima and J. Eichler, Phys. Rev. Lett. **60**, 573 (1988).

[ToE88b] N. Toshima and J. Eichler, Phys. Rev. A **38**, 2305 (1988).

[ToE89] N. Toshima and J. Eichler, Phys. Rev. A **40**, 125 (1989).

[ToE90a] N. Toshima and J. Eichler, Phys. Rev. A **41**, 5221 (1990).

[ToE90b] N. Toshima and J. Eichler, Phys. Rev. A **42**, 3896 (1990).

[ToE91] N. Toshima and J. Eichler, Phys. Rev. Lett. **66**, 1050 (1991).

[ToE92] N. Toshima and J. Eichler, Phys. Rev. A **46**, 2564 (1992).

[ToE95] N. Toshima and J. Eichler, Comments Ato. Mol. Phys. (1995), in press.

[ToI87] N. Toshima, T. Ishihara, and J. Eichler, Phys. Rev. A **36**, 2659 (1987).

[TrB83] D. Trautmann, G. Baur, and F. Rösel, J. Phys. B **16**, 3005 (1983).

[TrR80] D. Trautmann and F. Rösel, Nucl. Inst. Meth. **169**, 259 (1980).

[Tso83] N. Tsoulfanidis, *Measurement and Detection of Radiation* (McGraw-Hill, New York, 1983).

[UlD91] J. Ullrich, R. Dörner, S. Lencinas, O. Jagutzki, H. Schmidt-Böcking, and U. Buck, Nucl. Inst. Meth. Phys. Res. B **61**, 415 (1991).

[UlS88] J. Ullrich, H. Schmidt-Böcking, and C. Kelbch, Nucl. Inst. Meth. Phys. Res. A **268**, 216 (1988).

[UmW91] A.S. Umar, J.-S. Wu, M.R. Strayer, and C. Bottcher, J. Comput. Phys. **93**, 426 (1991).

[VaB84] S.R. Valluri, U. Becker, N. Grün, and W. Scheid J. Phys. B **17**, 4359 (1984).

[VaD92] C.R. Vane, S. Datz, P.F. Dittner, H.F. Krause, C. Bottcher, M. Strayer, R. Schuch, H. Gao, and R. Hutton, Phys. Rev. Lett. **69**, 1911 (1992).

[Val91] P. Valin, Phys. Rep. **203**, 233 (1991).

[ViG93] M. Vidoviĉ, M. Greiner, C. Best, and G. Soff, Phys. Rev. C **47**, 2308 (1993).

[WaC89] G. Waysand and C. Chardin, eds., *Superconducting and Low-Temperature Particle Detectors* (North-Holland, Amsterdam, 1989).

[Wei34] C.F. v. Weizsäcker, Z. Physik **88**, 612 (1934).

[WeO92] J.C. Wells, V.E. Oberacker, A.S. Umar, C. Bottcher, M.R. Strayer, J.-S. Wu, and G. Plunien, Phys. Rev. A **45**, 6296 (1992).

[Wil34] E.J. Williams, Phys. Rev. **45**, 729 (1934).

[Wil35] E.J. Williams, Dan. Vid. Selsk. Mat.-Fys. Medd. **13**, No. 4 (1935).

[Wil83] U. Wille, J. Phys. B **16**, L275 (1983).

[WiS87] K.-H. Wietschorke, P. Schlüter, G. Soff, K. Rumrich, and W. Greiner, Phys. Rev. A **36**, 377 (1987).

[WiW68] L. Wilets and S.J. Wallace, Phys. Rev. **169**, 84 (1968).

[Wiz79] J. L. Wiza, Nucl. Inst. Meth. **162**, 587 (1979).

[YaI81] A. Yamadera, K. Ishii, K. Sera, M. Sebata, and S. Morita, Phys. Rev. A **23**, 23 (1981).

# Index

ISBN 0-12-233675-5

90065

9 780122 336751